double stars
for small telescopes

double stars
for small telescopes

MORE THAN
2,100 STELLAR GEMS
for BACKYARD OBSERVERS

BY SISSY HAAS

Sky Publishing
Cambridge, Massachusetts

To my late sister Barb,
swept up to the stars but forever in my heart.

And to my steadfast *Sky & Telescope* editor,
Alan MacRobert, for believing that I could write.

Published by Sky Publishing Corporation
49 Bay State Road
Cambridge, MA 02138-1200, USA
SkyandTelescope.com

Library of Congress Cataloging-in-Publication Data

Haas, Sissy.
Double stars for small telescopes : more than 2,100 stellar gems for backyard observers / by Sissy Haas.
p. cm. -- (Stargazing series)
Includes bibliographical references and index.
ISBN 1-931559-32-5 (softcover)
1. Double stars--Observers' manuals. 2. Double stars--Catalogs. I. Title. II. Series.

QB821.H14 2006
523.8'41--dc22

2006007705

CONTENTS

INTRODUCING DOUBLE STARS

IMAGINE THAT a *pair* of stars rises in your morning sky: one yellow-orange, the other yellow-white. Or picture one of these suns setting while the other rises. What a sight that would be! If a planet orbits the double star Gamma Delphini, that's exactly what its inhabitants would see.

Of course, we don't know if such a planet (inhabited or not) exists. But with a small telescope on a clear night during mid-year, you can visit Gamma Delphini (in the constellation Delphinus, the Dolphin) and see its stunning visual beauty for yourself. Through my little 60-millimeter refractor, Gamma looks like a pretty pair of grapefruit-orange stars: brilliant, sparkling, and separated by just a small gap of black. The 19th-century astronomer John Ellard Gore called the colors "reddish yellow" and "greyish lilac."

Just to see Gamma Delphini is, I think, worth the price of a telescope. But it's only one of a myriad of beautiful double stars in our sky. A small scope will reveal that hundreds of bright stars are actually double or multiple suns. When fainter stars are included, tens of thousands turn out to be double. The *Washington Double Star Catalog (WDS)* lists more than 102,000 double or multiple stars, though the modest telescopes of most amateur astronomers cannot show them all.

I think that many double stars are more beautiful than any flower on Earth; some reveal subtle colors that no earthly object can duplicate. One star may show a fascinating blend of white, yellow, and orange, while its companion might reveal an unusual combination of white, silver, blue, and green. In his 1884 astronomy classic *A Cycle of Celestial Objects,* William Smyth detailed 280 fantastic color combinations such as bright white and pale purple; slight yellow and pale garnet; orange and smalt blue; light rose tint and dusty red. He must have had a keen eye for color — he described 16 variations of yellow and 15 of white!

Double stars are among the best-kept secrets in nature. It's a shame so few amateur astronomers observe them, because these stellar pairs make for fascinating viewing. Some show hauntingly beautiful colors that are often enhanced by their different brightnesses. Others are identical in color and brightness but are widely separated, so they look like a pair of celestial headlights approaching from the depths of space. Many are so close that only the tiniest sliver of black can be seen be-

DENNIS DI CICCO

At the crook of the Big Dipper's Handle are Mizar and Alcor, one of the sky's best-known stellar couples. While the two are likely not an orbiting pair, Mizar itself is a true binary and almost any telescope reveals it as double. (Dubhe, the upper-right star in the Big Dipper's Bowl, is also a double.)

AKIRA FUJII

tween them. Sometimes each star appears surrounded by a beautiful ring of light (an *Airy disk*), though this is a purely artificial image created by your telescope. There may be a startling contrast between a pair — a bright, vivid star touching a tiny globe of light so that it resembles a star with a planet. The permutations are almost infinite.

Occasionally there's a bonus sight in the same field of view — a star cluster, nebula, galaxy, or just a beautifully bright star. A number of multiple stars show interesting patterns, like a boomerang, a fishhook, or an arrow. And the motion of some binaries is so rapid that you can see their separation increase or decrease from year to year — it's subtle, but it's possible.

I encourage everyone to observe double stars. You don't need a large telescope to see some of the pairs in this catalog; even a little 60-mm refractor will show dozens of pretty double and multiple stars. And in my opinion, the best thing about these stellar pairs is their color. To the eye, much of the universe appears only as various shades of gray (planets being the major exception). But when you observe double or multiple stars, it's often the color contrast between them that takes your breath away!

GREEK LETTERS FOR BRIGHT STARS

The brightest stars in each constellation are assigned Greek letters. A constellation's most brilliant star is often called Alpha, the first letter of the Greek alphabet.

α	Alpha
β	Beta
γ	Gamma
δ	Delta
ε	Epsilon
ζ	Zeta
η	Eta
θ	Theta
ι	Iota
κ	Kappa
λ	Lambda
μ	Mu
ν	Nu
ξ	Xi
ο	Omicron
π	Pi
ρ	Rho
σ	Sigma
τ	Tau
υ	Upsilon
φ	Phi
χ	Chi
ψ	Psi
ω	Omega

A SLIDING SCALE

For those who have limited experience with the magnitude system, here's how I compare some of the numbers.

magnitude −1.5: Sirius, the brightest star in the night sky	**magnitudes 0.0 to 1.0:** the brightest naked-eye stars
magnitude 4.5: barely detectable from a typical (light-polluted) backyard	**magnitude 6.0:** the naked-eye limit for a dark, clear sky, but a bright object in any telescope
magnitude 8.0: just bright enough to be seen easily in good 7 × 50 binoculars or a small telescope	**magnitude 13.0:** visual limit of a 6-inch (150-mm) telescope

The magnitude scale continues on in each direction: the planet Venus is often magnitude −4, the full Moon shines at −13, and (going the other way) the Hubble Space Telescope can see stars as faint as magnitude 31.

STARS: THEIR NAMES AND BRIGHTNESSES

All stars have officially recognized designations. Many of the naked-eye stars have multiple labels: a number, a Greek letter, and a proper name, though for the most part, only the very brightest stars are known by a name. For example, the brilliant star HD 48915 is also known as Alpha Canis Majoris but is better known as Sirius.

When the Greeks of ancient days wanted to identify the naked-eye stars, they called the brightest one in a constellation Alpha (α) — the first letter of the Greek alphabet. The second-brightest star is Beta (β), the third-brightest Gamma (γ), and so on. (This is not a hard and fast rule, but it holds true for most of the constellations.) Today, the star's name is followed by the genitive (possessive) form of the constellation's name. So it's not called Alpha Canis Major; it's Alpha Canis Majoris. But after the telescope was invented, astronomers had to use other labeling schemes to identify the numerous faint stars they were discovering, because the Greek alphabet has only 24 letters.

The name of many faint double stars includes a letter or symbol (preceding a number), which specifies the identity of the discoverer. For example, if you glance through my catalog you'll find a lot of names that begin with the capitalized Greek letter sigma (Σ). This indicates that the star is listed in two catalogs (published in 1827 and 1837) compiled by the famous double-star observer Friedrich Georg Wilhelm Struve. Other common designations that you'll see include OΣ for his son Otto Wilhelm Struve, H for William Herschel, h for his son John Herschel, S for James South, and β for Sherburne W. Burnham. (Don't confuse this with Beta [β], the designation for the second-brightest star in a constellation.)

Astronomers describe the brightness of stars in terms of their *magnitude*. The magnitude system works like a golf score: the smaller the number, the better the score or, in the case of astronomy, the brighter the star. For example, a 1st-magnitude star is brighter than a 2nd-magnitude star, a 2nd-magnitude star is brighter than a 3rd-magnitude, and so on. Sadly, most urban stargazers can see stars only to about magnitude 4.5 due to light pollution. In a dark sky, stars of 6th magnitude abound.

For double stars, there are two specific bits of nomenclature you need to know. When the members of a binary system are unequal in brightness, the brighter star is called the *primary* and the fainter one is the *secondary* or companion.

STARS: THEIR TEMPERATURE AND LUMINOSITY

In my catalog I list a star's *spectral type*, which helps describe both its temperature and its luminosity. Astronomers use seven main *temperature classes* to organize stars. From hottest to coolest the sequence is *O, B, A, F, G, K,* and *M.* Astronomy students are often taught the mnemonic "Oh Be A Fine Girl (or Guy), Kiss Me" as a way to remember this non-alphabetical order. Each of these classes has 10 subdivisions from 0 to 9: an *O*2 star is cooler than an *O*1 but hotter than an *O*3 star. Our Sun is a *G*2 star.

Just because a star is hot doesn't necessarily mean it's bright. A cool, giant star can be more *luminous* than a hot, tiny star. The great majority of stars belong to what's called the main sequence, where hotter stars are usually brighter (and larger) than cooler stars. But then there are stars called dwarfs and giants, which are atypically luminous for their temperature class. The dwarfs are often fainter; the giants are usually brighter.

For a group of stars in the same temperature class, the luminosity class helps differentiate their sizes. Astronomers use eight luminosity classes — Ia and Ib are supergiant stars; II, III, and IV indicate giant suns; V is for main-sequence stars; and VI and VII are for dwarf stars.

So when you look at the "Spec." column of the catalog and see a star labeled *B*2III, it means it's a hot (*B*2) giant (III) sun.

STARS AND THEIR COLORS

Many factors influence the actual colors we see in starlight. The temperature of the star itself has an effect on its color; the table *(below)* gives some indication as to what to expect. But observational factors, including the Purkinje effect, dazzling, contrast, and analogous hues, can affect our perception of color.

STAR TEMPERATURES AND COLORS

Temperature class	Visible color
O	blue-white
B	blue-white
A	white
F	yellow-white
G	yellow
K	orange
M	reddish orange

Under low light, the eye isn't particularly sensitive to color; what it can detect tends to be blue and green. This causes dim objects, like faint companion stars, to look more blue or green than they actually are (the *Purkinje effect*).

Dazzling occurs when the intensity of starlight is so great that it causes the eye to see more white than is present (or to see only white instead of a color). It's probably because of dazzling that I see more white in bright stars when I use a large telescope, since a larger telescope gives brighter images.

The effect of *contrast* is to make a color seem more intense when it's seen against a contrasting color. It's a technique used by painters, and it certainly applies to double stars. A faint companion star will always look more intensely white, blue, or green than if seen without its brighter primary.

And when the color of a stellar pair is similar — in other words, they have *analogous hues* — it's hard to accurately describe their tints, particularly if they're close. Detecting the subtle difference between a tight pair of yellow and orange stars, especially if they're of nearly equal brightness, can be very challenging.

BINARY STARS AND OPTICAL DOUBLES

Astronomers refer to stars that are physically connected — held together by the force of gravity — as *binary stars.* In some cases, we can see (over time) the stars swing around each other. This orbital motion lets us calculate their *period,* the time it takes the pair to complete one full revolution.

Stars that look like pairs but are actually chance alignments are called *optical doubles.* Here one star is actually light-years away from the other, and any visual impression of closeness is a line-of-sight coincidence. (A light-year is the distance light travels in one year — 5.9 trillion miles, or 9.5 trillion kilometers. It's a measure of distance, not time.)

When both members of the pair are confidently known to be alike in distance and in proper motion (their motion through space), it's very likely that they're binary stars. On the other hand, when they are known to be different in distance and proper motion, there is little doubt that they're optical doubles.

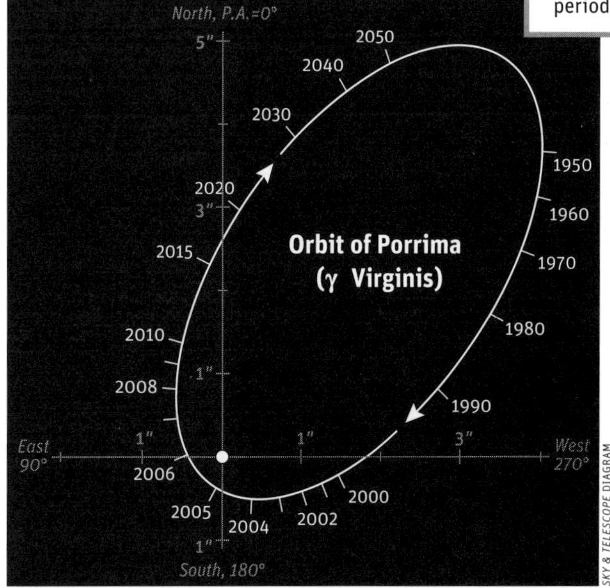

The apparent orbit of Porrima is plotted on the plane of the sky, with celestial north up. Ticks show the companion's location at the start of the years given. On the next page are a series of sketches that show the changing position of the two stars relative to each other during a six-year period.

The vast majority of binary stars are closer than 10 arcseconds (10″) apart, while the vast majority of optical doubles are wider than this. The smaller the separation, the greater the probability that the pair is a binary. In my catalog's "Status" column, I've listed as *binary* those stars for which there is a high degree of confidence that they're binary stars, *optical* for those stars that are very likely optical doubles, and *probable binary* for those for which the data are inconclusive but strongly tending toward the star being a binary. If the binary's period (p) is known, it's included. A blank entry means the status of the double is unknown.

For me, it's psychologically rewarding to know that the members of a double star are a real couple. But there are also a lot of wide optical pairs that are every bit as lovely as any binary star. Gamma Leporis — in the constellation Lepus, the Hare — is a perfect example of this; it's a bright yellow star beside a bright red star. The components are very wide apart but dominate the field and make a beautiful pair. James South cataloged a lot of similar, super-wide pairs, and William Smyth described many of these as brilliant shades of yellow or white with little companions of blue, green, plum, lilac, or red. I've included the best of these in my catalog.

The total number of binaries with a known period is not large. Of the more than 102,000 pairs in the *Washington Double Star Catalog,* only 1,745 have a published period at the present time. And among them are pairs with an orbital motion so rapid that we can actually see the change taking place in our lifetime. Castor (Alpha Geminorum), Xi Ursae Majoris, and Porrima (Gamma Virginis) are examples of bright pairs whose separation has decreased or increased visibly during the last decade. In 1990, Castor was a close pair and Xi UMa was too tight for me to resolve, but as of 2005 both were wide enough to be easily split in my 60-mm refractor. Porrima, on the other hand, went from an easy pair for my small scope to one that was too close for my 13-inch reflector.

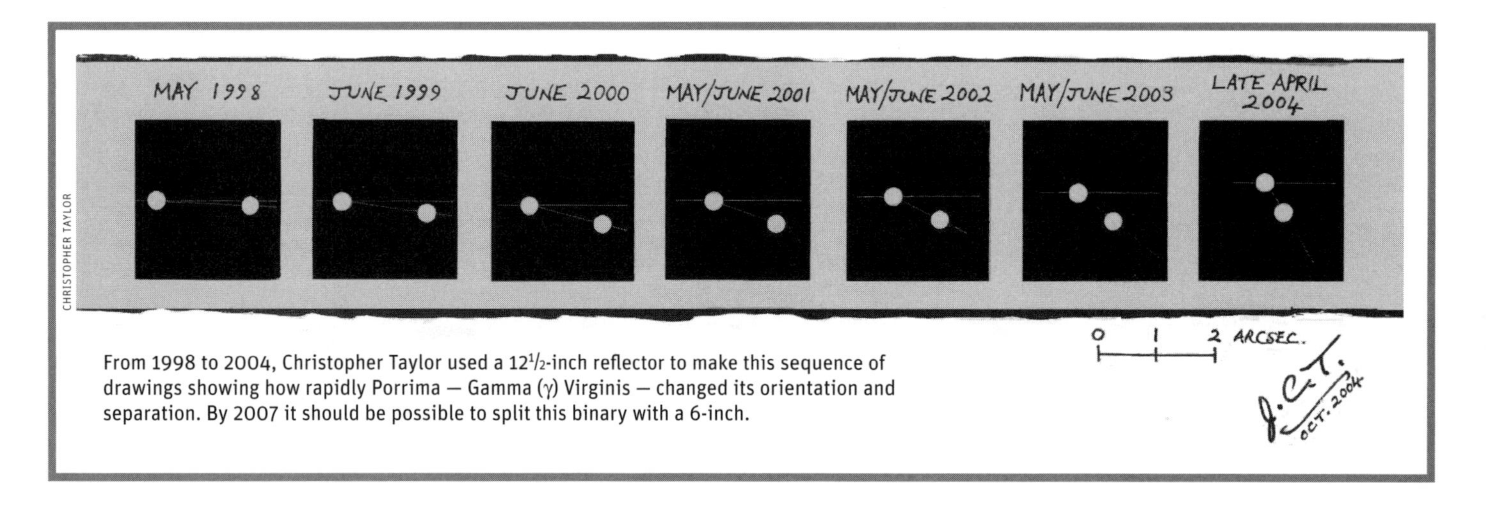

From 1998 to 2004, Christopher Taylor used a 12½-inch reflector to make this sequence of drawings showing how rapidly Porrima — Gamma (γ) Virginis — changed its orientation and separation. By 2007 it should be possible to split this binary with a 6-inch.

RESOLVING DOUBLE STARS

The visual description that accompanies each double-star entry generally includes the *aperture* of the telescope used — the diameter of its main mirror or lens in millimeters. If you need to convert to inches, divide the listed aperture by 25.4. So a 125-mm scope is the same as a 4.9-inch (call it a 5-inch) instrument.

Also in each listing is the *magnification* (or power) used. In simple terms, this is a measure of how many times the image has been enlarged (see the illustration opposite). Usually, double stars with small separations need to be observed at high power. But don't go too high. With each increase in magnification, the image definition suffers and the field of view (the amount of sky observed) grows smaller. When the pair is difficult to split primarily because the brightness of the stars is very unequal, a low or a medium power may give the best result.

If you flip through the catalog, you'll occasionally see the phrase *showcase pair.* These are the brightest and easiest doubles to observe; if you're just beginning, please start with these. (See the star charts on page 8 to locate several very easy and beautiful starter pairs.) Most are bright enough to spot with the naked eye, and the smallest telescope will show that they are doubles.

More and more telescopes now come equipped with a computer that locates celestial objects. They can find the showcase pairs for you. But many of these pairs are also easy to find with a simple star map — the kind that can be found in magazines such as *Night Sky, Sky & Telescope,* or *Astronomy.*

When the intended star is no brighter than the others in your low-power field of view, computer positioning of the telescope may not be good enough to show which one is the double. Here's where a more detailed star map will come in handy. One of the best is *Sky Atlas 2000.0* by Wil Tirion and Roger W. Sinnott — available from Sky Publishing.

As you gain experience, you'll probably want to tackle more challenging pairs. The most difficult doubles are the

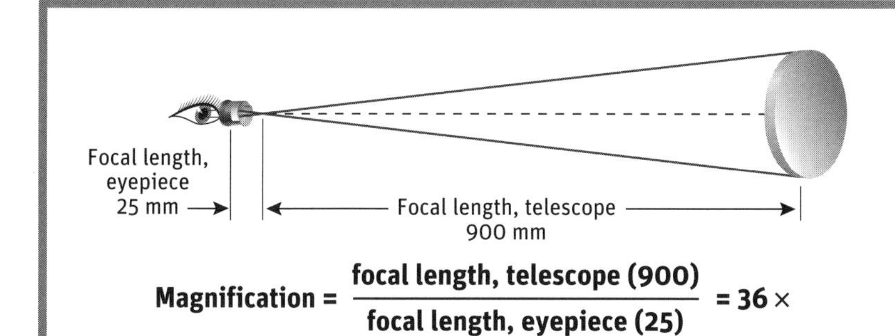

Determining a telescope's magnification requires a little arithmetic and the understanding that *focal length* is the distance (usually in millimeters) from a mirror or lens to the image that it forms. This number is usually inscribed somewhere on both the scope and the eyepiece. To change magnification, change eyepieces. A 10-mm eyepiece used in the setup depicted at left yields a magnification of 90 power (90×) — 900/10.

closest or most unequal in brightness — splitting a modestly close pair of very unequal stars is about as hard as a very close pair of equal stars. These require large telescopes, good weather, dark adaptation, averted vision, and patience.

Double-star enthusiasts will happily talk for hours about the ease (or not) of separating a particular double. So can you determine, in advance, whether or not your scope will separate any specific stellar pair? Numerous studies, using optical theory and actual observations, have attempted to provide an answer, but it's a complex problem. Given fine weather and good eyesight, and all else being equal, the variables are the gap between the stars, the difference in brightness between them, and the aperture of your telescope.

While looking for some guidance, I compared actual observations with published formulas for double-star resolution. The observations come from the great French work *Revue des Constellations,* and the formulas that consistently match them were developed by William Dawes, P. J. Treanor, and Christopher Lord. The resulting table *(below)* should be used only as a guide; as commercials often say, your actual results may vary.

In my catalog you'll see that I've included the magnitude of each member of a double-star system. The difference in brightness between the primary and secondary is Δm — the first column in the table. The table's top row lists a series of telescope apertures. To use the table, look down the aperture column that best matches your scope and across the Δm row of the stellar pair you wish to observe. Where the row and column meet is the smallest separation (in arcseconds) that your scope can resolve for that particular stellar-magnitude difference.

MINIMUM SEPARATION FOR AN APERTURE

Δm is the difference in magnitude between the primary and secondary stars. The estimated minimum separation for the different apertures is in arcseconds. These estimates are useful only when the primary star is magnitude 6.5 or brighter.

Magnitude difference	60-mm	100-mm	150-mm	200-mm	250-mm
Δm 0.0	2.0″	1.2″	0.8″	0.6″	0.5″
Δm 0.5	2.0″	1.2″	0.8″	0.6″	0.5″
Δm 1.0	2.2″	1.3″	0.9″	0.8″	0.6″
Δm 1.5	2.5″	1.4″	1.0″	0.9″	0.7″
Δm 2.0	3.2″	1.9″	1.2″	1.1″	0.9″
Δm 2.5	3.5″	2.0″	1.4″	1.3″	1.1″
Δm 3.0	3.7″	2.3″	1.6″	1.5″	1.3″
Δm 3.5	4.4″	2.4″	1.8″	1.6″	1.5″
Δm 4.0	4.5″	2.6″	2.0″	1.9″	1.6″

Lyra, the Lyre, is home to numerous double stars (see page 101), but only one is obvious in this image. Look carefully to the upper left of the bright star Vega. That little dumbbell-shaped star is Epsilon (ε) Lyrae, the famous Double-Double.

AKIRA FUJII

For example, can my 60-mm refractor split the brilliant burgundy star Antares in Scorpius, the Scorpion? Antares is listed on page 141: note there is about a four-magnitude difference between its two stars. On the table, if I look along the row marked Δm 4.0 and down the column marked 60-mm, I find that the stars need to be 4.5″ apart for my 60-mm scope to split them. In fact, they're 2.5″ apart, which means I can't separate them with my little scope. So I have to either look at a different star or dig out my 5-inch (125-mm) scope, which can split the pair.

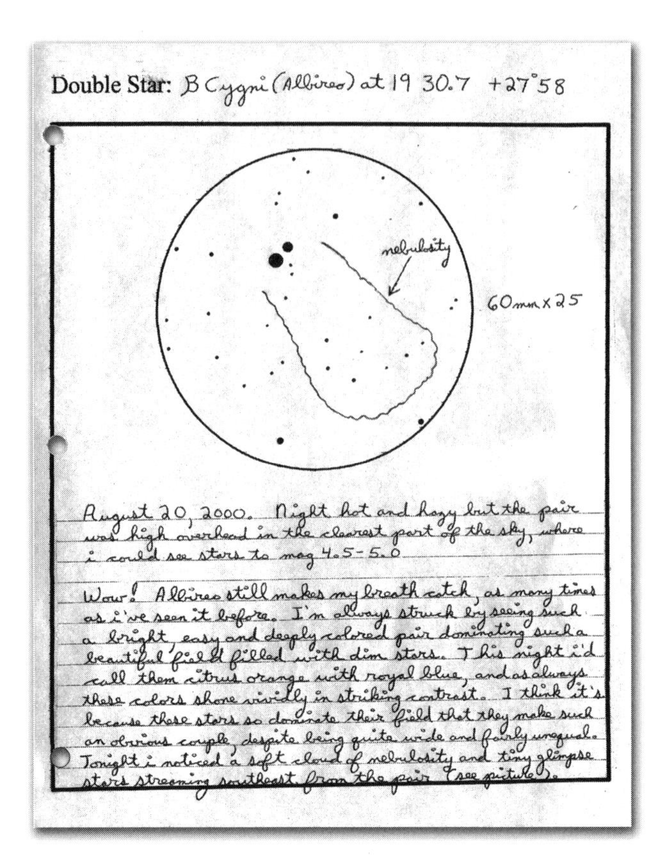
Double Star: β Cygni (Albireo) at 19 30.7 +27°58

August 20, 2000. Night hot and hazy but the pair was high overhead in the clearest part of the sky, where i could see stars to mag 4.5-5.0

Wow! Albireo still makes my breath catch, as many times as i've seen it before. I'm always struck by seeing such a bright, easy and deeply colored pair dominating such a beautiful field filled with dim stars. This night i'd call them citrus orange with royal blue, and as always these colors shone vividly in striking contrast. I think it's because these stars so dominate their field that they make such an obvious couple, despite being quite wide and fairly unequal. Tonight i noticed a soft cloud of nebulosity and tiny glimpse stars streaming southeast from the pair (see picture).

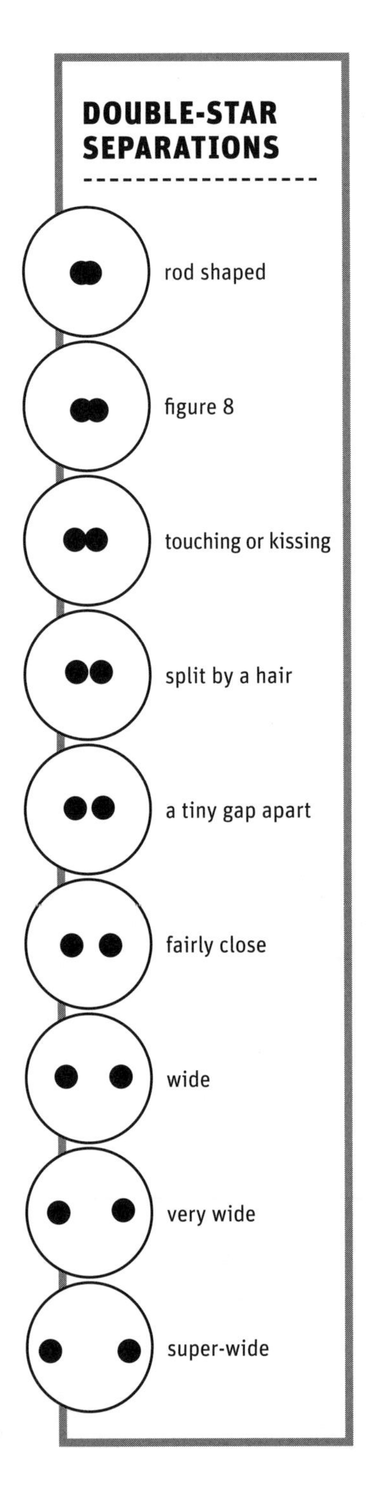

OBSERVING HINTS

It's beyond the realm of this book to describe how to observe. But here are a few comments that will help you make the most of your evenings under the stars.

Astronomers often use two words to describe the weather: transparency and seeing. *Transparency* refers to how clear the sky is, and *seeing* refers to how steady the sky is. A quick glance at the sky gives some indication of the weather at that moment: the stars don't twinkle when the seeing is good, and they seem particularly bright when the transparency is excellent. In an ideal sky, the air is both clear and steady, but this is a rare occurrence.

A transparent sky is always best for viewing very faint companion stars, assuming their separations are not minuscule. For close pairs, it's the seeing that counts. Poor seeing blurs the image of a star's disk so that it appears swollen. The swollen disks of a double take up the space that would otherwise be seen between them and make it difficult to split them cleanly.

Dark adaptation is a process that makes our eyes more sensitive to light. The pupils of our eyes constrict or widen according to the brightness of our surroundings. When you step into the dark, your pupils quickly widen and you become better able to see. If you remain in the dark, your eye produces a chemical that gradually makes you even more sensitive to light. Thus, we can see much fainter objects with dark-adapted eyes. Some of this adaptation takes place within 15 minutes, but noticeable improvement continues for another 30 to 45 minutes. If you need light to read a star chart, use a dim red-light flashlight; its light will have a minimal effect on your dark adaptation.

Averted vision refers to the practice of observing from the corner of your eye. The well-known British observer Patrick Moore explains it as "directing the eye slightly to one side, while the observer's attention is concentrated on the spot where the object is believed to be." Averted vision makes it possible to see faint or unequal companion stars that are invisible to direct vision.

RECORDING YOUR OBSERVATIONS

Few men have inspired amateur astronomers as much as the late Walter Scott Houston, a revered *Sky & Telescope* columnist and expert at describing the heavens. He would often tell his fellow stargazers, "You are all poets." While amateurs can make scientific measurements, most of us are indeed more like poets — we try to capture in words what we see in our telescopes. Many of us keep an *observing log*, which makes our time "infinitely more satisfactory" in the words of Patrick Moore. And if you don't, I urge you to start one tonight. For ideas and inspiration, I've included a page from mine *(above)*.

The best log includes an impression of the double, along with an accurate description. Is it an ordinary pair in a spectacular field? Are its stars ordinary in color but stunning in brightness? Is the companion feeble with barely enough separation to spot? These subjective impressions are what make us poets; they're what separate us from mechanical instruments.

Your observing log should include the aperture and

magnifications used, the weather, a description of colors, brightnesses, separation, surrounding field, and, if possible, a drawing. Your sketch should strive to accurately show the brightness difference and how much space between the two stars you actually saw. The illustration opposite shows a chart of my personal words for describing separations. I include it merely as a starting point. Use words that best suit you, as long as the drawings clearly show the meaning of your words.

THE CATALOG

This catalog contains 2,101 double and multiple stars whose primary is magnitude 8 or brighter. I consider these the prettiest or most interesting of the more than 102,000 pairs listed in the *Washington Double Star Catalog.*

I winnowed the *WDS* list on the basis of several factors: the magnitude of the secondary, the stars' separation, spectral type (which indicates color), orbital information, and the appearance of the surrounding field. Then I gathered visual descriptions of this reduced list from books and observations made by myself and others. I also included many of the pairs that caught the attention of William Smyth and Thomas William Webb. Their books — *A Cycle of Celestial Objects* and *Celestial Objects for Common Telescopes,* respectively — are classics that are well known to both amateur and professional astronomers.

The visual descriptions accompanying each entry are either mine or from an identified observer. I'm eternally grateful to all the observers listed below. Their contributions were made as a personal favor to me with no reward beyond my thanks.

Special recognition must go to Ross Gould of Canberra, Australia. Ross sent me eloquent notes for thousands of pairs that can be seen primarily from the Southern Hemisphere. Furthermore, his descriptions are those of a true double-star enthusiast with years of observing experience Without his tireless help, this catalog would have been suitable only for Northern Hemisphere observers. In addition to Smyth and Webb, other descriptions came from Ernst J. Hartung *(Astronomical Objects for Southern Telescopes)* and James Mullaney *(Celestial Harvest).*

Finally, I'd like to acknowledge the help of Brian Mason, the curator of the online *Washington Double Star Catalog* (http://ad.usno.navy.mil/wds), which is maintained at the U.S. Naval Observatory.

SISSY HAAS
Greensburg, Pennsylvania

THANK YOU

A special thanks to all the observers who helped make this catalog possible.

Kevin Barker, Christchurch, New Zealand: 100-mm refractor
Jeff Brydges, Tucson, Arizona: 60-mm refractor
Paul Castle, Rock Island, Illinois: 150-mm refractor
Joe DalSanto, Carol Stream, Illinois: 150-mm reflector
Jeff Dekanich, Sheboygan, Wisconsin: 200-mm Schmidt-Cassegrain
Tom Ferguson, Mimbres, New Mexico: 150-mm refractor
Bill Geertsen, Timonium, Maryland: 100-mm refractor
Ross Gould, Canberra, Australia: 175-mm refractor
Richard Harshaw, Kansas City, Missouri: 200-mm Schmidt-Cassegrain
Ernst J. Hartung, Victoria, Australia: several scopes from 75-mm to 300-mm
Randy Heckman, Kearney, Nebraska: 250-mm Schmidt-Cassegrain
Richard Jaworski, Melbourne, Australia: 100-mm refractor
Joseph Liu, Salinas, California: 200-mm refractor
James Mullaney, author, Celestial Harvest*: various scopes*
Jim Phillips, Charleston, South Carolina: 150-mm refractor
Darian Rachal, Alexandria, Louisiana: 70-mm refractor
Charles Sawyer, Pembroke, Maine: 80-mm refractor
Reed Smyth, Oxnard, California: 70-mm refractor

SOME EASILY FOUND SHOWCASE DOUBLES

IF YOU'VE NEVER SEEN a double star and want to know what all the fuss is about, check out the 15 doubles identified on these four star maps. They're perfect starter stars and are easy to find. More details about these stars can be found in their respective constellation listings.

LATE FEBRUARY *through* EARLY JUNE

Between late February and early June, Leo, the Lion, prowls the evening sky after sunset. His brightest star, Regulus, lies under an easily identifiable curve of fainter stars called the Sickle.

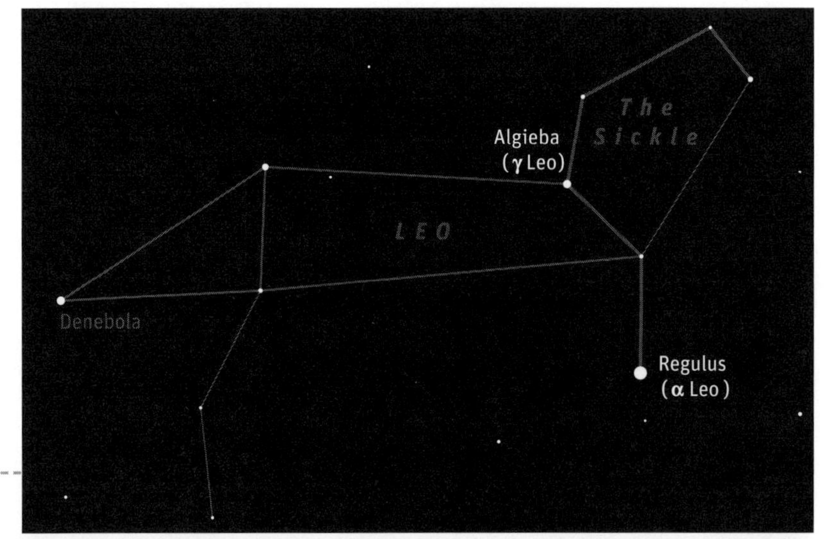

JULY *through* OCTOBER

After sunset between July and October, Northern Hemisphere observers should look high in the south for Cygnus, the Swan, and Lyra, the Lyre. Helping locate these two constellations are their bright stars: Deneb and Vega, respectively. (Below them — but not shown on the chart — is Altair, the third bright star in the Summer Triangle.)

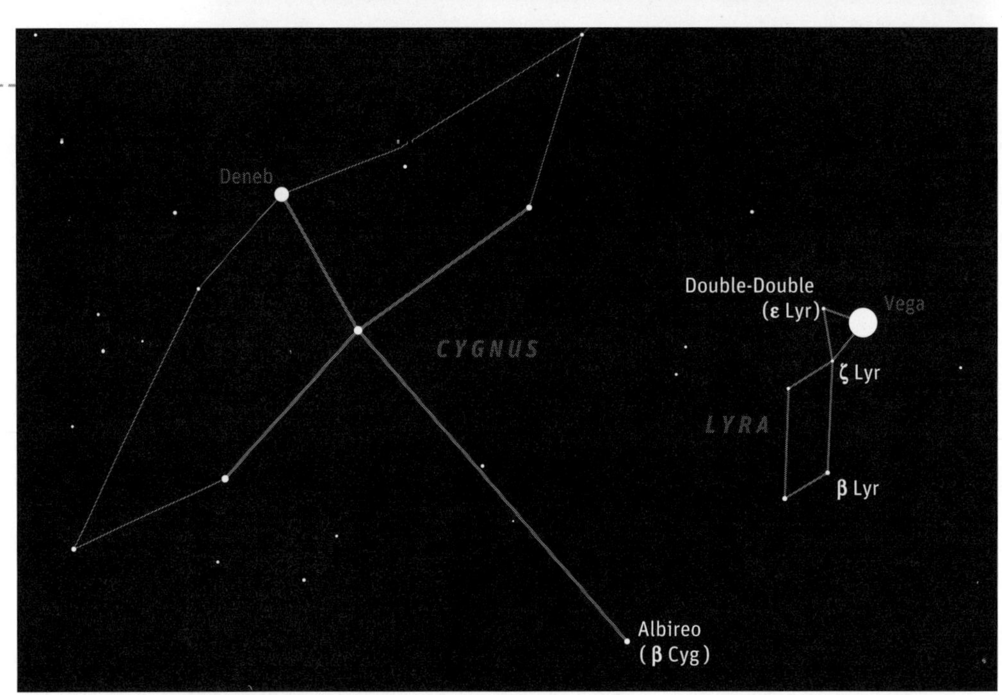

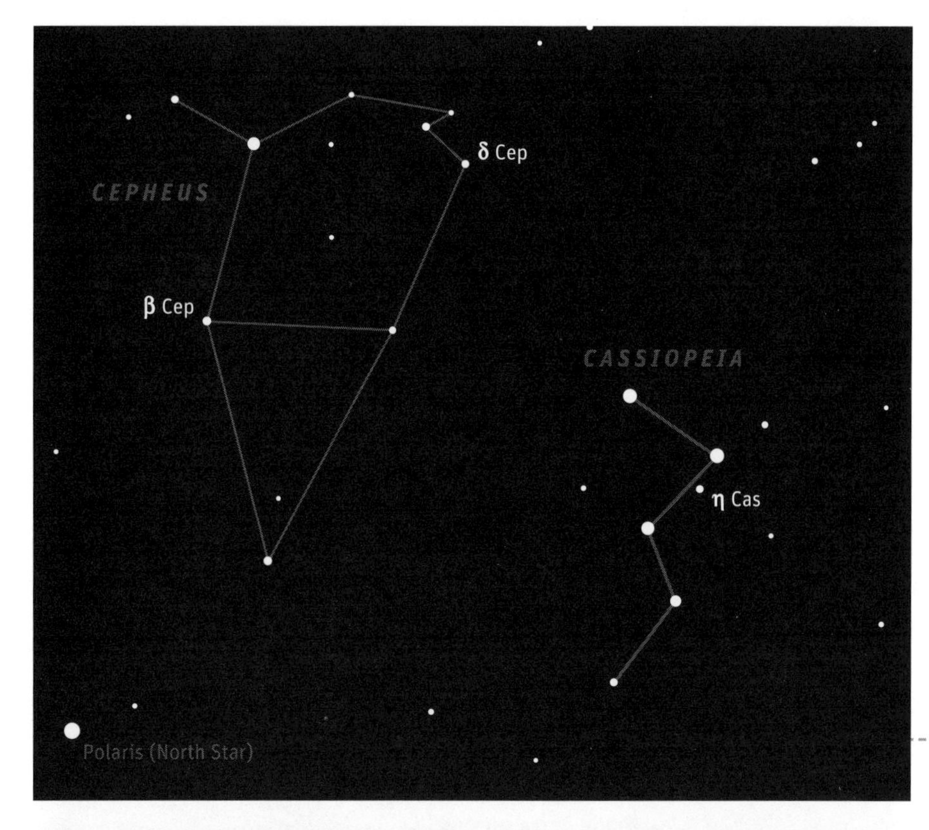

Northern Hemisphere observers can look high in the north after sunset between August and November for five bright stars in the shape of the number 3. These are the brightest stars of Cassiopeia, the Queen. To her left are the fainter stars of Cepheus, the King.

AUGUST *through* NOVEMBER

DECEMBER *through* APRIL

After sunset between December and April, Orion, the Hunter, looms large in the sky. Under his feet is the tiny constellation Lepus, the Hare.

λ Ori
Betelgeuse
ORION
δ Ori
θ Ori
Orion Nebula
σ Ori
ι Ori
Rigel
CANIS MAJOR
Sirius
LEPUS
γ Lep

CATALOG HEADINGS

Coordinates (RA and Dec.)
As seen from Earth, the night sky looks like a huge star-filled dome. To help you navigate the heavens, imagine Earth's lines of latitude and longitude extending out from the planet's surface and intersecting this great celestial sphere of stars. What we call latitude, astronomers call declination (Dec.); the equivalent of longitude is right ascension (RA).

Declination is measured in degrees (°), arcminutes (′), and arcseconds (″) from 90° north (+) to 90° south (–) of the celestial equator — an extension of Earth's equator into space. Right ascension is expressed in hours (^{h}), minutes (^{m}), and seconds (^{s}), from 0 to 24 hours. Together, RA and Dec. give the coordinates of a celestial object so you can find it on a star chart. You can also enter these coordinates into a computer-controlled mount and let it find the star for you.

Name and Year
See page 2 for more information about star names. The year is the date of the last satisfactory observation of the pair.

Position Angle (P.A.) and Separation (Sep.)
The compass direction, in degrees (°), of the fainter secondary star (B) as measured from the brighter primary star (A) is called the *position angle*. A P.A. of 0° means the secondary lies to the north of the primary, 90° is east, and so on.

The apparent space between a pair of stars is known as their *separation*; it's measured in arcseconds (″).

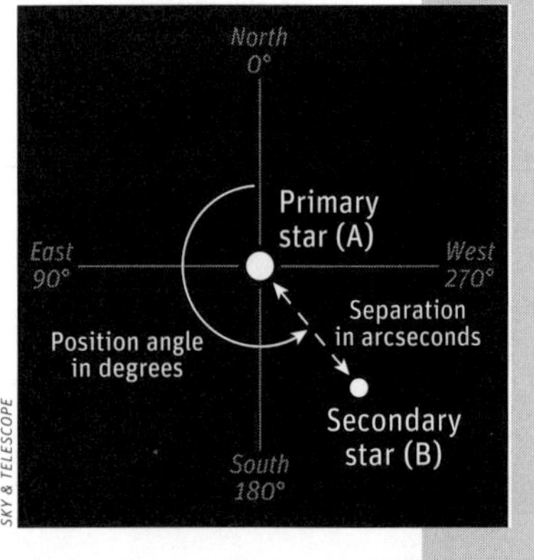

Magnitudes (m_1 and m_2)
Magnitude is a measure of the visual brightness of a star (see page 2 for more details). The primary star is m_1, while m_2 refers to the fainter star.

Spectral Type (Spec.)
The spectral type of a star includes both its temperature and luminosity class (see page 3). Usually only the primary star's class is listed. If there are two listings, the secondary's spectral type always follows that of the primary.

Status
In this column, *binary* refers to a star system for which there is a high degree of confidence that it's a gravitationally bound pair because both stars show the same parallax and have a common proper motion.

The term *optical* refers to a pair of stars, often widely separated, that is unlikely to be gravitationally bound and is, instead, the result of a chance alignment.

If the data are inconclusive but strongly tending toward the star being binary, it's called a probable binary and is labeled *p. binary*.

The period (listed as *p=xxx yr*) is the number of years required for the two stars to complete one orbit around their common center of mass.

If the entry is blank, the status of the double is unknown.

Comments
The visual description accompanying each entry generally includes the aperture of the telescope (in millimeters) and the magnification (×) used. Observers other than the author are identified by name.

CATALOG of DOUBLE STARS

Andromeda

RA	Dec.	Name	Year	P.A.	Sep.	m_1	m_2	Spec.	Status	Comments
23h 02.8m	+44° 04′	Σ2973	2002	40°	7.5″	6.4	10.1	*B*2V		125-mm, 160×: Nice color, contrast, and field. A bright whitish banana-yellow star with a little glimpse star next to it, in a conspicuous triangle of three stars.
23h 06.5m	+46° 55′	ΟΣΣ 242	1998	31°	79.9″	7.8	8.6	*B*3		125-mm, 50×: A conspicuous pair of white stars, super-wide apart, that are easily noticed in the field with a bright orange star (4 Andromedae).
23h 10.0m	+47° 58′	Σ2985	2001	256°	15.7″	7.2	8.0	*G*2IV-V		125-mm, 50×: A bright and attractively close pair without being hard. It's a pair of white stars, mildly unequal, that are split by a fairly small gap.
23h 10.0m	+36° 51′	S 825	1999	319°	67.2″	7.8	8.3	*K*4III		125-mm, 50×: Very nice view at the edge of cluster NGC 752. Coppery scarlet and a dim ash white.
23h 11.4m	+38° 13′	Ho 197								125-mm, 50×: This triple is a neat little arc of three stars stars in a sparse field — a gloss-white star with two dim companions.
		AB-C	2004	320°	36.5″	8.0	10.3			
		AB-D	2002	279°	55.3″	8.0	9.8	*F*2		
23h 13.1m	+40° 00′	Σ2992	2002	285°	13.8″	7.7	9.6	*A*7III	optical	125-mm: This is a conspicuous white star in a desolate field; 50× shows that it has a little dot of silver fairly close to it.
23h 19.1m	+48° 55′	Es 2725	2001	235°	53.3″	7.3	8.6	*A*2		125-mm, 50×: A pretty view. It shows a wide, conspicuous white pair in the field with a bright orange-red star (8 Andromedae) and a bright white star (11 Andromedae).
23h 20.7m	+44° 07′	Σ3004	2003	177°	13.5″	6.3	10.1	*A*5Vn		125-mm, 200×: Striking contrast and a pretty primary. A bright dusty gold star with a ghostly speck beside it, fairly wide apart.
23h 37.5m	+44° 26′	ΟΣ 500	2003	7°	0.5″	6.1	7.4	*B*8V	p=351 yr	This binary is 31′ east of Kappa (κ) Andromedae. The data suggest a bright pair of stars, very close together. Webb: "White, blue. A curious star group 15′ N."
23h 38.6m	+35° 02′	Σ3028	2003	199°	14.7″	7.1	10.0	*A*2	binary	125-mm, 50×: A straw-yellow star with a little misty speck next to it, attractively close for their contrast.
23h 40.4m	+44° 20′	κ And								The data suggest a very bright star with a very faint companion. Smyth: "A wide but delicate triple star . . . brilliant white; dusky; ash colored."
		AB	1998	202°	47.8″	4.1	11.3	*B*9IV		
		AC	1998	293°	113.3″	4.1	11.3			

Andromeda *(continued)*

RA	Dec.	Name	Year	P.A.	Sep.	m_1	m_2	Spec.	Status	Comments
$23^h 44.5^m$	+46° 23′	Σ3034	1998	104°	5.7″	7.7	9.9	*B*8V	binary	The data suggest a bright star with a faint companion. Smyth: "[In] a little [star] group, of which the principal is ψ [Psi Andromedae] . . . topaz yellow; blue."
$23^h 51.6^m$	+42° 05′	h 1911 AB-C	1998	347°	22.0″	7.4	9.4			Liu, 206-mm, 113×: An easy and wide uneven pair, "bluish white with deep yellow or orange."
$23^h 51.9^m$	+37° 53′	Σ3042	2003	86°	5.6″	7.6	7.8	*F*5V	binary	Liu, 206-mm, 113×: An "easy white pair," both stars nearly alike, beside an "orange star to the southeast, forming a pretty sight together."
$23^h 58.1^m$	+24° 20′	Σ3048	2003	312°	8.9″	7.9	9.5	*G*5		Webb: "Yellow white, no color given. Companion decidedly blue [in] 1850." Grover called the companion "beautiful blue."
$23^h 58.4^m$	+35° 01′	OΣ 513	2002	19°	3.8″	6.8	9.3	*A*3	binary	Liu, 206-mm: A white star with a close companion that's "easy and distinct" at 264× but hard at lower powers.
$23^h 59.5^m$	+33° 43′	Σ3050	2004	334°	2.0″	6.5	6.7	*F*8V	p=320 yr	Liu, 206-mm: A "lovely pair" of perfect twins, "quite close but easy at 264×, and they look like two approaching headlights." Webb: "Yellowish."
$00^h 04.6^m$	+42° 06′	OΣ 514 Aa-B	2002	170°	5.2″	6.2	9.7	*B*9III		Ferguson, 150-mm: "Attractive pair, both [stars] blue." Harshaw, 200-mm: "Yellow, red."
$00^h 04.7^m$	+34° 16′	Σ3056 AB-C	2003	3°	25.9″	7.7	10.1			325-mm, 54×: Nice view. A bright and obvious wide pair, very lemony white with silvery white, in a little group of a dozen stars.
$00^h 04.9^m$	+45° 40′	β 997	2000	340°	4.1″	7.6	9.4	*F*8IV-V	binary	125-mm: An attractively close pair of white stars, split by a hair at 83×, just beyond the field of 22 Andromedae.
$00^h 08.0^m$	+31° 23′	OΣΣ 256	1999	114°	110.0″	7.1	7.3	*A*5		125-mm, 50×: A bright pair that pops out of a vacant background — the stars are super-wide apart, but both look exactly alike. They're whitish banana yellow.
$00^h 08.4^m$	+29° 05′	α And	2000	284°	89.3″	2.2	11.1	*B*8IV		The data suggest a brilliant star with a faint companion. Smyth: "A standard Greenwich star with a minute companion . . . bright white; purplish."
$00^h 10.0^m$	+46° 23′	Σ3	2001	82°	5.1″	7.8	9.1	*A*4V	binary	125-mm, 83×: A dim but easy close pair, easily noticed in the field with 22 Andromedae. The colors are peach white and silvery blue.
$00^h 18.5^m$	+26° 08′	Σ24	2003	248°	5.2″	7.8	8.4	*A*2	binary	125-mm, 83×: An easy and attractively close binary. Nearly equal yellow-white and blue-white stars, only a small gap apart. Webb simply calls it "a pretty pair."
$00^h 18.7^m$	+43° 47′	26 And	2002	240°	6.4″	6.0	9.7	*B*8V		The data suggest a bright star with a close dim companion. Ferguson, 150-mm: "Bright blue and faint blue. Interesting pair." Harshaw, 200-mm: "White, red."

Andromeda *(continued)*

RA	Dec.	Name	Year	P.A.	Sep.	m_1	m_2	Spec.	Status	Comments
00ʰ 20.9ᵐ	+32° 59′	AC 1	2001	288°	1.8″	7.3	8.3	*F*5V	binary	Hartung: "In a field with a few widely scattered fainter stars this deep yellow star has a bright orange-red gem . . . 105-mm resolves it clearly."
00ʰ 31.4ᵐ	+33° 35′	h 5451	1998	85°	55.9″	6.0	9.3	*K*1III		125-mm, 50×: A conspicuous star with a distant companion, about 1° west of Pi (π) Andromedae. A bright Sun-yellow star with a little milky dot beside it. Forms a triangle with Pi (π) Andromedae and Σ33.
00ʰ 35.2ᵐ	+36° 50′	Σ40	2003	312°	11.7″	6.7	8.5	*K*4III	optical	125-mm: A pretty pair, though not bright. It's a goldish pumpkin star and a dim bluish silver, which are wide apart at 83×.
00ʰ 36.9ᵐ	+33° 43′	π And	2003	173°	36.0″	4.3	7.1	*B*5V		60-mm, 25×: A fine pair for low power. A bright gloss-white star with a little speck beside it; they look wide apart but much closer than the listed separation. Webb: "Yellowish, blue."
00ʰ 40.3ᵐ	+24° 03'	Σ47	2003	205°	16.8″	7.3	8.8	*A*4III	optical	125-mm, 50×: A dim but pretty pair, easily found about 1½° west of Zeta (ζ) Andromedae. It's a whitish banana-yellow star with a faint powder blue, attractively close with fine contrast.
00ʰ 44.2ᵐ	+46° 14′	Σ52	1998	5°	1.4″	7.9	9.0	*F*6V	binary	The data suggest a close binary. Webb: "Yellowish, no color given."
00ʰ 46.4ᵐ	+30° 57′	Σ I 1	2003	46°	47.1″	7.3	7.4	*K*1III		125-mm, 83×: A conspicuous pair of wide bright twins, and their color is an indescribable mix of ivory, yellow, and peach. Webb: "Yellow. Curious similarity."
00ʰ 55.0ᵐ	+23° 38′	36 And	2004	314°	1.0″	6.1	6.5	*K*1IV	p=168 yr	Smyth: "A miniature of [Eta] η CBr. This beautiful golden pair is very difficult." Webb: "Beautiful; strong yellow."
00ʰ 56.8ᵐ	+38° 30′	μ And								The data suggest a very bright star with very faint companions. Webb: "A light test . . . Ward has seen [C] with 4.7-inch [119-mm] achromat." Smyth: "Bright white; dusky gray."
		AB	1999	297°	49.8″	3.9	12.9	*A*5V		
		AC	1999	147°	29.2″	3.9	11.4			
01ʰ 00.1ᵐ	+44° 43′	Σ79	2004	193°	7.8″	6.0	6.8	*B*9.5V *A*2V	binary	60-mm: A bright, conspicuous pair, easily noticed between Phi (φ) Andromedae and M31 (the Andromeda Galaxy). It's a pearly white star and a pale blue-violet, mildly unequal and split by a hair at 25×.
01ʰ 03.0ᵐ	+47° 23'	OΣ 21	2003	176°	1.2″	6.8	8.1	*A*9IV	p=450 yr	A 150-mm should be able to split this binary; the gap is widening.
01ʰ 07.9ᵐ	+46° 50′	β 397	1998	142°	8.7″	7.5	10.3	*K*0II-III		125-mm: A dim pair with a pretty primary, just beyond the field of Phi (φ) Andromedae. It's a yellowish peach star and a glimpse star just separated at 50x.
01ʰ 09.5ᵐ	+47° 15′	φ And	2004	124°	0.5″	4.6	5.6	*B*7V	p=391 yr	This binary should now be resolvable with 150-mm; the separation is slowly increasing.

RA	Dec.	Name	Year	P.A.	Sep.	m_1	m_2	Spec.	Status	Comments
$01^{h}17.0^{m}$	+38° 28′	Σ104	2001	323°	13.8″	8.0	9.8	*G5*	optical	Smyth: "About 20 minutes [of right ascension from Mu (μ) Andromedae] . . . is [this] very neat double star . . . its colours, pale yellow and green, are very decided."
$01^{h}17.7^{m}$	+44° 38′	h 1077	1998	293°	40.9″	7.4	9.4	*K5*		125-mm, 50×: A conspicuous dusty gold star next to a dim smoky dot; the closeness gives a nice contrast.
$01^{h}17.8^{m}$	+49° 01′	Σ102 AB-C	2003	223°	10.0″	7.2	8.8	*A*	binary	125-mm: A modestly bright pair that pops out of a dim field. The stars look whitish fire yellow and dim silver. They're attractively close at 50×.
$01^{h}18.8^{m}$	+37° 24′	Σ108	2001	63°	6.5″	6.5	9.6	*A5*		125-mm: A difficult but striking pair. A Sun-yellow star with a tiny glimpse star split by hairs at 83×.
$01^{h}27.7^{m}$	+45° 24′	ω And AC	2003	113°	92.8″	5.0	10.4	*F4V*		60-mm, 25×: A bright white star with a wide dim pinpoint. It's a very wide pair, but the stars look much closer than their listed separation; a pretty couple in a small scope.
$01^{h}49.3^{m}$	+47° 54′	Σ162 AB	2004	200°	1.9″	6.5	7.2		p. binary	The data suggest this double star (AB) is easy and attractively close. Webb: Bird calls the B companion "dark blue."
		AC	2003	178°	20.3″	5.8	9.4			
$01^{h}53.2^{m}$	+37° 19′	Σ179	2000	160°	3.8″	7.6	8.1	*F2V*	binary	125-mm: A dim but attractively close pair, just beyond the field of 56 Andromedae. A plain white star and a blue-white star split by a hair at 83×. Harshaw, 200-mm: "Very fine. White, red."
$01^{h}53.3^{m}$	+40° 44′	55 And	1913	5°	59.6″	5.4	10.9	*K1III*		Smyth: "Yellow; bluish . . . [primary] sometimes had a blurred aspect . . . companion was only caught by . . . evanescent glimpses, being a *minimum visible* for my telescope."
$01^{h}56.2^{m}$	+37° 15′	56 And	2001	299°	200.5″	5.8	6.1	*K0III M0*		60-mm, 25×: Interesting. A bright pair of stars that are superwide apart but identical in brightness — they look whitish scarlet and whitish lemon. Smyth: "Both [stars] yellow."
$01^{h}58.1^{m}$	+41° 23′	S404	2003	83°	28.1″	7.6	9.7	*G5*		125-mm, 50×: A dim but easy wide pair, centered in a conspicuous line of seven stars just west of Gamma (γ) Andromedae. Lemon-tinted white and milky azure. Harshaw, 200-mm: "White, deep blue."
$02^{h}03.9^{m}$	+42° 20′	γ And A-BC	2004	63°	9.7″	2.3	5.0	*K3II*	binary	Showcase pair. 60-mm, 35×: A brilliant star touching a smaller one, in the beautiful combination of vivid citrus orange and whitish deep blue. Smyth: "Orange colour; emerald green."
		BC	2000	103°	0.4″	5.1	6.3	*B8V+A0V*	p=63.7 yr	
$02^{h}10.9^{m}$	+39° 02′	59 And	2003	36°	16.5″	6.1	6.7	*B9V A1V*		60-mm: A fine sight for low power. It's a pearly white star and a peach white, attractively close at 25×.
$02^{h}14.0^{m}$	+47° 29′	Σ228	2004	286°	0.9″	6.6	7.2	*F2V F7V*	p=144 yr	A 125-mm should be able to split this fast-moving binary, but the separation will close to 0.5″ by 2020.
$02^{h}18.6^{m}$	+40° 17′	Σ245 Aa-B	2001	294°	11.4″	7.3	8.0	*F3V F3V*	optical	125-mm, 50×: Lovely combination. A pale pumpkin-orange star paired with a fainter bluish turquoise. They are attractively close but wide enough to be easy.

Andromeda *(continued)*

RA	Dec.	Name	Year	P.A.	Sep.	m_1	m_2	Spec.	Status	Comments
$02^h 21.6^m$	+44° 36′	Σ249	2001	197°	2.4″	7.2	9.0	*A2*	binary	This star is easily spotted about 1½° east-northeast of 60 Andromedae. Webb, listed under Perseus: "White, ash."
$02^h 22.8^m$	+41° 24′	βpm 30								Webb, referring to it as (Sigma) σ 70: "Yellow, no color given. Field fine with low powers."
		AB	1917	2°	56.3″	5.8	10.9	*F0III-IV*		
		AC	2002	9°	303.2″	5.8	7.4	*F0III-IV*		
$02^h 35.6^m$	+37° 19′	Σ279								The data and location suggest a bright primary star, probably pretty in color, about 1¼° northeast of 14 Trianguli. Webb, listed under Perseus: "Gold, no color given."
		AB	2002	71°	18.1″	5.9	10.9	*K3III*		
		AC	1998	208°	44.9″	5.9	11.7			

Antlia

RA	Dec.	Name	Year	P.A.	Sep.	m_1	m_2	Spec.	Status	Comments
$09^h 30.8^m$	−31° 53′	ζ^1 Ant	1999	212°	8.1″	6.1	6.8	*A1V A1V*	binary	Gould, 175-mm, 50×: Good view! $\zeta^1 - \zeta^2$ is a "bright pair of pale yellow stars." ζ^1 has a companion of its own, and there is "a gathering of stars, some 20′ N, that includes some wide pairs."
$09^h 33.2^m$	−36° 24′	h 4218	1999	29°	5.7″	7.6	9.8	*A1IV*	binary	Gould, 175-mm, 100×: "Quite good. This is the brightest star in a neat arc of four others. It has a faint, fairly close companion."
$09^h 36.1^m$	−31° 14′	h 4224	1999	118°	7.4″	7.8	8.2	*A4V*	binary	Gould, 175-mm, 100×: An "easy but not bright pair, yellow and pale yellow, in a faint, moderately starry field."
$09^h 48.8^m$	−35° 01′	h 4249	1999	122°	4.2″	8.0	8.1	*A9IV/V*	p. binary	Gould, 175-mm, 100×: "Quite nice. An even and moderately bright easy pair; the field is fairly starry but not bright."
$10^h 20.2^m$	−33° 08′	h 4304	1999	286°	9.4″	7.6	9.8	*A3III*	binary	Gould, 175-mm, 100×: "Quite good." An "easy uneven pair, with a fairly bright cream-white primary. About 15′ west and south is a little grouping of moderate and faint stars."
$10^h 29.6^m$	−30° 36′	δ Ant	1999	226°	10.9″	5.6	9.8	*B9.5V*		Gould, 175-mm, 100×: This "bright pale yellow star dominates a moderately starry field; the companion is much less bright . . . but not difficult." Hartung: "Pale yellow and ashy."
$10^h 54.6^m$	−38° 45′	h 4381	1999	42°	25.8″	7.0	8.5	*B8III*		The data suggest an easily noticed pair 1.7° south-southwest of Iota (ι) Antliae. Gould, 175-mm, 100×: A "wide and moderately bright uneven pair, in a sparse field."

Apus

RA	Dec.	Name	Year	P.A.	Sep.	m_1	m_2	Spec.	Status	Comments
$14^h 22.6^m$	−73° 33′	h 4667	1991	138°	2.3″	8.2	8.6	*A0V*	binary	Jaworski, 100-mm: A "challenging tight pair, but worth the effort of tracking down. With 167× under steady seeing, it is a pair of white stars, only slightly contrasting in brightness."
$14^h 29.5^m$	−80° 06′	h 4671	2000	125°	4.2″	8.0	8.7	*F5V*	p. binary	Jaworski, 100-mm: A "fine better-than-average pair. It is a tight double . . . well shown at 167×, with a slight contrast in brightness. The stars are yellow white with bluish."

Apus *(continued)*

RA	Dec.	Name	Year	P.A.	Sep.	m_1	m_2	Spec.	Status	Comments
14h53.2m	−73° 11′	I 236	1991	123°	2.2″	5.9	7.6	*G*5III	p. binary	Hartung: "In a field sown profusely with small stars, this bright deep yellow star has a less bright companion very close [to it], quite clearly shown with 75-mm."
15h06.4m	−72° 10′	CapO 15	1991	42°	1.4″	7.2	8.5	*B*8V	binary	Hartung: "This fairly bright pale yellow star has a white companion very close [to it] which, though difficult . . . is quite clear with 105-mm."
15h59.5m	−71° 07′	BrsO 21 AC	2000	319°	37.1″	7.9	8.7	*G*5		Jaworski, 100-mm: An "attractive wide pair easily separated at 50× . . . with a pretty color contrast. It is a light yellow star with a deep blue companion."
16h20.3m	−78° 42′	δ Aps	2000	10°	103″	4.9	5.4	*M*5III		Gould: A "fairly good, easy wide pair . . . in a field only moderately starry." Both stars have an "orange-yellow mustard tone."
17h09.9m	−82° 19′	h 4884	2000	5°	34.8″	7.2	8.9	*B*6IV		Jaworski, 100-mm, 50×: An "easy wide pair, well shown with 50×. The stars are white and blue, with a moderate contrast in brightness."
17h10.3m	−75° 23′	h 4904	2000	187°	6.8″	7.6	9.1	*F*3V	binary	Jaworski, 100-mm: A "fine close pair . . . well seen with 80×. It is a yellow-white star with a dim companion, possibly bluish tinged."
17h44.3m	−72° 13′	HdO 275	1991	170°	0.4″	6.8	8.1	*F*7IV+*F*5V	p=101 yr	A 300-mm should probably resolve this fast-moving binary until at least 2025.
18h10.6m	−75° 11′	h 4999	2000	171°	13.2″	7.7	8.6	*B*9V *A*0/1	optical	This pair is in a broad arc of four conspicuous stars. Jaworski, 100-mm: A "fine . . . moderately wide pair that is well shown at 80×. The stars are white and bluish white."

Aquarius

RA	Dec.	Name	Year	P.A.	Sep.	m_1	m_2	Spec.	Status	Comments
20h51.4m	−05° 38′	4 Aqr	2002	21°	0.9″	6.4	7.4	*F*5IV-V	p=194 yr	Smyth: "A binary star . . . [that is] the middle one of three stars pretty close together [referring to 3, 4, and 5 Aquarii] . . . pale yellow, purple." Webb: "Yellow."
21h03.1m	+01° 32′	Σ2744	2003	117°	1.3″	6.8	7.3	*F*7IV	p=1,532 yr	125-mm: An easy, fairly bright, and attractively close binary. It's a pair of very lemony white stars, nearly equal, that are just apart with 200×.
21h04.1m	−05° 49′	12 Aqr	2003	196°	2.5″	5.8	7.5	*G*4III	binary	125-mm, 200×: Striking closeness and contrast — a dimmer Epsilon (ε) Bootis clone. A bright yellow star touched by a little globe of ocean blue.
21h21.8m	+02° 02′	Σ2787	2002	20°	22.7″	7.5	8.6	*A*2		125-mm, 50×: A dim pair with fine colors, wide in the field with a broad stream of conspicuous stars. It's a pure white star and a very blue star, quite widely apart.
21h23.8m	−06° 35′	S 788	2002	93°	55.3″	7.7	8.3	*F*8		60-mm, 25×: A dim but easy wide pair, easily noticed about 2° southwest of Beta (β) Aquarii. Both stars look pure white and nearly alike.

Aquarius *(continued)*

RA	Dec.	Name	Year	P.A.	Sep.	m_1	m_2	Spec.	Status	Comments
$21^h 31.6^m$	−05° 34′	β Aqr	1998	319°	36.9″	2.9	11.0	G0Ib		The data suggest a brilliant star with a faint companion. Webb: Ward calls it a "test for a 4.5- to 5-inch [114- to 127-mm] achromat." Smyth: "Pale yellow; blue."
$21^h 37.6^m$	−00° 23′	Σ2809	2003	163°	30.9″	6.2	9.4	A2V		This pair is about 1° east-northeast of globular cluster M2. Gould, 175-mm, 64×: This "bright pale yellow star has an easy wide companion. The field has a modest scatter of stars."
$21^h 54.6^m$	−03° 18′	Σ2838	2003	182°	16.1″	6.3	9.5	F5IV	binary	125-mm, 50×: An easy pair with lovely colors. It's a bright peach-orange star and a very bluish white star, modestly wide apart; the little companion is dim but easily seen.
$22^h 02.4^m$	−16° 58′	29 Aqr	1998	248°	3.5″	7.2	7.1	A2V+K0III		60-mm: Grand sight! A pair of perfect twins that seemed to shiver in different shades of yellow and orange, split by a hair at 42×. Hartung: "Yellow and white."
$22^h 07.1^m$	+00° 34′	Σ2862	1997	97°	2.5″	8.0	8.4	G0	binary	125-mm: An easy and attractively close binary. It's a pair of whitish stars that are nearly alike and almost, but not quite, fully apart at 83×. Webb: "Yellowish, yellow."
$22^h 14.3^m$	−21° 04′	41 Aqr	2002	112°	5.1″	5.6	6.7	K0III+F2V	p. binary	Showcase pair. 60-mm, 42×: This is a bright star almost touching an elusive little star, in the beautiful combination of yellowish peach and pale violet. Webb: "Reddish, blue."
$22^h 25.8^m$	−20° 14′	S 808	2002	153°	6.9″	7.1	8.0	G5III		125-mm, 50×: Good combination. A straw-yellow star paired with a fainter whitish green. This couple is attractively close while wide enough to be easy. About 1° northwest of the Helix Nebula.
$22^h 26.6^m$	−16° 45′	53 Aqr	2004	19°	1.5″	6.3	6.4	G0V G0V	p=3,500 yr	This is a beautifully tight binary, almost as bright as a showcase pair. It's a pair of lemony white stars, nearly alike — 60-mm shows a crisp rod, and 125-mm shows two hairsplit stars.
$22^h 28.8^m$	−00° 01′	ζ Aqr	2004	178°	2.0″	4.3	4.5	F3IV-V	p=587 yr	Showcase binary. 60-mm, 120×: A bright pair of identical stars in a figure-8; they have the pretty color of whitish citrus orange.
$22^h 30.5^m$	−08° 07′	Σ2913	2004	329°	8.6″	7.8	8.6	F0V	binary	125-mm: This binary is attractively close but easily apart at 50×; it pops out of a field without other bright stars. The stars are mildly unequal but both look yellow-tinted white.
$22^h 30.6^m$	−10° 41′	σ Aqr	1991	10°	3.7″	4.8	8.5	A0I		Gould, 175-mm: A "bright pale yellow star with two lesser [field] stars near it. At 180×, there is a hint of its [dim] close companion." Ferguson, 150-mm: "Bluish white and faint deep blue."
$22^h 38.6^m$	−14° 04′	h 5355								125-mm, 50×: Neat asterism! The stars of this broad triple make a nearly perfect triangle, and each looks sightly different in color — one is pure white, one beige, and one reddish silver.
		AB	2001	290°	82.0″	7.5	8.8	A5II/III		
		AC	2004	0°	107.6″	7.5	9.4			

Aquarius *(continued)*

RA	Dec.	Name	Year	P.A.	Sep.	m_1	m_2	Spec.	Status	Comments
$22^h 43.1^m$	−08° 19′	Σ2935	2003	307°	2.4″	6.8	7.9	*A5V*	binary	125-mm, 200×: A difficult but striking binary. It's a bright, very lemony white star with a speck of light at its edge. The companion was barely glimpsed.
$22^h 47.7^m$	−14° 03′	69 Aqr	2004	126°	21.7″	5.7	9.6	*A0V*		60-mm: A striking pair that's probably close to the limits for 60-mm. It's a very lemony white star paired with the faintest presence of light still possible to see. A fairly close pair at 25×.
$22^h 47.8^m$	−04° 14′	Σ2944								60-mm, 25×: The AC couple is a wide and pretty pair, beige white and dim arctic blue in color. 125-mm: The primary star becomes a pair of stars, split by a hair at 200×.
		AB	2004	297°	2.0″	7.3	7.7	*G2V G4*	p. binary	
		AC	2001	90°	58.5″	7.3	8.6	*G2*		
$23^h 12.0^m$	−11° 56′	Σ2988	1998	277°	2.6″	7.9	7.9	*G8III*		60-mm: A dim but attractively close pair. It's a pair of identical stars of uncertain color, split by a hair at 42×. Webb calls them yellowish.
$23^h 14.1^m$	−08° 55′	Σ2993	2002	176°	25.0″	7.6	8.2	*F8V G2V*		125-mm, 50×: This wide and easy but ordinary pair is easily noticed in the field with Psi[1] (ψ^1) Aquarii, which is about 1/2° to the east-southeast. It's a gloss-white star and a milky white; the stars look nearly equal.
$23^h 15.9^m$	−09° 05′	ψ^1 Aqr								The data suggest a very bright star with a wide companion. Webb: "Very yellow, blue." Smyth: This pair is "the first of three similar stars [referring to ψ^1, ψ^2, and ψ^3] . . . orange tint; sky blue."
		A-BC	2002	313°	49.4″	4.4	9.9	*K1III*		
$23^h 19.1^m$	−13° 28′	94 Aqr	2004	351°	12.3″	5.3	7.0	*G5IV*		Showcase pair. 60-mm, 25×: A whitish banana-yellow star almost touching a smaller globe of nebulous light. Webb: "Yellow, with perhaps reddish glare [and] greenish."
$23^h 20.9^m$	−18° 33′	h 3184	1998	282°	5.1″	7.3	8.4	*G8IV*	binary	125-mm, 83×: Splendid combination. A yellowish peach star and a little whitish green star, attractively close while wide enough to be easy.
$23^h 23.8^m$	−08° 28′	Σ3008	2004	151°	6.3″	7.2	7.7	*K0III-IV*	p. binary	125-mm: This pair is very bright, easy, and attractively close at 50×. It's a yellow-tinted white star paired with a plain white star.
$23^h 37.7^m$	−13° 04′	h 316	2003	98°	30.1″	5.7	9.6	*G9III*		The data suggest a bright primary star that is probably pretty colored with a wide dim companion.
$23^h 46.0^m$	−18° 41′	107 Aqr	2003	135°	6.8″	5.7	6.5	*A9IV F2V*	p. binary	Showcase pair. 60-mm: A bright pair of pure white stars, very nearly alike, that are almost but not quite apart at 25×. Webb: "White, or yellowish" and "blue." Hartung: "Pale and deep yellow."

Aquila

RA	Dec.	Name	Year	P.A.	Sep.	m_1	m_2	Spec.	Status	Comments
$18^h 46.5^m$	−00° 58′	5 Aql								60-mm, 25×: Grand! A bright pair of stars, whitish straw yellow and pure sapphire. They are attractively close while wide enough to be easy. BC was not seen.
		AB	2003	121°	12.5″	5.9	7.0	*A2V*	optical	
		BC	1999	171°	13.8″	7.9	11.3			

Aquila *(continued)*

RA	Dec.	Name	Year	P.A.	Sep.	m_1	m_2	Spec.	Status	Comments
$18^h 50.8^m$	+10° 59′	Σ 2404	2003	182°	3.6″	6.9	7.8		binary	Hartung: "In a field spangled with stars . . . this unequal orange pair shines like gems, making a fine sight."
$18^h 54.5^m$	+01° 54′	ΟΣΣ 176	2004	113°	94.8″	7.5	7.5	*F*5		60-mm, 25×: These fairly bright stars are super-wide apart but identical in brightness and color. They look greenish white.
$18^h 59.1^m$	+13° 38′	11 Aql	2003	300°	19.6″	5.3	9.3	*F*6IV		125-mm, 50×: A bright, colorful pair in a field dense with stars. It's a bright Sun-yellow star next to a dim ashy blue. These stars are very wide apart but look like a couple.
$19^h 00.0^m$	+12° 53′	Σ2426	2002	260°	16.8″	7.5	9.0	*A*5V *K*2III	optical	This pair is easily spotted in a low-power field with 11 Aquilae (above), which is about 3/4° to the north-northwest. Webb: "Reddish yellow, grey."
$19^h 00.6^m$	−08° 07′	Σ2425	2002	178°	29.6″	7.9	8.6	*G*5		Webb: "Yellowish, ash."
$19^h 05.0^m$	−04° 02′	15 Aql	2003	210°	39.1″	5.5	7.0	*K*0IV		Showcase pair. 60-mm, 25×: Grand! A bright wide pair with beautiful colors — an amber-yellow star and a bluish turquoise. Webb: "White, or yellow white [and] red lilac."
$19^h 05.8^m$	+06° 33′	Σ2446	2003	153°	9.4″	7.0	8.9	*F*5	binary	125-mm, 50×: An attractively close pair with pretty colors. It's a bright lemon star and a dim greenish sapphire with just a small gap between them. 19 Aquilae is about 1° east-southeast.
$19^h 06.4^m$	+07° 09′	Σ2449	2002	290°	8.1″	7.2	7.7	*F*2V	binary	60-mm, 25×: A modestly dim but easy pair, with attractive closeness. It's a lemon-white star and a silvery white star, nearly equal, that are split by a hair at 25×.
$19^h 06.6^m$	−01° 21′	Σ2447	2003	343°	13.9″	6.8	9.6	*B*5V		125-mm, 50×: A pretty pair for its contrast. It's a bright white star and a small blue dot, fairly wide apart.
$19^h 16.4^m$	+14° 33′	Σ2489	2002	347°	8.2″	5.7	9.3	*B*9.5V		125-mm, 50×: A dimmer Beta (β) Cephei clone. A bright white star almost touched by a little globe of silvery blue, and both stars are seen vividly despite their great contrast.
$19^h 18.1^m$	−05° 25′	h 881								This pair is in the low-power field with 26 Aquilae, about 30′ to its southwest. Webb: "Orange and two blue."
		AB	2003	340°	32.6″	7.8	9.8	*K*3II		
		BC	1999	307°	7.3″	10.1	10.1			
$19^h 18.5^m$	+01° 05′	23 Aql	1997	4°	3.0″	5.3	8.3	*G*9III	binary	Webb: "Yellow, blue. Herschel and Smyth found the companion increasingly visible with higher powers, which struck me independently." Smyth: "Light orange; grey."
$19^h 18.8^m$	+00° 20′	24 Aql	2001	317°	427″	6.5	6.8			60-mm, 25×: This is a pair of bright twins, super-wide apart, in a field without other bright stars. They make a fine couple, goldish white in color.
$19^h 19.7^m$	+12° 22′	28 Aql	2002	176°	59.7″	5.5	9.0	*F*0III		125-mm, 50×: Pretty colors! A bright straw-yellow star beside a dim ocean blue. These stars are very wide apart, but they look like a couple.

Aquila *(continued)*

RA	Dec.	Name	Year	P.A.	Sep.	m_1	m_2	Spec.	Status	Comments
$19^h 20.0^m$	+05° 35′	Σ2497	2002	356°	30.0″	7.7	8.5	G5		125-mm, 50×: Nice effect — this ordinary wide pair, lemon white and pearly white in color, form a neat triangle with a dimmer third star.
$19^h 20.0^m$	−06° 38′	Σ2494	2003	89°	25.6″	7.0	10.4	K0		Webb: "Ruddy orange, blue. Fine colours."
$19^h 29.5^m$	+03° 05′	Σ2531								At low power, these two pairs will be in the same field. Webb calls Σ2531 "a pretty pair, very unequal" and calls Σ2532 "orange and green" with a companion "readily visible."
		AC	1999	29°	31.4″	8.1	10.2			
$19^h 30.2^m$	+02° 54′	Σ2532	1983	3°	32.7″	6.1	10.6	M1III		
$19^h 30.1^m$	−00° 27′	Σ2533	2002	212°	21.7″	7.4	9.2	A3V		125-mm, 50×: An easy wide pair, easily noticed about halfway between Nu (ν)and Iota (ι) Aquilae. It's a bright gloss-white star and a milky bluish olive. The companion is dim but easily seen.
$19^h 36.8^m$	−10° 27′	Σ2541	1999	326°	5.4″	8.7	10.1	K2V	optical	A low-power field, just northeast of 37 Aquilae, will show all three primary stars. 125-mm: A hairsplit pair (Σ2545), a slightly wider pair (Σ2541), and a broad triple that form a straight line (Σ2547).
$19^h 38.7^m$	−10° 09′	Σ2545	2003	326°	3.8″	6.8	8.5	A9III	binary	
$19^h 38.9^m$	−10° 20′	Σ2547								
		AB	2003	331°	20.7″	8.1	9.5	A0		
		AC	1999	142°	49.6″	8.1	11.1			
$19^h 42.8^m$	+08° 23′	Σ2562	2002	251°	27.1″	6.9	8.7	F8V G0V		125-mm, 50×: A conspicuous white pair, wide and unequal, that's inside dark cloud B339. It's just beyond the field of Altair, which is 2° to the east-northeast.
$19^h 48.7^m$	+11° 49′	π Aql	2003	106°	1.4″	6.3	6.8	A3V F9III	binary	60-mm, 120×: A difficult pair for 60-mm, but beautifully tight. It's a pair of lemon-white stars in a small figure-8 — one is a little brighter than the other. Smyth: "A miniature (Alpha [α] Geminorum) Castor."
$19^h 50.8^m$	+08° 52′	α Aql	2003	287°	192″	0.9	9.8	A7V		*(Altair)* This is a brilliant star with a companion. Smyth: "Pale yellow, violet tint."
$19^h 51.4^m$	+04° 05′	Σ2587	2003	101°	4.3″	6.7	9.4	A5V K2III	p. binary	Hartung: "A very pretty pair in a field sown with stars, and an easy object for small apertures."
$19^h 52.3^m$	+10° 21′	Σ2590	2003	308°	13.4″	6.5	10.3	B7V		Webb: "Pale orange, blue . . . [a] light test for moderate apertures," and Sadler calls it a "very white pair."
$19^h 54.0^m$	+15° 18′	Σ2596	2004	301°	2.0″	7.3	8.7	F8V	binary	Webb calls this binary "yellowish, ash."
$19^h 54.6^m$	−08° 14′	57 Aql	2004	171°	35.9″	5.7	6.3	B7V B8V		Showcase pair. 60-mm, 25×: A wide pair of bright stars, mildly unequal, that are pure white and scarlet-tinted white. Webb calls them "pale yellow, pale lilac" and "distinctly contrasted."
$19^h 55.3^m$	−06° 44′	Σ2597	2000	104°	0.5″	6.9	8.0	F2V	binary	Hartung calls this binary a "yellow pair," which "I have found difficult."
$20^h 01.4^m$	+10° 45′	Σ2613	2004	354°	3.7″	7.5	8.0	F5V F5V	p=2,356 yr	125-mm: Grand sight at 50×— a bright pair of touching stars. They are deep white with a vibrant sparkle.

Aquila *(continued)*

RA	Dec.	Name	Year	P.A.	Sep.	m_1	m_2	Spec.	Status	Comments
20h 01.7m	−00° 12′	H I 93	1997	299°	1.8″	7.7	8.4	*A0*	binary	125-mm: A dim but attractively close binary. It's a pair of yellow-white stars, just kissing at 125×. One is slightly brighter than the other.
20h 02.8m	+14° 35′	Σ2616	2003	266°	3.5″	6.8	9.6	*K0*	binary	Hartung: "A fine pair in a field sown with stars, quite clearly seen with 75-mm."
20h 06.6m	+07° 35′	OΣ 198	2003	185°	65.2″	7.1	7.6	*A2V A3V*		125-mm, 50×: A bright wide pair of stars, lemon white and azure white, with bright yellow Tau (τ) Aquilae also in the field.
20h 07.8m	+09° 24′	Σ2628	2003	340°	3.3″	6.6	8.7	*F3V*	binary	The data suggest a fairly bright star with a close companion. Webb: "Very white, purple." Hartung: "Both stars are yellow and small apertures show them well."
20h 12.6m	+00° 52′	Σ2644	2004	208°	2.6″	6.9	7.1	*B9*	binary	60-mm: Grand sight! A bright pair of gloss-white stars, exactly alike, locked in a deep kiss at 120x. Hartung: It lies "in a black field."
20h 12.8m	−03° 00′	Σ2643	1998	80°	3.0″	7.1	9.4	*A0V*	binary	The data suggest a close, unequal pair for a modest aperture, but the companion was not seen with 125-mm on a clear night.
20h 14.2m	+06° 35′	S 740	2002	193°	43.5″	7.8	8.1	*G4IV*		60-mm, 25×: A wide easy pair with a lovely color. It's a pair of grapefruit-orange stars, exactly alike, that look like a pair of eyes in a dark field. Among the best cataloged by James South.
20h 14.4m	−06° 03′	Σ2646	2003	40°	18.4″	7.5	9.3	*F0*		125-mm, 50×: An ordinary but conspicuous wide pair in a blank field. It's a lemon-white star and a smaller tan-pink; the companion is dim but visible at a glance.
20h 15.2m	−03° 30′	Σ2654	2003	233°	14.3″	7.0	8.1	*F2V F4V*	optical	125-mm, 50×: Nice effect. This easy pair seems to jump out of a black background, and the stars look attractively close while wide enough to be easy. Bright gold and pearly white.
20h 19.9m	−02° 15′	Σ2661	1998	340°	24.4″	7.9	9.2	*A0*		125-mm, 50×: A pretty view — an easy wide pair in a neat curved line of four field stars. Very lemony white star and milky silver. Webb: Franks calls the companion "lilac."
20h 24.6m	+01° 04′	Σ2677	2000	30°	31.4″	6.1	10.6	*B9V*		The data suggest a bright star with a dim companion. Webb: The companion was "steady [seen] with 3.7-inch [94-mm], a good test." Smyth, listed under Antinous: "White; grey."
20h 27.5m	−02° 06′	S 749	2002	189°	59.7″	6.8	7.5			60-mm, 25×: Grand! A bright wide pair with lovely colors, just northwest of 69 Aquilae. It's a lemon-yellow star and a subtle tangerine, super-wide apart. Very slightly unequal.
20h 33.4m	−06° 13′	h 1529	1996	110°	32.0″	7.4	9.6	*M3III*		125-mm, 50×: This pair shows beautiful contrast. It's a bright red-orange star with a little speck beside it. These stars look wide but much closer than their listed separation.

Ara

RA	Dec.	Name	Year	P.A.	Sep.	m_1	m_2	Spec.	Status	Comments
16ʰ39.7ᵐ	−57° 00′	R Ara	1999	120°	3.3″	7.2	7.8	*B9IV/V*		Hartung: An "elegant yellow close pair," next to a red star, "in a field sown profusely with stars."
16ʰ39.9ᵐ	−47° 47′	Slr 12	1991	159°	1.4″	8.0	8.1	*F5IV/V*	binary	Gould, 175-mm: "CorO 198 is the easiest of three doubles in the same field . . . it's a nice little pair, though not bright, and just split at 100×." The other two pairs were also resolved.
16ʰ40.7ᵐ	−47° 40′	CorO 198	1991	99°	2.5″	8.6	9.4	*A*		
16ʰ41.1ᵐ	−47° 45′	Slr 21	1991	321°	1.7″	7.4	9.4	*B5III*	binary	
16ʰ41.3ᵐ	−48° 46′	Δ 206								This bright pair is part of the little star cluster NGC 6193. Jaworski, 100-mm: A "beautiful close pair set in a rich field of stars, well shown at 80×. The stars are yellowish white and light blue."
		AC	2002	265°	9.5″	5.7	6.8	*O6.5V*		
16ʰ47.5ᵐ	−48° 19′	Δ 211								Gould, 175-mm, 100×: This multiple is "three fairly bright stars in a bent line, whose third star (CD) is really a close pair of stars. The brightest star is deep yellow."
		AB	1999	125°	106.4″	7.4	8.1	*M0III*		
		BC	1999	193°	45.3″	8.1	8.2	*F2/3IV/V*		
		h 4885								
		CD	1999	240°	3.5″	8.4	9.7			
16ʰ50.6ᵐ	−50° 03′	CorO 201	1991	42°	3.1″	7.2	7.3	*A5III*	binary	Hartung: A "dainty equal pair, cleanly shown with 75-mm."
16ʰ54.0ᵐ	−46° 55′	h 4890	1999	321°	30.4″	7.9	8.1	*B3/4III*		Gould, 175-mm, 100×: An "obvious wide white pair in a rather bare patch of sky."
16ʰ56.2ᵐ	−46° 51′	h 4896	1999	23°	3.9″	7.8	8.0	*B7V*	binary	Gould, 175-mm, 100×: A "fairly close but easy uneven pair, in a field of scattered stars."
17ʰ00.4ᵐ	−48° 39′	See 316	1991	173°	1.0″	6.3	7.7	*G8/K0III+G*	binary	Gould, 175-mm: A "bright deep yellow star in a thin field, that becomes elongated with 230×." Hartung calls it "two star disks clearly apart with 150-mm."
17ʰ02.9ᵐ	−50° 10′	CorO 206	1999	233°	7.9″	7.3	8.1	*A0V*	binary	Gould, 175-mm, 100×: An "easy pair, slightly uneven, with two fainter wide companions. The field is moderately starry."
17ʰ10.3ᵐ	−46° 44′	Δ 213	2000	168°	8.2″	7.0	8.3	*B1I B1.5V*	binary	Gould, 175-mm, 100×: An "easy, fairly bright and attractive pair. The field is sparse near the pair, with more stars out wide."
17ʰ13.0ᵐ	−58° 36′	h 4920	1991	323°	3.0″	7.0	9.2	*F4IV-V*	binary	Jaworski, 100-mm, 167×: A "superb tight double with a marked contrast in brightness. It is a yellowish white star with a companion of uncertain color, and is well shown at 167×."
17ʰ13.3ᵐ	−67° 12′	Δ 214	2000	14°	37.4″	6.0	8.8	*K0-1III*		This pair is about 1° northwest of Zeta (ζ) Apodis. Gould, 175-mm, 100×: This "bright orange star has a wide companion, blue in contrast; two faint stars are wide nearby."
17ʰ19.1ᵐ	−46° 38′	BrsO 13	1999	253°	8.7″	5.6	8.9	*G8V M0V*		Gould, 175-mm,100×: "Fine object. A very attractive, fairly close pair, bright yellow and dull orange-brown in color. There is a faint scattered asterism 10′ to the east."
17ʰ22.8ᵐ	−58° 28′	CorO 213	2000	284°	9.2″	6.9	9.3	*A5V*	binary	Gould, 175-mm, 100×: This "easy uneven white pair stands out in a middling starry field."

Ara *(continued)*

RA	Dec.	Name	Year	P.A.	Sep.	m_1	m_2	Spec.	Status	Comments
$17^h 25.4^m$	−56° 23′	γ Ara	2000	327°	18.4″	3.3	10.2	*B*1I		Hartung: "105-mm will just show the companion of this brilliant white star, which dominates a well sprinkled field."
$17^h 26.9^m$	−45° 51′	h 4949								Showcase pair. Gould, 175-mm, 100×: A "beautiful tight bright pair . . . with a very wide third companion and two faint companions wide. White."
		AB	1991	253°	2.1″	5.6	6.5	*B*7V+*B*9.5V	p. binary	
		Δ216								
		AC	1999	312°	102.4″	5.6	7.1	*A*0V		
$17^h 31.8^m$	−46° 02′	I 40	1999	209°	20.1″	6.0	10.5	*F*8-*G*0I		Hartung: A "bright yellow star . . . [with] a wide companion, which 105-mm will show with care." A "less bright orange star [is in the field], 4′ to the southeast."
$17^h 41.5^m$	−53° 28′	Pol 4	1999	296°	10.8″	7.8	10.4	*A*8III		Gould, 175-mm, 100×: An "easy uneven pair, 15″ from globular NGC 6397."
$17^h 42.2^m$	−48° 39′	h 4970								Gould, 175-mm, 100×: "Good combination though not bright. It is a neat easy pair of stars and two less bright stars [that form a broad quadruple]."
		AB	1999	69°	8.0″	8.0	9.1	*F*5V	binary	
		AC	1999	234°	20.2″	8.0	10.5			
$17^h 45.1^m$	−54° 08′	R 303	1999	101°	2.7″	8.0	9.1	*A*		Gould, 175-mm, 100×: This "uneven fairly close pair is not bright, but it is fairly attractive in its field of faint stars."
$17^h 50.5^m$	−53° 37′	h 4978	2000	268°	12.4″	5.7	9.2	*B*2V+*B*3V		Gould, 175-mm, 100×: "An easy uneven pair with a nice contrast in brightness, in a thin field."
$17^h 57.2^m$	−55° 23′	Rmk 22	1991	95°	2.4″	7.0	7.9	*K*0III+*A*2	binary	Jaworski, 100-mm: A "grand tight pair of stars . . . clearly split with 167×." Gould, 175-mm, 100×: A "tight little yellow pair."
$18^h 08.5^m$	−45° 46′	h 5015	1999	259°	3.5″	6.2	9.6	*B*7-8II	binary	Gould, 175-mm: A "bright white star with a tiny close companion, in a pleasant field of mixed stars [about 1/2° west-northwest of Epsilon (ε) Telescopii]. Easy at 180×."

Aries

RA	Dec.	Name	Year	P.A.	Sep.	m_1	m_2	Spec.	Status	Comments
$01^h 50.1^m$	+22° 17′	1 Ari	2003	165°	2.9″	6.3	7.2	*G*3III	binary	125-mm: A lovely pair of stars! They're bright Sun yellow and pale ocean blue and are split by a hair at 83×. Webb: "Gold, very blue." Smyth: "Topaz yellow, smalt blue."
$01^h 52.0^m$	+10° 49′	Σ178	1999	204°	3.4″	8.0	8.0	*F*1V		125-mm, 83×: A miniature Gamma (γ) Arietis! It's a pair of identical stars, very white in color, that are attractively close while wide enough to be easy. Smyth: "Both [stars] lucid white."
$01^h 53.5^m$	+19° 18′	γ Ari	2004	0°	7.5″	4.5	4.6	*A*1 *B*9V	p. binary	Showcase double called the "Ram's Eyes." 60-mm, 25×: A pair of identical stars, radiantly white and brilliant, that are attractively close without being hard.
$01^h 57.9^m$	+23° 36′	λ Ari	2003	47°	36.7″	4.8	6.7	*F*0IV *F*7V		Showcase pair. 60-mm, 25×: Very nice contrast. A bright white star with a little silvery companion, wide apart. Webb: "Yellow, greenish." Smyth: "Yellowish white; blue."

RA	Dec.	Name	Year	P.A.	Sep.	m_1	m_2	Spec.	Status	Comments
$01^{h}59.3^{m}$	+24° 50′	Σ194	1999	276°	1.2″	7.6	9.5	*A3*	binary	This binary is about 1¼° north-northeast of Lambda (λ) Arietis, and Webb included it in his catalog. He called the stars "yellow white."
$02^{h}03.7^{m}$	+25° 56′	10 Ari	2003	336°	1.2″	5.8	7.9	*F8IV*	p=325 yr	Smyth: "A close double star . . . yellow; pale gray . . . so beautiful an object that William Herschel calls it a miniature of ε Boo [Epsilon Bootis]." Webb: "Yellow, ash."
$02^{h}09.4^{m}$	+25° 56′	14 Ari								60-mm, 25×: Grand! A bright yellow star with a super-wide little companion. The pair jumps out at you in a barren field. Smyth: "A wide triple . . . white; blue; lilac."
		AB	1983	34°	93.2″	5.0	8.0	*F2III*		
		AC	2003	278°	106.7″	5.0	8.0			
$02^{h}09.7^{m}$	+20° 21′	Σ221	2003	145°	8.4″	8.0	9.5	*A8IV*	binary	125-mm, 83×: This pair is easily noticed in a blank field, about 1° north-northwest of 15 Arietis. It's an amber-yellow star and a little silvery star, wide apart.
$02^{h}26.8^{m}$	+10° 34′	OΣΣ 27	2001	32°	72.9″	6.7	8.3	*A3*		125-mm, 50×: Nice view — a bright, conspicuous pair in the field with Xi (ξ) Arietis. Its stars are whitish lemon and beige white, and they're super-wide apart.
$02^{h}37.0^{m}$	+24° 39′	30 Ari	2004	274°	38.7″	6.5	7.0	*F5V F7V*		60-mm, 25×: Nice! This is a fairly bright pair of stars in the lovely combination of Sun yellow and blue-white. They're wide apart and nearly equal.
$02^{h}40.7^{m}$	+27° 04′	33 Ari	2002	2°	28.6″	5.3	9.6	*A3V*		Smyth calls 33 Arietis "a fine double star . . . pale topaz; light blue." Webb notes that the binary OΣ 43 is easily spotted in the same field with it, 27′ to the south.
$02^{h}40.7^{m}$	+26° 37′	OΣ 43	2004	352°	0.7″	7.9	9.0	*F7V*	p=289 yr	
$02^{h}41.1^{m}$	+18° 48′	Σ291								125-mm, 83×: Nice effect. A cone shape formed by kissing white twins and a dim third member.
		AB	2000	118°	3.1″	7.7	7.5	*B9.5V*	binary	
		AC	2000	242°	65.7″	7.7	9.5	*A8V*		
$02^{h}44.6^{m}$	+29° 28′	Σ300	1998	314°	3.2″	7.9	8.1	*F0IV*	binary	125-mm. 83×: Good view. A tight pair of bluish white stars, split by only a small gap, in the same field with amber-yellow 39 Arietis. They're very slightly unequal.
$02^{h}47.5^{m}$	+19° 22′	Σ305	2003	308°	3.6″	7.5	8.3	*G0V*	p=720 yr	125-mm, 83×: An attractively close pair of mildly colored stars. They're modestly unequal but both look peach white; they're split by only a small gap.
$02^{h}49.3^{m}$	+17° 28′	π Ari								125-mm: Striking contrast! A bright yellow-white star with a little glimpse star beside it, split by a modest gap. B not seen. Smyth: "Pale yellow; flushed; dusky . . . superb trio."
		AB	2002	118°	3.5″	5.3	8.0	*B6V*	binary	
		AC	2000	109°	25.1″	5.3	10.7			
$02^{h}50.0^{m}$	+27° 16′	41 Ari								The data suggest a very bright star with three faint companions. Smyth: "A coarse quadruple star . . . white; deep blue; lurid; pale gray."
		Aa-B	1997	292°	32.1″	3.6	10.8	*B8V*		
		Aa-C	1997	231°	27.9″	3.6	10.6	*B8*		
		Aa-D	2002	236°	122.5″	3.6	8.8	*B8*		
$02^{h}59.2^{m}$	+21° 20′	ε Ari	2004	208°	1.4″	5.2	5.6	*A2V A2V*	binary	125-mm: Grand! A bright pair of touching stars, nearly equal in brightness but different in color — one is pearly white and one is vaguely blue. Smyth: "Pale yellow; whitish."

Auriga

RA	Dec.	Name	Year	P.A.	Sep.	m_1	m_2	Spec.	Status	Comments
$04^h 59.3^m$	+37° 53′	ω Aur	2004	3°	4.7″	5.0	8.2	*A*1V	binary	125-mm, 83×: Grand! A bright fire-yellow star almost touched by a little blue speck. The companion is easily seen despite its great contrast. Smyth: "Pale red; light blue."
$05^h 00.3^m$	+39° 24′	5 Aur	2002	281°	4.1″	6.0	9.5			The data suggest a bright star with a close companion. Webb: "Yellowish."
$05^h 10.3^m$	+37° 18′	Σ644	2003	224°	1.6″	7.0	6.8	*B*2II+*K*3	binary	125-mm: Grand! A beautifully bright pair of stars just apart at 200×. They're both yellow-white and nearly twins. Open cluster NGC 1778 is about 1/2° southwest.
$05^h 11.0^m$	+32° 03′	Σ648	2003	62°	4.9″	7.9	8.9	*G*5	p. binary	325-mm, 98×: A nicely colored close double. It's a kissing pair of tangerine-orange stars, quite unequal but identical in color. 14 Aurigae (below) is just 1° northeast.
$05^h 15.4^m$	+32° 41′	14 Aur								125-mm, 83×: Beautiful color contrast! The stars are bright straw yellow and royal blue, and they're wide enough to be easy while attractively close. Member B was not seen.
		AB	2002	10°	10.0″	5.0	11.1	*A*9IV		
		AC	2004	225°	14.1″	5.0	7.3	*F*3V		
$05^h 20.7^m$	+46° 58′	Σ681	2002	182°	23.1″	6.6	9.2	*A*2V+*G*III		125-mm, 50×: A conspicuous star with a wide little companion just outside the field of brilliant yellow Capella. Its colors are bright whitish banana yellow and dim ocean blue.
$05^h 24.4^m$	+42° 37′	Es 576								125-mm, 50×: This is a dim but easy wide pair in an attractive field of bright and dim stars. It's a pair of bluish white stars, nearly equal in brightness.
		AC	2002	237°	42.5″	8.0	8.9	*A*2		
$05^h 25.2^m$	+34° 51′	Σ698	2003	347°	31.5″	6.7	8.3	*K*2III		325-mm, 54×: Splendid view! It shows a bright yellow star with a little blue companion. It's in the field with a bright white star and the bright orange star Phi (φ) Aurigae.
$05^h 25.6^m$	+38° 03′	Σ699	2003	345°	8.9″	7.9	8.6	*A*1V	binary	325-mm: A conspicuously bright pair in the field with Sigma (σ) Aurigae. It's a pair of hot-looking, very bluish white stars split by a modest gap. They're moderately unequal.
$05^h 30.1^m$	+29° 33′	Σ719								125-mm, 50×: A modestly dim but easy wide pair that is easily spotted about 1 1/2° northeast of Beta (β) Aurigae. It's a peach-white star with a dim nebulous companion.
		AC	2002	353°	14.8″	7.5	9.4	*G*5	binary	
$05^h 30.8^m$	+39° 50′	Σ63	1998	277°	75.9″	6.5	7.7	*G*9III		125-mm, 50×: Nice view. It shows a pretty pair of citrus-orange stars, super-wide apart, in a little group of dim stars.
$05^h 32.3^m$	+49° 24′	Σ718	1998	73°	7.8″	7.5	7.5	*F*5	p. binary	125-mm: An easy and attractively close double star. It's a bright pair of azure-white stars, exactly alike, split by only a small gap at 50×.

Auriga *(continued)*

RA	Dec.	Name	Year	P.A.	Sep.	m_1	m_2	Spec.	Status	Comments
$05^h 37.1^m$	+41° 50′	Σ736	1995	358°	2.5″	7.5	8.6	*F8*	binary	125-mm, 200×: Striking contrast for the separation! It's a pure lemon-yellow star with a dim little speck near its edge. It's surprising that Webb didn't include this binary in his catalog.
$05^h 38.6^m$	+30° 30′	26 Aur								125-mm, 50×: Grand! A bright straw-yellow star and a dim Atlantic blue, modestly wide apart; both stars are seen vividly in striking contrast. Smyth: "Pale white; violet."
		AB-C	2004	268°	12.1″	5.5	8.4	*A2*		
$05^h 41.3^m$	+29° 29′	Σ764	2004	15°	26.1″	6.4	7.1	*B8IV*		60-mm, 25×: A fine sight for low power. It's a wide, easy, and fairly bright pair, lemon white and azure white in color. They're modestly unequal.
$05^h 47.4^m$	+29° 39′	β 560	1999	129°	1.6″	7.8	8.2	*F8*	p. binary	125-mm, 200×: This double is difficult but attractively tight. It's a pair of white stars in a figure-8, slightly unequal, and the fainter star is just barely visible.
$05^h 49.9^m$	+31° 47′	Σ796	2002	63°	4.0″	7.2	8.2	*A3*	binary	125-mm, 83×: An attractively close binary just outside the view field from cluster M37. The stars are blue-white, mildly unequal, and split by only a small gap.
$05^h 51.5^m$	+39° 09′	ν Aur	2003	206°	55.9″	4.0	11.4	*K0III*		The data suggest a bright star with a wide faint companion. Smyth: "A coarse double star . . . rich yellow; dusky red . . . I have not found difficulty seeing [the companion]."
$05^h 52.4^m$	+46° 49′	Es 1321	1998	349°	4.8″	7.9	9.7	*K0*		125-mm, 200×: Striking contrast for the separation! It's a gloss-white star with a little glimpse star next to it, just outside its glow. The primary is easily spotted about 1½° northwest of Pi (π) Aurigae.
$05^h 54.2^m$	+30° 29′	Σ811	2000	233°	5.0″	8.0	9.3	*B5*		125-mm: A dim but attractively close pair in a rich patch of stars. It's a deep white star and a nebulous globe split only by hairs at 50×.
$05^h 55.4^m$	+29° 58′	Mil 2	1995	230°	10.7″	7.3	9.5	*F0*		125-mm, 83×: A difficult pair with arresting contrast. It's a gloss-white star with a little speck next to it. This companion is the faintest ember of light still possible to see.
$05^h 55.6^m$	+53° 28′	ΟΣ 120	2003	142°	48.4″	7.6	8.8	*M0*		125-mm, 50×: A super-wide pair of stars with lovely colors, just outside the field of Delta (δ) Aurigae. It's a vivid apricot-orange star and a whitish Atlantic blue.
$05^h 59.7^m$	+37° 13′	θ Aur								125-mm, 200×: Fantastic but difficult. A brilliant white star with a greenish spur on its edge, and a silvery speck that's wide away. Smyth: "Brilliant lilac; pale violet."
		AB	2004	308°	3.8″	2.7	7.2	*A0*		
		AD	2002	351°	135.3″	2.7	10.1			
$06^h 00.3^m$	+44° 36′	H VI 91								125-mm, 50×: A wide pair with a pretty primary, in the field with brilliant white Beta (β) Aurigae. It's a bright amber-yellow star with a little speck next to it.
		AC	2000	345°	35.6″	6.4	10.3	*K2III*		

Auriga *(continued)*

RA	Dec.	Name	Year	P.A.	Sep.	m_1	m_2	Spec.	Status	Comments
$06^h04.5^m$	+51° 34′	35 Cam								This star is still called 35 Camelopardalis, though it is now in Auriga. Smyth, listed under Camelopardalis: "A small double star . . . white; lilac." Webb, listed under Camelopardalis: "Yellow white [and] bluish, or purplish."
		A-BC	2003	14°	40.0″	6.4	9.3	*A7III*		
$06^h11.6^m$	+48° 43′	41 Aur	2002	357°	7.6″	6.2	6.9	*A1V A6V*	binary	Showcase pair. 60-mm: A bright, easy, and attractively close binary. It's a pair of yellow-white stars, nearly alike, that are split by a hair at 25×. Smyth: "Silvery white; pale violet."
$06^h15.6^m$	+36° 09′	Σ872	2004	216°	11.2″	6.9	7.4	*F4IV*	optical	325-mm, 54×: A bright pair of close stars, very nice. It's a pair of white stars, nearly equal, that are split by only a tiny gap of space.
$06^h20.0^m$	+28° 26′	Σ888								Webb included this pair in his catalog. The data suggest a dim pair of close stars. Webb: "Very white, ash."
		AB-C	1999	260°	2.8″	7.5	9.6	*A6V*	binary	
$06^h34.0^m$	+52° 28′	Σ918	1998	335°	4.8″	7.3	8.2	*A3*	binary	125-mm, 83×: An attractively close pair that dominates a little gathering of dim stars. It's a pure white star and a turquoise white split by only a small gap. Moderately unequal.
$06^h34.3^m$	+38° 05′	OΣ 147								325-mm, 54×: Nice effect — a broad triple (OΣ 147) and a tight binary (Σ928) that are in the same low-power view. The primary star of OΣ 147 is bright orange and quite pretty.
		AB	2003	75°	43.5″	6.8	8.7	*K0*		
		A-CD	2004	119°	46.4″	6.8	9.9			
$06^h34.7^m$	+38° 32′	Σ928	1998	133°	3.4″	7.9	8.6	*F5*	binary	
$06^h35.3^m$	+37° 43′	Σ929	1998	24°	6.2″	7.4	8.4	*G5*	binary	325-mm, 98×: Striking contrast for the separation. A citrus-orange star almost touching a little puff of silvery blue amid a rich field of faint stars.
$06^h36.8^m$	+41° 08′	Σ933	2004	75°	25.7″	8.6	9.0	*A2*		125-mm: A tight binary (Σ941) and wide dim pair (Σ933) that are in the same low-power field. The binary is lovely — it's a pair of white stars, split by hairs at 200×.
$06^h38.7^m$	+41° 35′	Σ941	1997	83°	1.9″	7.3	8.2	*B9*	binary	
$06^h39.6^m$	+28° 16′	54 Aur	1995	35°	0.9″	6.2	7.9	*B7III*	binary	The data suggest a bright star with a very close little companion. Ferguson, 150-mm: A "fine pair, both stars blue."
$06^h44.3^m$	+40° 37′	OΣ 154	1998	90°	23.0″	7.1	9.7	*M4III*		325-mm, 73×: A wide pair with a pretty primary star just beyond the field of cluster NGC 2281. It's an orangy scarlet star with a little gray dot next to it.
$06^h46.7^m$	+43° 35′	56 Aur	1998	38°	31.1″	5.3	8.7	*G0V*		125-mm, 50×: An easy wide pair with lovely contrast. It's a bright lemon star with a little silvery white companion, and both stars are seen vividly. Smyth and Webb call the companion "lilac."
$06^h53.0^m$	+38° 52′	59 Aur	1998	224°	22.2″	6.1	10.2	*F2V*		325-mm, 56×: Grand view! It shows a cone of three bright stars (59, 60, and 61 Aurigae), and the brightest of these stars (59 Aur) has a fairly close little companion. Smyth: "Pale yellow, livid."

RA	Dec.	Name	Year	P.A.	Sep.	m_1	m_2	Spec.	Status	Comments
$06^h 59.5^m$	+37° 06′	Σ994	1999	55°	26.8″	7.9	8.1	*B*		125-mm, 50×: A pretty pair of stars easily noticed about 1° south of 62 Aurigae. They're both blue-white and nearly alike. They seem attractively close despite their listed separation.
$07^h 28.5^m$	+42° 45′	Σ1086	2003	103°	12.3″	8.0	10.1	*K0*		125-mm, 50×: A rather ordinary wide pair with a pretty primary star. It's a yellowish terra-cotta star with a pale speck of light next to it.

Boötes

RA	Dec.	Name	Year	P.A.	Sep.	m_1	m_2	Spec.	Status	Comments
$13^h 40.7^m$	+19° 57′	1 Boo	2000	133°	4.5″	5.8	9.6	*A1V*	binary	The data suggest a bright star with a close, dim companion. Webb: "Bluish white, very blue . . . a 7th-magnitude bluish [star] . . . in [the] field."
$13^h 49.1^m$	+26° 59′	Σ1785	2004	176°	3.5″	7.4	8.2	*K4V K6V*	p=156 yr	125-mm: An easy and attractively close binary that forms a triangle with 3 Bootis and M3. It's a pair of grapefruit-orange stars, exactly alike, that are almost kissing.
$13^h 50.4^m$	+21° 17′	S 656	2003	208°	85.9″	6.9	7.4	*G0*		Webb, referring to the pair as OΣΣ 126: "White, yellow . . . some difference in colors but with very little in magnitude." R. Smith, 70-mm: "In a nice star field; both stars are white."
$13^h 54.7^m$	+18° 24′	η Boo	2002	89°	110.6″	2.7	10.0	*G0IV*		125-mm, 50×: A brilliant Sun-yellow star with a misty little dot beside it, super-wide apart; these stars are alone in their field and look like a pair. R. Smith, 70-mm: "Orange and white."
$13^h 59.1^m$	+25° 49′	Σ1793	2000	243°	4.7″	7.5	8.4	*A5V*	p. binary	125-mm: An easy and attractively close double star. It's a pair of gloss-white stars, mildly unequal, that are split by a hair at 50×.
$14^h 12.4^m$	+28° 43′	Σ1812 AB-C	2002	109°	14.1″	7.9	9.5	*F2V*		R. Smith, 70-mm: A pair for a small scope, but "it takes high power to keep the split in view. The stars are white and gray."
$14^h 13.5^m$	+51° 47′	κ Boo	2003	235°	13.5″	4.5	6.6	*A7V F1V*	optical	Showcase pair. 60-mm, 25×: A bright and attractively close optical double in the same field with Iota (ι) Bootis (below). It's a bright white star and a smaller silver star split by a fairly small gap.
$14^h 13.9^m$	+29° 06′	Σ1816	2001	94°	0.6″	7.4	7.8	*F0+A2*	p. binary	Webb measured 1.9″ separation when he cataloged this pair, but it now probably needs at least 300-mm. He calls both stars yellowish.
$14^h 16.2^m$	+51° 22′	ι Boo	2003	32°	38.7″	4.8	7.4	*A7IV K0V*		60-mm, 25×: An easy, very conspicuous pair in the field with Kappa (κ) Bootis (above) — a nice bonus. It's a gloss-white star and a little nebulous star, quite wide apart. Smyth: "Pale yellow, creamy white."

Boötes *(continued)*

RA	Dec.	Name	Year	P.A.	Sep.	m_1	m_2	Spec.	Status	Comments
14ʰ 16.5ᵐ	+20° 07′	Σ1825	2003	156°	4.4″	6.5	8.4	F6V	p. binary	125-mm, 83×: A fine pair that's nicely placed. It's a bright amber-yellow star almost touching a little silvery globe. Easily spotted just outside the viewfield of Arcturus, which is about 1° to the south-southwest.
14ʰ 23.4ᵐ	+08° 27′	Σ1835 A-BC	2004	194°	6.2″	5.0	6.8	A0V F2V	binary	Showcase pair. 60-mm, 45×: Splendid brightness, closeness, and contrast. A bright goldish white star almost touching a powder-blue star, and both are seen vividly.
14ʰ 24.1ᵐ	+11° 15′	Σ1838	2000	333°	9.1″	7.5	7.7	F8V G1V	binary	125-mm: A bright, easy, and attractively close binary. It's a pair of tangerine-orange stars, exactly alike, split by only a small gap.
14ʰ 28.6ᵐ	+28° 17′	Σ1850	2002	261°	25.5″	7.1	7.6	A1V A1V		60-mm: A fine sight with 60-mm. It's a wide and easy pair of stars, just mildly unequal, that are strikingly pure white.
14ʰ 40.7ᵐ	+16° 25′	π Boo	2003	111°	5.5″	4.9	5.8	B9	p. binary	Showcase pair. 60-mm: Two bright stars, just mildly unequal, that are split by a hair at 35×. Both are strikingly pure white. Webb also calls them "very white."
14ʰ 41.1ᵐ	+13° 44′	ζ Boo	2004	298°	0.7″	4.5	4.6	A0V + A0V	p=123 yr	This is the most deeply white binary that most amateurs can resolve, as well as one of the fastest moving. 125-mm, 500×: When the separation was 0.9″, it was a pair of twins in a figure-8.
14ʰ 44.8ᵐ	+07° 42′	Σ1873	2001	93°	6.8″	8.0	8.4	G5III	binary	Webb calls this binary "yellow, bluish."
14ʰ 45.0ᵐ	+27° 04′	ε Boo	2004	343°	2.9″	2.6	4.8	K0II-III	binary	Showcase pair. 60-mm, 120×: A brilliant amber-yellow star with a deep blue spur on its edge; both stars are vivid in glorious contrast. Smyth: "Pale orange; sea green"
14ʰ 46.3ᵐ	+09° 39′	Σ1879	2004	85°	1.7″	7.8	8.5	G2V	p=253 yr	R. Smith: "It took a 150-mm telescope to split this beautiful pair. The stars are white and yellow-orange."
14ʰ 48.4ᵐ	+24° 22′	Σ1884	2004	55°	2.1″	6.6	7.5	F8IV-V	binary	125-mm: Beautiful brightness, closeness and color. A bright pair of amber-yellow stars, mildly unequal, that are split by hairs at 200×. Webb: "Yellowish, bluish."
14ʰ 49.5ᵐ	+51° 22′	Σ1889	2003	92°	15.0″	6.5	9.6	F3V	binary	The data suggest a bright star with a dim companion. Ferguson 150-mm: "Bluish white and faint blue."
14ʰ 49.7ᵐ	+48° 43′	39 Boo	2004	46°	2.7″	6.3	6.7	F6V+F5V	p. binary	Showcase pair. 60-mm, 120×: A pair of touching twins, whitish gold in color, that are just bright enough for a 60-mm to show them sharply. Smyth: "White; lilac." Webb: "Both [stars] white."
14ʰ 51.0ᵐ	+09° 43′	Σ1886	2002	225°	7.3″	7.6	9.7	K0	binary	This pair, and three 7th-magnitude stars, form an L-shaped asterism (about 1° in size). R. Smith, 70-mm: "White and yellow."
14ʰ 51.4ᵐ	+19° 06′	ξ Boo	2004	315°	6.3″	4.8	7.0	G8V K5V	p=152 yr	Showcase binary. 60-mm, 25×: A bright white star touching a vivid little gray star. Smyth: "Orange; purple." Webb: "Yellow, purplish red." Hartung: "Yellow and deep orange."

Boötes *(continued)*

RA	Dec.	Name	Year	P.A.	Sep.	m_1	m_2	Spec.	Status	Comments
14h 53.4m	+15° 42′	OΣ 288	2004	164°	1.1″	6.9	7.6	*F9V*	p=313 yr	This binary is probably resolvable with a 200-mm until at least 2025.
14h 56.0m	+32° 18′	OΣ 289	2000	112°	4.8″	6.2	10.2	*A2V*		The data suggest a bright star with a companion. Ferguson, 150-mm: "Blue and a very faint companion."
15h 00.6m	+47° 17′	OΣ 291	2004	156°	35.6″	6.3	9.6	*B9*		150-mm, 36×: The contrast is attractive for this wide pair. It's a brilliant white star with a little speck of light beside it, and the companion takes effort to see.
15h 03.8m	+47° 39′	44 Boo	2004	57°	1.9″	5.2	6.1	*F7V+K4V*	p=206 yr	125-mm: A bright, pretty, and attractively close binary. It's a lovely pair of grapefruit-orange stars, mildly unequal, that are just kissing at 200×. Smyth: "Pale white; lucid grey."
15h 07.5m	+09° 14′	Σ1910	2000	212°	3.8″	7.4	7.5	*G2V G3V*	binary	The data suggest a close pair of fairly bright stars. R. Smith, 70-mm, 150×: "Jewel-like yellow twins." Smyth: "Each [star] pale white."
15h 14.1m	+31° 47′	OΣ 292	2002	158°	119.0″	6.2	9.4	*K5*		The data suggest a bright and deeply colored star with a very wide companion. R. Smith, 70-mm: "A pretty pair. Pale yellow and medium blue."
15h 15.5m	+33° 19′	δ Boo	2003	78°	103.8″	3.6	7.9	*G8III*		125-mm, 50×: This super-wide pair is pretty. It's a brilliant citrus-orange star and a silvery greenish star nearly alone in their field. Webb: "Bright yellow, fine blue."
15h 15.8m	+50° 56′	OΣΣ 137	2004	102°	68.1″	6.6	8.9	*G*		60-mm, 25×: A bright and pretty amber-yellow star with a little silvery white companion. They stand alone in a black field and look much closer than the data suggest.
15h 24.5m	+37° 23′	μ Boo								125-mm, 200×: Splendid sight. It's a bright single star and a pair of hair-split stars that form a super-wide couple. The super-wide Aa-BC pair is a fine sight for low power.
		Aa-BC	2002	170°	107.1″	4.3	7.1	F2IV *G0V*		
		BC	2004	8°	2.2″	7.1	7.6	*G0V*	p=257 yr	
15h 36.0m	+39° 48′	OΣ 298								60-mm, 25×: Splendid view! A very wide pair of pale-orange twins in a large asterism of bright orange stars (ν^1 Boo, ν^2 Boo, φ Boo, and μ CrB). R. Smith, 70-mm: "An equal yellow pair."
		AB	2004	171°	0.8″	7.2	8.4	*K2V*	p=55.6 yr	
		AB-C	2002	328°	121.5″	6.9	7.8	*K3V*		

Caelum

RA	Dec.	Name	Year	P.A.	Sep.	m_1	m_2	Spec.	Status	Comments
04h 21.9m	−41° 13′	h 3646	1998	138°	38.1″	8.0	10.1	*F5/6V*		This pair is included because it is a double star for amateurs in a constellation without many others. Gould, 175-mm, 100×: "A wide uneven pair with a pale yellow primary."
04h 26.6m	−40° 32′	h 3650	1993	183°	2.9″	7.0	8.2	*A1V*	binary	Gould, 175-mm: "A nice double star, easy though fairly close, set in a rambling line of middling and fainter stars running across the field. It is a moderately uneven white pair."
04h 33.0m	−38° 17′	β 747	1993	222°	3.2″	7.9	9.8	*A4V*	binary	Gould, 175-mm, 100×: An "easy though not bright pair, fairly close and uneven — a delicate effect. The field is faint."

Caelum *(continued)*

RA	Dec.	Name	Year	P.A.	Sep.	m_1	m_2	Spec.	Status	Comments
$04^h 40.6^m$	−41° 52′	α Cae	1933	120°	6.3″	4.5	12.5	F2V		Hartung: "The faint companion of this bright yellow star is difficult and needs 300-mm on a clear dark night of good definition to show it."
$04^h 50.3^m$	−41° 19′	h 3697	1998	258°	13.0″	6.1	10.4	F2/3V		Gould, 175-mm, 100×: This "bright, light yellow star has a well separated and easy but faintish companion. Southwest is a scatter of 7th- to 10th-magnitude stars."
$05^h 04.4^m$	−35° 29′	γ Cae	1999	306°	3.2″	4.7	8.2	K3III	binary	Hartung: A "fine unequal pair, orange and white . . . 75-mm shows the fainter star clearly." Gould: A "test for a 4-inch [100-mm]."

Camelopardalis

RA	Dec.	Name	Year	P.A.	Sep.	m_1	m_2	Spec.	Status	Comments
$03^h 22.1^m$	+62° 44′	OΣ 373 AB	2004	120°	20.1″	7.7	10.0	F8	optical	125-mm, 50×: This triple is three stars in a neat line — a super-wide pair of identical stars with a little companion next to one of them. Gloss white, peach white, bluish khaki.
		OΣΣ 33 AC	2004	112°	115.5″	7.7	7.8			
$03^h 24.2^m$	+67° 28′	Σ374	1999	296°	11.1″	7.8	9.0	F8	optical	125-mm: A dim but attractively close pair next to a little star group. It's a peach-white star and a dim bluish white split by a fairly small gap at 50×.
$03^h 28.5^m$	+59° 54′	Σ384	1998	274°	2.0″	8.1	8.9	F8	binary	A "double double" — two pairs just 2′ apart. Webb calls Σ384 "gold, blue" and calls Σ385 "yellow, no color given."
$03^h 29.1^m$	+59° 56′	Σ385	1991	159°	2.5″	4.2	7.8	B9I		
$03^h 30.0^m$	+55° 27′	Σ390	1999	160°	14.6″	5.1	10.0	A1V		The data suggest a bright star with a companion. Webb: "Greenish white, no color given . . . a pale ruby [star] is south [east]."
$03^h 30.2^m$	+59° 22′	Σ389	2004	71°	2.6″	6.4	7.9	A2V	binary	125-mm, 200×: An attractively close binary. It's a bright white star and a smaller pure silver split by just a tiny gap.
$03^h 32.0^m$	+67° 35′	OΣ 54	1991	0°	22.6″	7.7	9.0	F0		125-mm, 50×: Nice effect — this wide pair forms an arrow-shaped asterism with three field stars. The colors are pure white and bluish khaki.
$03^h 33.5^m$	+58° 46′	Σ396	2004	244°	20.1″	6.4	7.7		optical	DalSanto, 150-mm: "A nice low-power pair. A bright orange star with a deep orange companion." Webb: Franks calls it "white, blue."
$03^h 35.0^m$	+60° 02′	Σ400	2004	266°	1.4″	6.8	8.0	F3V	p=244 yr	125-mm, 200×: An easy, attractively close, and colorful binary. It's a bright lemon-yellow star and a smaller silvery star split by only a couple of hairs. Webb: "Yellow white, bluish white."
$03^h 40.0^m$	+63° 52′	OΣΣ 36 A-BC	1999	71°	45.9″	6.9	8.3	F5V G8V		125-mm, 31×: Nice view! It shows a wide pair of stars, lemon white and beige, in the field with a bright red star and a bright yellow star.
$03^h 42.7^m$	+69° 50′	Σ419	1998	74°	2.9″	7.8	7.8	A5IV	binary	125-mm: A very attractively close binary. It's a pair of identical stars, azure white in color, that are split by a hair at 83×. Looks like a pair of eyes staring back at you.

Camelopardalis *(continued)*

RA	Dec.	Name	Year	P.A.	Sep.	m_1	m_2	Spec.	Status	Comments
$03^h 42.7^m$	+59° 58′	Webb 2 AD	2002	36°	55.3″	5.9	8.5	*K*4II		125-mm, 50×: A super-wide pair with lovely colors. It's a bright citrus-orange star and a small bluish turquoise, and these colors are seen vividly in strong contrast.
$03^h 51.7^m$	+70° 30′	h 1139	1999	177°	47.3″	7.5	9.6	*A*3		125-mm, 31×: This wide pair forms a broad boomerang shape with Gamma (γ) Camelopardalis and another bright star, making a pleasant view. Both stars look white.
$04^h 07.8^m$	+62° 20′	Σ485 AE	2004	305°	17.7″	6.9	6.9	*B*0II *B*0II		60-mm, 25×: Σ485 is a multiple star that dominates the cluster NGC 1502. It looks like a pair of sparkling gems in a swarm of tiny fireflies.
$04^h 19.2^m$	+61° 35′	Σ513 A-BC	1999	60°	5.4″	7.9	10.5	*A*3		125-mm, 83×: A difficult pair with striking contrast for the separation. It's a blue-fringed white star with a ghostly presence touching its edge.
$04^h 21.1^m$	+55° 32′	ΟΣΣ 46	1999	160°	97.8″	7.7	8.0	*A*3		125-mm, 50×: Grand sight! A bright pair of identical stars, super-wide apart, that seem to jump out of a thin field. Both stars are white, and they look like a pair of eyes.
$04^h 23.0^m$	+59° 37′	Arg 100 AB-C	2002	60°	32.1″	6.2	9.3	*A*4V		125-mm, 50×: A wide pair with a pretty primary star. It's a strikingly pure white star with a little blue companion; it dominates a rich starry field.
$04^h 32.0^m$	+53° 55′	1 Cam	2003	309°	10.5″	5.8	6.8	*B*0III	binary	Showcase pair. 60-mm, 25×: A bright, obvious pair that jumps out of a barren field. The stars are pearly white and dim silver, and they're attractively close while wide enough to be easy.
$04^h 33.3^m$	+52° 48′	Es 2607	2002	130°	47.4″	7.5	9.5	*K*5		125-mm, 50×: This pair is included because of its pretty primary star. It's an apricot-orange star with a little white companion, very wide apart. It's easily noticed about 1° west-southwest of 3 Camelopardalis.
$04^h 57.3^m$	+53° 45′	7 Cam AB-C	1999	240°	25.8″	4.5	11.3	*A*1V		The data suggest a bright star with a faint companion. Smyth: "A delicate and very difficult double star . . . white; orange." Webb: Burnham calls it "easy [with] 6-inch."
$05^h 03.4^m$	+60° 27′	β Cam	2003	210°	83.0″	4.1	7.4	*G*0I		125-mm, 50×: Grand sight! A brilliant citrus-orange star and a bright sapphire, super-wide apart but so vivid in the field that they look like a beautiful couple.
$05^h 06.1^m$	+58° 58′	11-12 Cam	2002	9°	178.7″	5.2	6.2	*B*2.5V		125-mm, 50×: Grand sight! A bright pair of stars, super-wide apart, in the pretty combination of whitish lemon and citrus orange. Mildly unequal. Webb: "Yellowish, pale red. Fine field."
$05^h 14.3^m$	+69° 49′	Σ638	2000	219°	5.0″	7.5	9.1	*K*1IV	binary	The data and location suggest fairly close pair beside a little group of stars. Webb: "Yellow, very blue."
$05^h 22.6^m$	+79° 14′	19 Cam	2001	135°	25.8″	5.0	9.2	*F*6V		Smyth: "A fine double star . . . white, lilac." Webb: "Yellowish, white [and later] yellow, violet."

Camelopardalis *(continued)*

RA	Dec.	Name	Year	P.A.	Sep.	m_1	m_2	Spec.	Status	Comments
$05^h 24.7^m$	+63° 23′	Σ677	2003	128°	1.0″	7.9	8.5	G0	p=340 yr	A 200-mm should be able to split this binary; its separation is slowly widening.
$05^h 49.1^m$	+62° 48′	Σ3115	2003	342°	1.0″	6.6	7.5	A4V	p. binary	Webb measured the separation as 1.7″ when he included this double star in his catalog. He calls it "white, ashy white."
$05^h 50.6^m$	+56° 55′	29 Cam	2001	128°	26.1″	6.5	10.4	A4IV-V		The data suggest a bright star with a companion. DalSanto, 150-mm, 98×: "A white star, with an extremely faint companion."
$05^h 51.1^m$	+65° 45′	Σ780								125-mm, 83×: Pretty combination. A close pair of stars and a wide third companion in a neat cone shape. Bright yellow, dim arctic blue, and a glimpse star.
		AB	2002	105°	3.8″	7.0	8.2	F8		
		AC	2000	149°	12.5″	7.0	10.2			
$07^h 04.1^m$	+75° 14′	Σ973	2003	31°	12.8″	7.2	8.2	G0	optical	DalSanto, 150-mm, 76×: "A nice white pair with a nice difference in magnitude. It seemed wider than its listed separation."
$07^h 26.6^m$	+73° 05′	Σ1051								125-mm, 50×: A bright, wide, and conspicuous pair in a field without other bright stars. Peach white and pearly white.
		AC	1999	84°	31.7″	7.6	7.8	F2IV+F0IV		
$07^h 45.9^m$	+65° 09′	Σ1122	1999	6°	15.2″	7.8	7.8	F2	optical	125-mm, 50×: Nice view! An easy pair of white twins, moderately wide apart, in the field with bright yellow 51 Camelopardalis. They look like a pair of headlights.
$07^h 47.0^m$	+64° 03′	Σ1127								The primary is a conspicuous star about 1½° south of 51 Camelopardalis. DalSanto, 150-mm, 100x: "Neat triple! The B companion is faint, and C is very faint."
		AB	1999	340°	5.4″	7.0	8.5	A2	binary	
		AC	1999	177°	11.7″	7.0	9.7		binary	
$08^h 02.5^m$	+63° 05′	SHJ 86	1999	81°	51.1″	6.2	7.5	G1III		125-mm, 50×: Lovely primary. It's a bright banana-yellow star that seems to jump out of a field without other bright stars. The little companion looks closer than its listed separation.
$08^h 06.0^m$	+71° 47′	Σ1159	1999	95°	34.1″	7.4	9.7	K0		125-mm, 50×: Striking contrast. An amber-yellow star with a little glimpse star next to it, fairly wide apart but close enough to be attractive.
$12^h 11.0^m$	+81° 43′	SHJ 136	2000	74°	70.8″	6.2	8.3	K5		125-mm: This pair is included because of its pretty primary star. It's a bright apricot-orange star with an arctic-blue companion super-wide apart. Easily spotted about 2° southwest of 32 Camelopardalis.
$12^h 16.2^m$	+80° 08′	Σ1625	1999	218°	14.7″	7.2	7.8	F1V F3V		125-mm, 50×: A fairly close pair of stars, nearly alike, that seem to jump out of a field without bright stars. They look white with a touch of yellow.
$12^h 49.2^m$	+83° 25′	32 Cam	2003	328°	21.5″	5.3	5.7	A1III		Showcase pair. 60-mm, 25×: A bright, wide, and easy pair that's visible all year (in the Northern Hemisphere). It's a pair of lucid-white stars, slightly unequal. Webb: "Pale yellow, pale violet."

Cancer

RA	Dec.	Name	Year	P.A.	Sep.	m_1	m_2	Spec.	Status	Comments
07h 57.4m	+13° 12′	Σ1162	1997	327°	9.1″	8.0	10.2	G5		125-mm, 83×: Nice effect. Two prominent stars dominate the field, and the one listed here has a small companion. It's a pale lilac star and a nebulous dot, wide apart.
08h 05.6m	+27° 32′	Σ1177	2003	350°	3.5″	6.7	7.4	B9V	binary	125-mm, 50×: A bright and interesting pair in the field with Chi (χ) Geminorum. A yellow-white star touched by a blue-white star — they're nearly alike except for these colors.
08h 08.8m	+27° 29′	11 Cnc	1991	216°	3.2″	7.1	10.1	G8III-IV	binary	Smyth: "A close double star . . . pale yellow; lilac."
08h 09.5m	+32° 13′	Σ1187								125-mm: A bright, pretty, and attractively close binary.
		Aa-B	2004	23°	3.0″	7.2	8.0	F5V	p=1,385 yr	A pair of very pale orange stars, nearly equal, that are just kissing at 50×.
08h 12.2m	+17° 39′	ζ Cnc								Showcase pair. 60-mm, 45×: A bright pair of lemon-yellow stars, modestly unequal, split by only a small gap. Hartung resolved the tight AB with 150-mm when the separation was 1.2″.
		AB	2004	61°	0.9″	5.3	6.3	F8V	p=59.5 yr	
		AB-C	2004	72°	5.9″	5.1	6.2	G0V	p=1,115 yr	
08h 24.8m	+20° 09′	OΣ 191	1999	191°	37.5″	7.4	8.6	A5		125-mm: A conspicuous wide pair easily noticed in the thin field about 3½° west of the Beehive Cluster (M 44). The colors are peach white and azure white.
08h 26.7m	+24° 32′	24 Cnc								60-mm: A bright, easy, and attractively close binary.
		A-BC	2003	51°	5.7″	6.9	7.5	F0V F7V	binary	An amber-yellow star and a smaller silvery star split by hairs at 45×.
08h 26.8m	+26° 56′	φ² Cnc	2003	218°	5.2″	6.2	6.2	A3V A6V	binary	60-mm: A bright, easy, and attractively close binary. A pair of goldish white stars, nearly alike, that are split by hairs at 45×. Smyth: "Both [stars] silvery white."
08h 31.6m	+18° 06′	θ Cnc	1988	62°	70.4″	5.4	10.0	K5III		The data suggest a bright star with a wide, dim companion. Smyth: "A star with a distant companion, in the middle of the Crab's body . . . yellow; gray."
08h 35.8m	+06° 37′	Σ1245	2004	25°	10.0″	6.0	7.2	F8V G5V	optical	Grand! This is a bright yellow star whose companion has been called yellowish olive (125-mm, 50×); "yellow red" (Webb); "purple" (Franks); and "rose tint" (Smyth).
08h 39.9m	+19° 33′	S 571								Smyth: "A coarse triple . . . pale yellow; dusky; lucid white . . . forms a very fair scalene triangle."
		AC	2002	157°	45.2″	7.3	7.5	A		
		AD	2002	242°	92.7″	7.3	6.7			
08h 40.1m	+20° 00′	39-40 Cnc	2002	153°	150.1″	6.5	6.6	K0III		125-mm, 50×: This is a bright pair of identical stars, super-wide apart, that are outer members of the Beehive Cluster (M44). It's a part of the cluster worth noticing.
08h 44.7m	+18° 09′	δ Cnc	1998	77°	39.6″	3.9	12.2			The data suggest a very bright star with a very dim companion. Smyth: "A very delicate double star . . . straw color; blue . . . [companion] only seen by glimpses."
08h 46.7m	+28° 46′	ι Cnc	2003	308°	30.7″	4.1	6.0	G7.5III		Showcase pair. 60-mm, 25×: A bright wide pair with striking colors — it's a Sun-yellow star and a royal-blue star. These colors are seen vividly in dramatic contrast.

Cancer *(continued)*

RA	Dec.	Name	Year	P.A.	Sep.	m_1	m_2	Spec.	Status	Comments
$08^h 49.9^m$	+14° 50′	Σ1283	2002	123°	16.5″	7.7	8.5	*F0*	optical	125-mm, 50×: A conspicuous wide pair, easily noticed in the field with 54 Cancri. It's a straw-yellow star and a small silvery nebulous one.
$08^h 52.0^m$	+25° 43′	OΣΣ 96	1999	313°	49.0″	7.7	8.5	*K2III*		125-mm, 50×: This pair has a pretty primary star. It's a bright citrus-orange star with a small white companion very wide apart.
$08^h 52.6^m$	+32° 28′	σ^1 Cnc AC	2003	24°	77.0″	5.7	10.2	*A8*		125-mm, 50×: A broad triple star with a very pretty primary. It's three stars that form a cone shape — a bright banana-yellow star with two little gray companions.
$08^h 54.0^m$	+08° 25′	OΣ 195	2004	140°	9.9″	7.7	8.3	*F8*	p. binary	125-mm, 83×: An easy and attractively close binary. It's a pair of bright stars, yellow-white and blue-white, that are split by a fairly small gap. Modestly unequal.
$08^h 54.2^m$	+30° 35′	57 Cnc	2004	311°	1.5″	6.1	6.4	*G7III*	binary	125-mm, 200×: Grand! A bright pair of identical stars, citrus-orange in color, that are just fully apart. Webb: Bird calls it "one of the loveliest pairs, both [stars] crocus yellow."
$08^h 58.5^m$	+11° 51′	α Cnc	2000	321°	10.6″	4.3	11.8	*A5*		325-mm, 98×: Profound contrast, fantastic. A stunning gold star, brilliantly bright, with a tiny ghostly presence within its glow.
$09^h 01.4^m$	+32° 15′	σ^4 Cnc	2003	134°	4.5″	6.0	8.6	*A2V*	binary	125-mm, 50×: Lovely contrast! A bright peach-white star with a little blue speck next to it, and the companion is just bright enough to show color. Smyth: "Lucid white; sky blue."
$09^h 01.8^m$	+27° 54′	67 Cnc	1998	329°	103.6″	6.1	9.2	*A8V*		125-mm, 50×: This super-wide pair makes a fine couple. It's a bright Sun-yellow star and a small pale scarlet that stand alone in a blank field.
$09^h 07.4^m$	+22° 59′	Σ1311	2003	198°	7.4″	6.9	7.1	*F4V F5V*	binary	60-mm, 25×: Nice effect. This binary is a pair of twins that seem to flicker between orange and red. They're attractively close while wide enough to be easy.
$09^h 17.4^m$	+23° 39′	Σ1332	2003	28°	5.9″	7.9	8.1	*F6V F7V*	binary	60-mm: An easy and attractively close binary. It's a pair of white stars, nearly equal, that are modestly bright and split by a hair at 45×.
$09^h 17.9^m$	+28° 34′	Σ3121	2003	200°	0.8″	7.9	8.0	*K0*	p=34.2 yr	A 300-mm should be able to split this very fast-moving binary, but the separation is decreasing.

Canes Venatici

RA	Dec.	Name	Year	P.A.	Sep.	m_1	m_2	Spec.	Status	Comments
$12^h 16.1^m$	+40° 40′	2 CVn	2003	260°	11.3″	5.9	8.7	*M1III+F7V*	optical	125-mm, 50×: Lovely contrast and colors. A bright brick-red star with a tiny dot of silvery sapphire next to it, just beyond its glow.
$12^h 20.2^m$	+37° 54′	Σ1632	2003	192°	10.0″	6.8	10.0	*K0III F9V*	binary	Webb included this binary in his catalog and calls it "yellow, no color given."

Canes Venatici *(continued)*

RA	Dec.	Name	Year	P.A.	Sep.	m_1	m_2	Spec.	Status	Comments
$12^h28.1^m$	+44° 48′	Σ1645	2004	159°	9.6″	7.5	8.1	*F9V*	optical	125-mm, 50×: An easy and pleasingly close pair. It's a pair of peach-white stars, nearly alike, that are split by a fairly small gap.
$12^h56.0^m$	+38° 19′	α CVn	2004	229°	19.3″	2.9	5.5	*A0*	optical	Showcase pair (*Cor Caroli*). 60-mm, 25×: A wide, bright, and easy pair. It's a bright white star and a bluish sea green; the color contrast is quite vivid. Webb: "Pale yellow, pale copper."
$13^h10.1^m$	+38° 30′	16-17 CVn	2004	296°	277.5″	6.0	6.3	*A9IV*		60-mm, 25×: Very nice! Two beautifully bright and pretty stars, pearly white and sapphire white, that form a super-wide pair.
$13^h12.0^m$	+32° 05′	Σ261	2004	340°	2.6″	7.4	7.6	*F6V*	binary	125-mm: A bright and close binary at a barren location. It's a bright pair of bluish white stars, exactly alike, that are split by a hair at 83×.
$13^h32.4^m$	+36° 49′	Σ1755	2000	132°	4.2″	7.3	8.1	*G5III*	binary	125-mm, 50×: An easy pair that's fairly bright and attractively close. It's a yellow-white star and an arctic-blue star, modestly unequal, that are split by only a tiny gap.
$13^h37.5^m$	+36° 18′	25 CVn	2004	98°	1.8″	5.0	7.0	*A7IV*	p=228 yr	125-mm, 200×: Striking contrast for the separation! A bright lemon-white star with a small shadow touching its edge. Webb: "White, blue."
$13^h47.0^m$	+38° 33′	S 654	2004	236°	70.1″	5.6	8.9	*K0III+F8V*		60-mm, 25×: A bright star with a very wide companion, easy with a small aperture. It's a peach-white star with a little misty speck beside it.

Canis Major

RA	Dec.	Name	Year	P.A.	Sep.	m_1	m_2	Spec.	Status	Comments
$06^h19.3^m$	−24° 58′	S 516								The data and location suggest a broad, conspicuous little arc of three stars. Gould: "An 8 × 50 [finder] shows a pair (AC) with a faint third component (B)."
		AB	1999	8°	60.2″	7.3	8.3	*A1V*		
		AC	1991	243°	300.5″	7.3	7.0	*B9*		
$06^h21.4^m$	−11° 46′	Σ3116	1999	21°	3.8″	5.6	9.7	*B1V*		Hartung: "75-mm shows clearly the faint, ashy companion close [to] this bright pale yellow star," and there's a "bright orange red star" close by in the field.
$06^h24.4^m$	−16° 13′	S 518	2002	87°	16.3″	7.0	8.4	*A6V*	optical	Gould, 175-mm, 100×: An "easy white pair sandwiched between a few fainter stars near to it."
$06^h28.7^m$	−32° 22′	β 753	1999	43°	1.2″	5.9	7.6	*B4V*	binary	Hartung: "200-mm will resolve [this binary] in good conditions." [Said when the separation was 1.3″.]
$06^h32.6^m$	−32° 02′	h 3869	1999	258°	24.7″	5.7	7.9	*B2IV*		Gould, 175-mm: A "wide and easy uneven pair, with a bright white primary; the field is fairly starry but not bright."
$06^h34.1^m$	−29° 38′	h 3871	1999	354°	7.8″	7.1	8.2	*A1V*	binary	Gould, 175-mm, 100×: A "bright and easy white double star, in an attractive setting – it is inside a line of stars that converges with another line of stars."

Canis Major *(continued)*

RA	Dec.	Name	Year	P.A.	Sep.	m_1	m_2	Spec.	Status	Comments
06^{h}35.8^m	−16° 06′	Howe 13	1991	116°	12.6″	7.5	7.4	*B8/9III*		Gould, 350-mm, 110×: A "bright easy white pair, with two wide pairs also in the field."
06^{h}36.4^m	−18° 40′	ν^1 CMa	2002	264°	17.8″	5.8	7.4	*F3IV-V*	optical	125-mm, 50×: Splendid view — a bright yellow-white star with a wide companion and a brilliant yellow star (ν^2) in the field. Hartung: "Orange yellow and bluish."
06^{h}36.7^m	−22° 37′	H II 60	1999	336°	9.2″	6.4	9.3	*B5V*	binary	125-mm, 50×: Grand! A bright banana-yellow star with a little blue-green companion split only by a small gap; the contrast is striking but the companion is seen easily.
06^{h}42.0^m	−16° 00′	β 19	1998	168°	3.0″	7.1	9.0	*B8/9III*	binary	Gould, 350-mm,120×: A "fairly bright, cream white star with a fainter close companion; the field is moderately starry."
06^{h}42.8^m	−22° 27′	S 534	1999	144°	18.2″	6.3	8.3	*F2V*	optical	Hartung: "The field is beautiful, sown profusely with stars dominated by this pair."
06^{h}45.0^m	−30° 35′	CorO 44	1999	221°	4.4″	6.5	9.8	*B8IV*	binary	This pair is just $^1/_2$° north-northeast of 10 Canis Majoris. Gould, 350-mm, 120×: A "bright yellow star (despite spectral type) with a small companion fairly close [to it] — not difficult."
06^{h}45.1^m	−16° 43′	α CMa	1997	190°	3.7″	−1.5	8.5	*A1V*	p= 50.9 yr	*(Sirius)* This binary is famous for how unequal its members are. The separation should reach 9.0″ by 2012, when it will be easier for amateurs with small telescopes to spot the tiny companion.
06^{h}45.5^m	−30° 57′	h 3891	1999	223°	4.9″	5.7	8.2	*B2III*	binary	Gould, 350-mm, 120×: An "easy and beautiful unequal pair with a pale yellow primary (despite the spectral type). The field is mostly thin."
06^{h}49.6^m	−24° 09′	S 538	1999	4°	27.0″	7.2	8.2	*A2*		150-mm, 36×: A splendid "double triple" — two easy wide pairs just arcminutes apart. The primary star of S 537 (known separately as β 324) should split with more aperture.
06^{h}49.7^m	−24° 05′	β 324	1991	209°	1.8″	6.6	7.9	*A0/1V*	binary	
		S 537								
		AC	1999	283°	30.1″	6.6	8.3	*A0/1V*		
06^{h}50.4^m	−31° 42′	H V 108								Gould, 350-mm, 120×: A "bright pale-yellow star with a wide bluish companion — a striking and beautiful combination. Various small faint pairs are about in the field."
		A-BC	1999	66°	42.7″	5.8	7.7	*B6V*		
06^{h}55.0^m	−20° 24′	17 CMa								150-mm, 36×: A pretty and unusual view. This triple looks like a bright yellow star with a little pair of silvery twins beside it.
		AB	2002	148°	43.5″	5.8	8.7	*A3IV*		
		AC	2002	186°	49.4″	5.8	9.0	*A3IV*		
06^{h}55.6^m	−20° 08′	π CMa	1999	13°	11.7″	4.6	9.6	*F2*		325-mm, 100×: Fantastic contrast! A brilliant Sun-yellow star with a tiny ash-white companion just beyond its glow. It forms a neat triangle with 15 and 17 Canis Majoris.
06^{h}56.1^m	−14° 03′	μ Cma	2004	345°	3.2″	5.3	7.1	*G5III+A2*		125-mm, 200×: Superb contrast for the separation. A bright grapefruit-orange star with a little shadow on its first diffraction ring. Webb: "Yellow, blue." Hartung: "A beautiful orange pair."

Canis Major *(continued)*

RA	Dec.	Name	Year	P.A.	Sep.	m_1	m_2	Spec.	Status	Comments
06ʰ56.6ᵐ	−22° 39′	S 541	1999	45°	23.3″	7.5	8.5	*K*0/1III		Gould, 175-mm, 100×: "An easy double star, deep and pale yellow, in a starry field. 30′ to the south is a very bright star."
06ʰ58.6ᵐ	−28° 58′	ε CMa	2000	161°	7.0″	1.5	7.5	*B*2II		*(Adhara)* Hartung: "This brilliant white star has a deep yellow companion . . . which 75-mm will show . . . with many scattered stars [in the field]."
07ʰ04.6ᵐ	−11° 31′	Σ1016	1999	149°	5.3″	7.4	9.5	*B*0V	binary	Gould, 350-mm: An "easy uneven pair in a good field of fairly bright stars and some nebulosity." [It's inside the large nebula IC 2177.]
07ʰ06.7ᵐ	−11° 18′	FN CMa								Hartung, referring to it as β 328: "This bright white star has a wide companion . . . which looks fainter than magnitude 9 to me." Gould: It lies "in a pleasant field."
		AB	2003	111°	0.6″	5.7	6.9	*B*0.5IV	binary	
		AB-C	1999	350°	17.5″	5.4	9.0	*B*0.5IV		
07ʰ08.4ᵐ	−26° 24′	δ CMa	1991	225°	265.1″	1.9	8.8	*F*8I		The data suggest a very bright star with a very wide companion. Ferguson, 150-mm: "White, blue."
07ʰ11.3ᵐ	−21° 48′	h 3934								150-mm, 36×: Nice view. A bright, easy pair at the end of a curved line of dim stars. It's a pure white star and an ash white, split by only a small gap.
		A-BC	1999	237°	13.7″	6.9	8.5	*B*3V		
07ʰ13.8ᵐ	−22° 54′	h 3938	1999	251°	19.4″	6.3	9.1	*B*2IV+*G*5IV		125-mm, 50×: Lovely combination. A bright white star with a small nebulous companion, split just wide enough for the companion to be easily seen while striking in contrast.
07ʰ16.6ᵐ	−23° 19′	h 3945	1999	52°	26.8″	5.0	5.8	*K*3I *F*0		Showcase pair. 150-mm, 36×: A bright, wide, and easy pair with deep colors. The stars are bright citrus orange and royal blue; these colors are seen vividly and in strong contrast.
07ʰ17.0ᵐ	−30° 54′	BrsO 2	1999	183°	37.7″	6.3	7.8	*A*9II		Gould, 175-mm, 100×: A "wide and easy bright pair, pale yellow and dull ash, in a field moderately starry."
07ʰ18.6ᵐ	−30° 48′	h 3949	1993	77°	3.0″	7.7	7.9	*B*2V	binary	Gould, 175-mm, 100×: "Easy and attractive. A slightly uneven, fairly close pair, just 20′ east and north from BrsO 2 [above]."
07ʰ19.3ᵐ	−22° 03′	Lal 53	2000	346°	3.8″	7.7	7.6	*A*4V	binary	Gould, 350-mm, 120×: "Attractive pair. A fine, even, and fairly close pale yellow pair; 20′ NE in a good field is the little cluster NGC 2367."
07ʰ24.1ᵐ	−29° 18′	η CMa	1999	286°	179.0″	2.5	6.8	*B*5I		Smyth: "A star with a distant companion . . . pale red, dull grey." Gould : "Easy in an 8 × 50 finder."
07ʰ24.7ᵐ	−31° 49′	Δ 47								This pair is an outer member of the open cluster Cr 140. Gould: A "very wide uneven pair of white stars . . . visible in an 8 × 50 finder."
		A-CD	1999	343°	98.5″	5.4	7.6	*B*8V		
07ʰ25.1ᵐ	−21° 10′	β 199	2000	24°	1.7″	7.2	8.1	*B*1I	binary	Gould, 350-mm: This pair is in a low-power field with clusters NGC 2383 and NGC 2384. It is a "yellowish pair, just split with 120×."

Canis Major *(continued)*

RA	Dec.	Name	Year	P.A.	Sep.	m_1	m_2	Spec.	Status	Comments
07ʰ 26.1ᵐ	−18° 22′	Stone 17	1999	74°	4.8″	7.6	9.6	*A0IV/V*	binary	Gould, 350-mm, 120×: An "easy uneven pair, fairly close, with a pale yellow primary. A little line of [dim] stars arcs southeast from it."
07ʰ 27.9ᵐ	−11° 33′	β 332								Gould, 350-mm, 110×: A "bright yellow star with two easy fainter companions . . . the AB pair can be split with the Celestron 14 [350-mm] at high power. Cluster NGC 2396 is in the field."
		AB	2003	171°	0.7″	6.2	7.4		binary	
		Σ 1097								
		AC	2003	313°	19.8″	6.0	8.5			
		β 332								
		AD	1998	157°	22.8″	6.0	9.5			

Canis Minor

RA	Dec.	Name	Year	P.A.	Sep.	m_1	m_2	Spec.	Status	Comments
07ʰ 21.0ᵐ	+10° 12′	Σ1073	1999	67°	8.9″	7.9	10.2	*A1IV*	binary	125-mm, 50×: A binary with attractive contrast for the separation. It's a pearly white star with a tiny wisp of light next to it, split by only a modest gap.
07ʰ 30.6ᵐ	+05° 15′	Σ1103	1999	246°	3.8″	7.1	8.6	*B9*	p. binary	125-mm, 83×: Beautiful contrast for the separation. This straw-yellow star is almost touched by a sharp little globe of nebulous light.
07ʰ 39.3ᵐ	+05° 14′	α CMi								*(Procyon)* Smyth: "Yellowish white, orange tinge . . . several small stars in the field. [The primary] is a splendid star."
		AC	1984	17°	148.4″	0.4	11.7	*F5IV-V*		
07ʰ 40.1ᵐ	+05° 14′	Σ1126	2003	171°	0.9″	6.6	7.0	*A0III*		125-mm: A very tight pair that's very easily spotted — it's so close to Alpha (α) Canis Minoris (Procyon) that it's almost like a super-wide companion of it; 200× shows an elongated white star.
07ʰ 46.6ᵐ	+04° 08′	Σ1137	2000	130°	2.5″	8.0	9.1	*F5*	binary	125-mm, 83×: A dim but attractively close binary. It's an amber-yellow star and a smaller gray dot, split by only a hair.
07ʰ 49.5ᵐ	+03° 13′	Σ1149	2003	41°	21.7″	7.8	9.2	*G0*	optical	125-mm, 50×: This pair is included because of its pretty primary star. It's a grapefruit-orange star with a silvery yellow companion quite wide apart.
07ʰ 58.3ᵐ	+02° 13′	14 CMi								125-mm, 50×: A pretty and interesting triple. It's three stars that form a nearly perfect triangle — a bright grapefruit-orange star with two dim companions.
		AB	2004	84°	98.8″	5.4	9.4	*K0III*		
		AC	2004	148°	133.3″	5.4	9.9			
08ʰ 02.4ᵐ	+04° 09′	Σ1175	2003	280°	1.4″	7.9	9.1	*G5*	p=718 yr	160-mm should split this binary; its separation is slightly widening.
08ʰ 05.4ᵐ	+05° 50′	Σ1182	2002	73°	4.4″	7.5	8.8	*B9*	binary	125-mm: An attractively close binary. It's a white star and an almond brown, mildly unequal, that are split by only a hair at 50×. It's part of a neat cone-shaped asterism of four stars.

Capricornus

RA	Dec.	Name	Year	P.A.	Sep.	m_1	m_2	Spec.	Status	Comments
$20^h 17.6^m$	−12° 30′	α^1 Cap	2000	221°	46.0″	4.2	9.6	*G2I*		Showcase pair. 60-mm, 25×: A stunning pair of whitish gold stars, beautifully bright, that are super-wide apart but remarkably matching in appearence. One of them has a small companion.
$20^h 18.1^m$	−12° 33′	α^1 - α^2	2002	292°	381.2″	3.7	4.3			
$20^h 19.4^m$	−19° 07′	σ Cap	2002	180°	56.3″	5.4	9.4	*K3III*		60-mm, 25×: A very wide pair with a pretty primary star. It's a bright reddish yellow star with a silvery nebulous companion.
$20^h 21.0^m$	−14° 47′	β^1 - β^2 Cap	2001	267°	207.0″	3.2	6.1	*F8V+A0*		Showcase pair. 60-mm, 25×: Beautiful colors — a bright grapefruit-orange star and vivid whitish blue. They're super-wide apart. Webb: There is a "minute pair between [the members]."
$20^h 27.3^m$	−18° 13′	π Cap								Hartung: "This bright star has a white close companion, which
		Aa-B	2002	146°	3.2″	5.1	8.5	*B8II-III*	binary	75-mm shows well, a fine pair in a field of scattered stars."
$20^h 28.9^m$	−17° 49′	ρ Cap								60-mm, 25×: The AD pair is extremely wide but pretty —
		AB	2004	194°	1.3″	5.0	6.9	*F3V*	p=278 yr	it's a bright Sun-yellow star and a pale rose red. A 250-mm
		AD	2001	150°	258.7″	5.0	6.7			should now be able to separate the tight AB pair, and the gap is widening rapidly.
$20^h 29.9^m$	−18° 35′	o Cap	2003	238°	21.6″	5.9	6.7	*A3V A7V*	optical	60-mm: Easy and pretty. A bright pair of stars, whitish gold in color, split by only a modest gap at 25×. Despite their listed magnitudes, they look almost equal. Webb: "White, bluish."
$20^h 32.2^m$	−22° 09′	h 2973	1998	129°	39.5″	7.8	8.1	*F0V*		125-mm, 50×: A very wide but pretty pair. It's a pair of amber-yellow stars, exactly alike, that look like a pair of headlights deep in outer space.
$20^h 36.9^m$	−12° 44′	Σ2699	2002	196°	9.4″	8.0	9.2	*F2IV/V*	binary	125-mm: An attractively close pair with a pretty color. It's a pair of citrus-orange stars, mildly unequal, that are split by only a modest gap at 50×.
$20^h 46.5^m$	−26° 52′	h 5220	1991	10°	13.3″	7.1	10.5	*F3V*	optical	125-mm, 50×: A wide easy pair with pretty color contrast. It's a bright white star next to a dim opal green.
$20^h 48.4^m$	−18° 12′	S 763	2002	294°	15.8″	7.2	7.8	*G8III-IV*		60-mm, 25×: Nice combination. A pair of grapefruit-orange stars, nearly equal, that are attractively close while wide enough to be easy.
$20^h 50.1^m$	−27° 22′	h 5226	2000	67°	18.5″	7.3	8.8	*K0III*	optical	125-mm, 50×: An easy but rather ordinary wide pair, pale amber yellow with bluish silver, that's easily noticed in the field with bright peach Omega (ω) Capricornus.
$21^h 19.8^m$	−26° 21′	β 271	1991	269°	1.9″	6.7	9.8	*G5V*	p=262 yr	Hartung: "This attractive pair, deep yellow and whitish, is well shown by 75-mm; it lies in a field of scattered stars which it far outshines."
$21^h 26.7^m$	−22° 25′	ζ Cap	1998	12°	17.3″	3.7	12.5	*G4Ib*		Hartung: "Dominating a field of scattered stars, this brilliant golden-yellow star has a faint companion wide [beside it]."

Carina

RA	Dec.	Name	Year	P.A.	Sep.	m_1	m_2	Spec.	Status	Comments
$07^h 03.3^m$	−59° 11′	Δ 39	1997	86°	1.4″	5.8	6.8	B9IV	p. binary	Gould, 175-mm: A “very tight and moderately uneven pair, in a moderately starry field. White.”
$07^h 10.4^m$	−55° 36′	Rmk 5	1999	226°	6.9″	7.7	7.8		binary	Gould, 175-mm, 100×: “Elegant pair. A nearly even pair, pale yellow in color, in a scattered field of stars.”
$07^h 20.4^m$	−52° 19′	Rmk 6	2000	26°	9.1″	6.0	6.5	F0-2IV-V	p. binary	Gould, 175-mm, 100×: “The field shows three bright stars in a row — h 3958 is the middle star and Rmk 6 is the southernmost . . . Rmk 6 is a pale bright yellow pair.”
$07^h 20.7^m$	−52° 12′	h 3958	2000	279°	28.5″	6.9	9.1	F3/5V		
$07^h 48.6^m$	−63° 41′	h 4014	2000	156°	11.0″	7.9	9.1	B9V	optical	Gould, 175-mm: “Good effect. A white pair, easy and good, that connects a wide pair in the field with a line of three stars.”
$07^h 49.6^m$	−55° 05′	CorO 60	2000	55°	3.9″	7.5	9.1	A1V	binary	Gould, 175-mm, 100×: A “bright and fairly close uneven pair, yellow and ashy in color; three other pairs are in the field [all of these are dim pairs].”
$07^h 52.2^m$	−59° 37′	h 4018	1999	327°	5.0″	7.6	9.6	A	binary	Gould, 175-mm, 100×: A “fairly close uneven pair with a wide dim companion, in an ordinary field.”
$07^h 55.4^m$	−53° 37′	Gli 77	1999	341°	35.0″	7.9	9.4	G5I		Gould, 175-mm, 100×: This “bright orange star” has a “wide easy companion; 6′ north is a faint little pair, just west of a moderately bright star.”
$07^h 58.4^m$	−60° 51′	h 4031	2000	357°	5.5″	7.1	7.7	B8.5III	binary	Gould, 175-mm, 100×: “This pair is in the heart of the large bright open cluster NGC 2516. A fairly close, fine white pair, with two obvious pairs [CorO 66 and I 1104] nearby.”
$08^h 01.4^m$	−54° 31′	Δ 60	1999	162°	40.2″	6.1	7.9	B2IV-V		Gould, 175-mm, 100×: “An easy and very wide uneven pair, which has two fairly wide companions.”
$08^h 08.2^m$	−61° 05′	h 4053								Gould, 175-mm, 100×: “Good triple: a bright orange star with two companions, in a field dominated by a bright yellow star to the southeast.”
		AB	2000	99°	11.6″	6.9	8.5	G8III		
		AC	2000	323°	20.3″	6.9	10.5			
$08^h 15.3^m$	−62° 55′	Rmk 8	2000	67°	3.6″	5.3	7.6	A2V	binary	Showcase pair. Hartung: A “beautiful easy object, whitish and deep yellow, in a field of scattered stars.” Gould, 175-mm, 100×: A “beautiful close uneven pair, with a yellow primary.”
$08^h 17.9^m$	−59° 10′	h 4084								Gould, 175-mm: “Nice object. A bright yellow star with a double companion . . . easy at 180×.”
		AB	2000	150°	42.2″	6.5	9.8	F5V		
		BC	2000	87°	3.1″	9.8	9.9	A2V		
$08^h 37.3^m$	−62° 51′	h 4125	2000	237°	5.8″	5.5	11.0	K0III		Gould: 175-mm, 180×: A “bright deep yellow star, in a field with a few other fairly bright stars nearby. The faint companion is visible at 100×, better at 280×.”
$08^h 39.2^m$	−60° 19′	h 4128	1991	202°	1.2″	6.8	7.5	A0V	p. binary	Gould, 175-mm: “In a moderately starry field is this cream/pale yellow star; it is elongated at 100×, and a neatly split close pair at 180×.”

Carina *(continued)*

RA	Dec.	Name	Year	P.A.	Sep.	m_1	m_2	Spec.	Status	Comments
08ʰ40.7ᵐ	−57° 33′	h 4130	1998	245°	3.3″	6.5	8.3	*A*3V	binary	Hartung: A "fine easy unequal pair, pale yellow and reddish . . . the field is sown thickly with stars."
08ʰ45.1ᵐ	−58° 43′	Rmk 9	1998	293°	3.7″	6.9	6.9	*B*7III	binary	Gould, 175-mm 100×: A "fine even pair, nicely split, with a fainter wide companion N and another SW." Hartung: "The two white stars . . . decorate a field sprinkled with stars."
08ʰ57.0ᵐ	−59° 14′	Δ 74	2000	76°	40.1″	4.9	6.6	*B*2IV-V		Gould: A "wide white pair, better than many other very wide pairs. It is in a moderately starry field and visible as a double with an 8 × 50 finder."
09ʰ17.4ᵐ	−74° 54′	h 4206								Gould, 175-mm, 100× [He describes an interesting triple.]: A "bright yellow star with a wide companion and a close one, nearly on the same line."
		AB-C	2000	345°	7.0″	5.3	9.6			
		AB-D	2000	354°	46.0″	5.5	10.3			
09ʰ25.5ᵐ	−61° 57′	h 4213	2000	330°	8.8″	5.8	9.6	*A*4V		Gould, 175-mm, 100×: "Nice contrast. A bright off-white star with a small companion moderately separated. There are several faint pairs nearby, and two of these are in a patch of stars."
09ʰ27.9ᵐ	−57° 20′	I 830	2000	59°	17.8″	7.5	9.5	*M*2III		Gould, 175-mm, 100×: This "fairly bright orange star has an easy companion, and [another] companion wide northeast. Some 15′ north is the cluster IC 2488." [The 3rd-magnitude star N Velorum is also in the field.]
09ʰ33.3ᵐ	−57° 58′	R 123	1991	34°	1.9″	7.5	7.6	*B*8III	p. binary	Gould, 175-mm, 100×: An "even, close and attractive white pair, in a moderately starry field that includes some fainter pairs."
09ʰ38.4ᵐ	−57° 32′	h 4232	2000	303°	10.8″	7.7	8.2	*A*1/2V		Gould, 175-mm, 100×: An "easy white pair, with two faint wide companions; it lies in a pleasantly starry field."
09ʰ43.3ᵐ	−60° 02′	h 4240	2000	58°	12.4″	7.6	9.9	*B*5V	optical	Gould, 175-mm, 100×: An "easy uneven pair of white stars in a good field, which includes an S-shaped asterism 20′ to the south."
09ʰ47.1ᵐ	−65° 04′	υ Car	2000	129°	5.0″	3.0	6.0	*A*8Ib	binary	Showcase pair. Gould, 175-mm, 100×: A "brilliant and fairly close uneven pair that dominates a quite starry field."
09ʰ55.1ᵐ	−69° 11′	Rmk 12	2000	213°	9.2″	6.9	8.9	*B*9V	binary	Gould, 175-mm, 100×: An "easy uneven pair with a white primary, in a moderately starry field. There are three 12th-magnitude companions."
10ʰ03.6ᵐ	−61° 53′	Hrg 47	1991	352°	1.2″	6.3	7.9	*B*7I	binary	Gould, 175-mm, 100×: A "very tight uneven pair, in a fine fairly rich starry field; white."
10ʰ15.1ᵐ	−67° 17′	CPO 48								Gould, 150-mm, 75×: This triple is a "fairly bright yellow star with a close companion, and a wide fainter one. 150× makes the close AB pair easy."
		AB	1991	341°	1.9″	7.7	9.1	*K*1III/IV	binary	
		AB-C	2000	236°	62.4″	7.7	8.4			
10ʰ19.1ᵐ	−64° 41′	h 4306	1997	133°	2.4″	6.5	6.3	*A*1V	p. binary	Gould, 150-mm, 112×: "Good effect. A beautiful pale yellow pair, quite close, with a curving line of faint stars beside it."

Carina *(continued)*

RA	Dec.	Name	Year	P.A.	Sep.	m_1	m_2	Spec.	Status	Comments
$10^h 20.1^m$	−67° 10′	R 141	1991	44°	1.8″	7.5	8.3	*B6V*	binary	Gould, 150-mm, 112×: "These two pairs are just 20′ apart. R 141 is a delicate, moderately uneven, close pair. h 4314 is a wide and easy but ordinary pair."
$10^h 20.8^m$	−67° 32′	h 4314	2000	13°	19.1″	8.3	8.8	*B9/A1*	optical	
$10^h 30.7^m$	−61° 21′	Δ 87	2000	331°	82.3″	6.6	7.6	*M2III*		Gould, 175-mm, 100×: A "bright but extremely wide pair . . . visible in an 8 × 50 finderscope. Orange and dull cream."
$10^h 34.9^m$	−64° 08′	Δ 93								Gould, 175-mm: An "easy wide pair" with a double companion — the fainter member becomes "a faint, uneven close pair with 100×."
		AB	2000	39°	24.2″	7.5	8.4	*A0V*		
		I 74								
		BC	1993	228°	2.7″	8.4	9.7	*F0*		
$10^h 35.1^m$	−57° 41′	CorO 107	1991	238°	5.1″	7.0	9.9	*G8/K0III*	binary	Gould, 175-mm, 100×: An "easy and attractive uneven pair next to a bright pale yellow star; at 65× the cluster NGC 3293 is [also] in the field."
$10^h 38.7^m$	−59° 11′	Δ 94	1991	21°	14.5″	4.9	7.5	*K4-5III*	optical	Showcase pair. Gould, 175-mm, 100×: A "bright orange star with an obvious, fairly bright companion . . . in a wonderful region not far from the η Car [Eta Carinae] Nebula."
$10^h 39.0^m$	−58° 49′	Gli 152	2000	80°	26.3″	6.2	8.0	*M1III*	optical	Hartung: "A fine wide unequal pair, orange and white, in a beautiful field sown profusely with stars."
$10^h 43.2^m$	−61° 10′	Δ 97	1998	175°	12.4″	6.6	7.9	*B3III*	optical	Gould, 175-mm, 100×: "This easy white pair is at one corner of a triangle of brighter stars."
$10^h 44.3^m$	−70° 52′	Δ 99								Gould, 175-mm, 100×: This triple is a "broad [bright] pair and a fainter 3rd star, in a field not very starry."
		AB	2000	75°	62.8″	6.3	6.5	*A5IV-V*		
		AC	2000	44°	42.2″	6.3	9.6	*A3*		
$10^h 48.9^m$	−59° 27′	h 4374	2000	120°	13.8″	7.8	10.6	*O*		Gould, 175-mm: "These two pairs are just 8′ apart, not far from the η Car area. R 161 is a neat and very close uneven pair, just split with 180×; h 4374 is a faint wide pair."
$10^h 49.4^m$	−59° 19′	R 161	1991	290°	1.0″	6.1	7.4	*B9.5IV-V*	p. binary	
$10^h 53.0^m$	−63° 05′	I 418	1991	203°	2.2″	8.0	8.2	*A8IV*	binary	Gould, 175-mm, 100×: "A tiny pale yellow pair that is neat, close, and even in brightness."
$10^h 53.5^m$	−58° 51′	u Car								Gould, 175-mm, 100×: This triple is "three stars nearly in a line — a bright orange star between two white companions. The AB pair is easy with an 8 × 50, which shows η Car in the field."
		AB	2000	204°	159.4″	3.9	6.2	*K1III*		
		AC	2000	7°	56.4″	3.9	7.8			
$10^h 53.7^m$	−70° 43′	h 4383	1991	288°	1.5″	6.4	7.1	*B6V B6V*	binary	Gould, 175-mm, 100×: A "bright and attractive close pair, light golden in color. . . . 180× makes a neat, easy split." Despite the spectral class, Hartung also calls it a yellow pair.
$10^h 57.3^m$	−69° 02′	h 4393	2000	132°	8.7″	6.6	8.7	*B7V*	binary	Gould, 175-mm, 100×: An "easy uneven pair, white and dull yellow, in an average field."
$10^h 59.2^m$	−61° 19′	R 164	2000	78°	3.3″	6.2	9.7	*B8IV*		Hartung: "In an attractive field sown profusely with stars . . . is this bright yellow star with a faint white companion close [to it] . . . 105-mm will show [the companion] with care."

Carina *(continued)*

RA	Dec.	Name	Year	P.A.	Sep.	m_1	m_2	Spec.	Status	Comments
$11^h04.9^m$	−61° 03′	Δ 105	2000	222°	23.9″	7.6	9.8	*B*4V		Gould, 175-mm, 100×: A "wide easy pair . . . in a loose gathering of stars within a good starry field."
$11^h05.1^m$	−59° 43′	Gli 159	2000	275°	17.5″	7.7	9.2	*B*1Ib/II	optical	Gould, 175-mm, 100×: An "easy and fairly wide uneven pair; it lies in a fine field, with lines and patterns of stars in it."
$11^h17.5^m$	−59° 06′	R 163								Gould, 175-mm, 100×: "Quite good. A close and nearly even pale yellow pair in a field that includes some bright stars."
		A-BC	1991	57°	1.6″	7.2	7.6			

Cassiopeia

RA	Dec.	Name	Year	P.A.	Sep.	m_1	m_2	Spec.	Status	Comments
$23^h02.7^m$	+55° 14′	Σ485	2003	48°	18.7″	6.5	10.2	*B*9III		125-mm, 83×: An easy wide pair with strong contrast. It's a bright white star with a dim spot of bluish silver beside it.
$23^h30.0^m$	+58° 33′	SHJ 355								125-mm, 200×: Grand! This triple is a perfect triangle — a wide pair of stars (AC) forms one side, and a close dim pair (FG) forms the other side. All the stars look white.
		AC	2004	268°	75.8″	4.9	7.2	*B*3IV		
		h 1887								
		FG	2001	73°	10.8″	8.9	9.1			
$23^h33.2^m$	+57° 24′	ΟΣ 499								125-mm, 50×: An easy, tight binary in a rich little patch of stars within a spectacular field full of stars. It's an amber-yellow star and a dim blue-green separated by only a small gap.
		A-BC	2002	77°	9.9″	7.6	9.5	*G*5	binary	
$23^h33.6^m$	+60° 28′	Σ I 60	2003	211°	234.5″	7.5	7.3	*G*1IV		8 × 50: A fine sight in a finderscope, which shows a wide, conspicuous pair of nearly equal stars in a beautiful starry field. The stars look pinkish yellow and white.
$23^h46.1^m$	+60° 28′	Σ3037								125-mm, 200×: A neat little multiple. The members form an attractive triangle — a close pair of stars (AB) form one side, and two wide dim stars form the other side.
		AB	1999	212°	2.6″	7.4	9.2	*K*0		
		AC	1996	189°	39.5″	7.4	9.5			
		AD	1991	232°	52.3″	7.4	10.9			
$23^h48.7^m$	+64° 53′	ΟΣ 507	2003	316°	0.7″	6.8	7.8	*A*0	p=566 yr	This binary is now probably resolvable with 250-mm until at least 2025.
$23^h48.8^m$	+62° 13′	6 Cas	1994	197°	1.4″	5.7	8.0	*A*3I	binary	The star can likely be split by a 150-mm. Webb: "Very yellow, orange."
$23^h59.0^m$	+55° 45′	σ Cas	2004	327°	3.2″	5.0	7.2	*B*1V	binary	125-mm, 200×: Striking contrast for the separation. A bright yellow star almost touching a dim ash white, and both are seen vividly with strong contrast. Smyth: "Flushed white; smalt blue."
$00^h02.6^m$	+66° 06′	Σ3053	2003	70°	15.0″	6.0	7.2	*G*9III *A*1V	optical	125-mm, 50×: A wide unequal pair with beautiful colors. The stars look bright amber yellow and royal blue; the colors are seen boldly.
$00^h04.9^m$	+58° 32′	Σ3057	2002	299°	3.8″	6.7	9.3	*B*3V	binary	125-mm, 200×: Very nice view. A bright straw-yellow star with a little speck of light next to it, in the field with another bright yellow star.

Cassiopeia *(continued)*

RA	Dec.	Name	Year	P.A.	Sep.	m_1	m_2	Spec.	Status	Comments
00ʰ 06.3ᵐ	+58° 26′	Σ3062	2004	336°	1.6″	6.4	7.3	*G3V*	p=107 yr	125-mm: Grand! A very fast-moving binary that's attractively close but not difficult. It's a bright pair of yellowish peach stars, modestly unequal, that are just kissing at 200×.
00ʰ 14.8ᵐ	+62° 50′	Σ10	2000	176°	17.6″	8.0	8.6	*A2V*		125-mm, 50×: A dim but easy and conspicuous pair in a rich starry field. A peach-white star and a fainter bluish silver split widely apart.
00ʰ 16.7ᵐ	+54° 39′	Σ16	2000	39°	5.9″	7.7	8.8	*A3*	binary	125-mm: A dim but attractively close binary. A gloss-white star and a fainter whitish blue split by only a small gap at 83×.
00ʰ 27.2ᵐ	+49° 59′	Σ30	2003	313°	13.8″	7.0	8.9	*B9III*	optical	125-mm, 50×: A wide unequal pair with pretty colors in a gloriously starry field. It's a bright banana-yellow star with an ash-white companion.
00ʰ 38.7ᵐ	+46° 57′	Σ45	2003	90°	18.3″	6.9	9.9	*G9III*		125-mm, 50×: A colorful unequal pair just west of Pi (π) Cassiopeiae. A pure Sun-yellow star and a delicate sky blue split by a modest gap.
00ʰ 39.2ᵐ	+49° 21′	OΣ 16	1998	22°	12.8″	5.7	10.2	*K4III*		The data suggest a bright star with a companion. Ferguson, 150-mm: "Yellowish white and faint blue."
00ʰ 40.5ᵐ	+56° 32′	α Cas AD	1998	282°	69.5″	2.4	9.0	*K0III*		The data suggest a brilliantly bright star with a wide little companion. Smyth: "Pale rose tint; smalt blue." Webb: "Fine yellow; bluish."
00ʰ 42.6ᵐ	+71° 22′	Σ48	1999	334°	5.4″	7.8	8.0	*A*	binary	125-mm, 83×: A miniature Gamma (γ) Arietis. It's a pair of white twins, split by only a small gap, inside a cloud of scattered faint stars.
00ʰ 45.7ᵐ	+74° 59′	21 Cas	1999	160°	35.9″	5.7	10.6	*A2IV*		150-mm, 60×: Very nice view. Two bright white stars (21 and 23 Cassiopeiae) are in the field, and one of these (21) has a colorless smudge next to it. It's an easier pair than the data suggest.
00ʰ 45.8ᵐ	+54° 59′	Arg 2	2003	72°	2.7″	7.9	9.8	*F8V*	p. binary	125-mm, 200×: A difficult but gratifying pair. It's a gloss-white star whose disk is brighter in one spot. This much resolution is gratifying for such a dim, close pair.
00ʰ 47.4ᵐ 00ʰ 48.0ᵐ	+51° 06′ +51° 27′	H V 82 Σ59	2003 1997	75° 146°	56.3″ 2.3″	8.0 7.2	8.4 8.1	*K2* *B9.5IV*	 binary	125-mm: These two couples are in the field with bright white 25 Cassiopeiae. H V 82 is a pretty pair of reddish white twins; Σ59 is a straw-yellow star and a little gray star split by a hair at 200×.
00ʰ 49.1ᵐ	+57° 49′	η Cas AB	2003	319°	13.0″	3.5	7.4	*G0V M0*	p=480 yr	Showcase pair. It's a bright yellow star with a vivid little companion; Webb and Smyth call the companion "purple"; Franks calls it "pale garnet." To the author, it looks almond brown.

Cassiopeia *(continued)*

RA	Dec.	Name	Year	P.A.	Sep.	m_1	m_2	Spec.	Status	Comments
00h 52.7m	+68° 52′	Σ65	1997	39°	3.0″	8.0	8.0	*A2*	binary	125-mm: A dim but attractively close binary. It's a pair of peach-white stars, exactly alike, that are split by a hair at 83×. Easily spotted 1.6° west of 31 Cassiopeiae.
00h 53.8m	+52° 42′	Σ70	2002	246°	8.2″	6.3	9.5	*A0*	binary	125-mm, 83×: Striking contrast for the separation. A bright whitish lemon star with a ghostly wisp of light close beside it, and this companion tests the threshold of visibility.
00h 56.7m	+60° 43′	γ Cas	1961	252°	2.3″	2.2	10.9	*B0IV*		Webb included this pair in his catalog and gave the same magnitudes and separation listed here; did he actually resolve it? He calls it "beautifully contrasted, with minute surrounding stars."
01h 16.6m	+74° 02′	h 2028	2001	205°	60.8″	7.1	7.9	*B9*		60-mm, 25×: A good pair for low power. It's a nearly equal super-wide pair of stars that seem to jump out of their thin field. The colors are straw yellow and ash white.
01h 20.1m	+58° 14′	φ Cas	2001	233°	135.3″	5.1	7.0	*B5*		This wide pair is probably a fine object for binoculars, but the companion looks more like a field star with 125-mm at 50×.
01h 21.1m	+64° 39′	35 Cas	2001	342°	57.0″	6.3	8.6	*A2V*		125-mm, 50×: An extremely wide pair with lovely colors. The stars look bright lemon yellow and rose red.
01h 25.9m	+68° 08′	ψ Cas								Smyth: "A fine triple . . . orange tint; blue; reddish."
		AC	1998	123°	22.1″	4.7	9.2	*K0III*		
		AD	1983	120°	22.4″	4.7	10.0			
		CD	2001	252°	2.6″	9.4	10.0			
01h 31.7m	+61° 03′	Σ128	1999	309°	11.6″	7.8	9.5	*B7III*		125-mm, 83×: Very nice view. This is a wide, rather ordinary pair in the rich starry field with cluster M103. It's a white star with a ghostly phantom next to it.
01h 33.2m	+60° 41′	Σ131	1999	143°	13.8″	7.3	9.9	*B5I*	binary	This is another pair on the edge of cluster M103. Webb: Franks calls it "yellow, blue."
01h 37.4m	+58° 38′	OΣ 33	2001	77°	26.7″	7.3	9.0	*B3IV*		125-mm, 50×: A conspicuous wide pair in a beautiful starry field, easily noticed about 3/4° southeast of Chi (χ) Cassiopeiae. The colors are pure white and coppery white.
01h 42.3m	+58° 38′	h 1088	2003	168°	19.4″	6.3	9.8	*B7III A*		125-mm, 50×: This pair is included because of its pretty primary star. A bright, very lemony white star with a wide little speck beside it in a glorious starry field.
01h 51.3m	+64° 51′	Σ163	2003	36°	33.9″	6.8	9.1	*K4+I/II*		125-mm, 50×: A conspicuous wide pair with pretty colors. The stars look pumpkin orange and pale sky blue, and the pair is easily spotted 1 1/4° north-northwest of Epsilon (ε) Cassiopeiae. Webb: "Red gold, blue."
01h 55.4m	+76° 13′	Σ170	1998	244°	3.2″	7.5	8.2	*A5*	binary	125-mm: A dim but very pretty binary. A peach-white star and smaller blue-green split by a hair at 83×. It's just beyond the field of 49 Cassiopeiae, which lies to the east.

Cassiopeia *(continued)*

RA	Dec.	Name	Year	P.A.	Sep.	m_1	m_2	Spec.	Status	Comments
$02^h 03.2^m$	+73° 51′	Σ191	2002	196°	5.3″	6.2	9.1	*A5III*	binary	125-mm, 83×: Beautiful contrast for the separation! This is a bright yellow-white star almost touching a faint misty smudge of light. It's 1.4° north of 50 Cassiopeiae.
$02^h 15.7^m$	+67° 40′	ENG 10	2003	334°	19.9″	7.3	9.4	*K2V*		125-mm: An easy but rather ordinary wide pair with a pretty primary. It's an apricot-orange star with a dim blue-green companion.
$02^h 19.7^m$	+60° 02′	OΣΣ 26	2002	202°	63.2″	7.0	7.3	*A2V*		60-mm, 25×: A good view for low power. This is a very wide and bright pair, almost equal, just beyond the edge of the giant open cluster Stock 2. The colors are blue-white with orange-white.
$02^h 29.1^m$	+67° 24′	ι Cas								125-mm, 200×: Grand triple — a bright lemon-yellow star touching a little blue star, with a wide third companion that's only a tiny speck. Webb, Smyth: "[A] Yellow, [B] lilac, [C] blue."
		Aa-B	2004	230°	2.9″	4.6	6.9	*A5*	p=620 yr	
		Aa-C	2004	115°	7.3″	4.6	9.0	*A5*		
$02^h 51.1^m$	+60° 25′	Σ306	2001	94°	2.0″	7.4	9.1	*O6*	binary	This binary is inside the cluster IC 1848; Webb included it in his catalog. He says there are three other pairs in the field, but these refer to members of the cluster.
$03^h 10.8^m$	+63° 47′	Σ349	1999	322°	6.0″	7.9	8.6	*F5V*	binary	125-mm, 50×: A dim but attractively close binary. A straw-yellow star and a silvery white, split by only a small gap.

Centaurus

RA	Dec.	Name	Year	P.A.	Sep.	m_1	m_2	Spec.	Status	Comments
$11^h 13.1^m$	−47° 03′	R 165	1999	73°	2.8″	7.5	7.5	*F8/G0IV*	p. binary	Gould, 175-mm, 100×: A "bright and easy yellow pair, even in brightness, in a scattered field with some fairly bright stars."
$11^h 16.5^m$	−45° 53′	h 4423	1997	276°	2.5″	7.0	7.3	*F3V*	binary	Gould, 175-mm, 100×: A "bright close pair, nearly even and pale yellow in color." Hartung: A "dainty object in a black field."
$11^h 24.7^m$	−61° 39′	BrsO 5	1996	242°	4.4″	7.7	8.8	*K5/M0V*	p=400 yr	Gould, 175-mm, 100×: "Neat pair in a pleasant field. A fairly bright orange star with an easy companion . . . partly enclosed by a circlet of faint stars."
$11^h 28.6^m$	−42° 40′	BrsO 6	1999	168°	13.1″	5.1	7.4	*B9V*	optical	Showcase pair. Gould, 175-mm, 100×: An "easy uneven pair . . . in a thin faint field. Bright pale yellow and dull yellowish."
$11^h 33.6^m$	−40° 35′	I 78	1998	98°	0.7″	6.1	6.2	*A2IV-V*	p. binary	Gould, 175-mm: "Quite good. A bright yellow star in a thin field . . . with 180× it becomes a pair of equal stars that are barely separated."
$11^h 35.8^m$	−63° 01′	λ Cen	1944	317°	16.0″	3.1	11.5	*B9III*		The data suggest a brilliant star with a companion. It's probably easy to resolve with 160-mm on a clear night.
$11^h 39.2^m$	−57° 44′	h 4460	2000	176°	8.6″	7.2	8.2	*A0V*	binary	Gould, 175-mm, 100×: "Quite good. An easy uneven pair, both stars white, in a fairly rich field of mostly faint stars."

Centaurus *(continued)*

RA	Dec.	Name	Year	P.A.	Sep.	m_1	m_2	Spec.	Status	Comments
11h 40.0m	−38° 06′	Δ 114	1999	95°	17.1″	6.7	8.0	G8III	optical	Gould, 175-mm, 100×: A "wide and bright uneven pair with a yellow primary. The field is sparse."
11h 40.6m	−62° 34′	CapO 11	1991	220°	2.6″	6.9	7.4	O9.5	binary	Gould, 175-mm, 100×: A "close pair of bright stars, nearly even in brightness . . . in a beautiful and varied Milky Way region, with a little clustering of dim stars 10′ to the NE."
11h 41.2m	−61° 08′	HdO 212	1991	323°	1.0″	7.5	7.9	K1III	binary	Gould, 175-mm: A "deep yellow orange star in a pleasing field that becomes a very close, nearly even pair at 180×. This is a test for a 4-inch [100-mm]."
11h 51.8m	−64° 36′	Gli 169	2000	226°	4.3″	7.5	9.0	B4V	binary	Gould, 175-mm, 100×: A "moderately bright and easy white pair in a fairly attractive field. The members are moderately unequal."
11h 55.0m	−56° 06′	Hld 114	2000	172°	3.3″	7.4	7.8	G3IV/V	p. binary	Gould, 175-mm, 100×: "Quite good little yellow pair, in a thin field with more stars to the south."
11h 55.0m	−62° 35′	BrsO 20	2000	266°	18.8″	7.7	8.9	M3III		Gould, 350-mm, 120×: "An easy wide pair of orange stars in a quite rich, attractive starry field."
11h 58.3m	−40° 57′	h 4484	1991	313°	3.0″	7.0	9.7	K1III	binary	Gould, 175-mm: "Good pair. A bright orange star with a small, close companion that's glimpsed at 100× and clear at 180×. The field is thin."
12h 03.6m	−39° 01′	89 Cen	2001	61°	0.8″	7.1	7.7	F7V	p= 110 yr	A 270-mm should be able to split this fast-moving binary until 2012.
12h 08.4m	−50° 43′	δ Cen								Jaworski, 100-mm, 50×: A "pleasing, very wide group of three contrasting stars . . . blue white, white, bluish green." Gould: "A 10 × 70 finderscope shows a bright triangle of stars."
		AB	1992	325°	268.9″	2.5	4.4	B2I		
		AC	1999	227°	216.9″	2.5	6.3	B9V		
12h 14.0m	−45° 43′	D Cen	1994	243°	2.9″	5.8	7.0	K3III	binary	Showcase pair. Jaworski, 100-mm: A "beautiful tight pair . . . just split at 80×. The stars are yellowish-orange and white, with a moderate degree of [brightness] contrast."
12h 24.7m	−41° 23′	h 4518	1999	208°	10.0″	6.5	8.4	K2-3III	optical	Barker: "The primary star is lovely bright orange and 100-mm easily shows the companion."
12h 41.5m	−48° 58′	γ Cen	2000	347°	0.9″	2.8	2.9	A1IV	p=84.5 yr	Barker, 100-mm: In 1999, this very fast-moving binary was seen as a "tight, yellow-white peanut, clearly not a single star."
12h 56.7m	−47° 41′	I 83	1997	230°	0.8″	7.4	7.7	F5IV/V	p=191 yr	A 200-mm can probably split this binary until at least 2025.
12h 58.2m	−54° 11′	CorO 143	1999	112°	16.5″	7.4	8.7	A0V		Barker, 100-mm, 100×: A "lovely white and orange pair, centered in a trapezium shaped asterism that looks like the constellation Corvus."
13h 00.3m	−48° 36′	CapO 13	2000	68°	4.8″	7.2	9.2	G8IV	binary	Barker, 100-mm: A "lovely faint pair, yellow orange in color."

Centaurus *(continued)*

RA	Dec.	Name	Year	P.A.	Sep.	m_1	m_2	Spec.	Status	Comments
13h 01.1m	−33° 37′	h 4563	1999	237°	6.4″	7.0	8.2	*G*6III		Barker, 100-mm: An "easy pair at the apex of a triangle of stars, with two prominent stars also in the field."
13h 06.3m	−48° 28′	f Cen	1999	78°	11.5″	4.7	10.8	*B*5V		Gould, 350-mm, 120×: An "ordinary pair with a very bright primary. It is a blue-white star in a moderately starry field, which has a small companion, moderately separated."
13h 06.9m	−49° 54′	ξ² Cen	1999	99°	25.0″	4.2	10.1	*B*1.5V		Gould, 350-mm, 120×: "This very bright white star has an obvious faint companion, well separated."
13h 07.4m	−59° 52′	R 213	1997	22°	0.7″	7.0	7.0	*B*9IV	binary	Hartung: A "close deep yellow pair . . . in 1961 [when the separation was 0.9″], the stars were just clear of one another with 200-mm." Russell: "Both [stars] alike and orange."
13h 08.0m	−56° 41′	h 4569	1999	243°	4.6″	7.5	9.1	*A*1V		Barker, 100-mm: "Gorgeous. A tight uneven pair that is white and orange in color."
13h 11.5m	−35° 08′	h 4571	1999	266°	23.7″	6.8	9.1	*K*1III		Barker, 100-mm: A "wide and very uneven pair, white and orange in color, in an interesting field. The companion seems fainter than its listed magnitude."
13h 12.3m	−59° 55′	I 424								Hartung: "These two pairs are in the same viewfield."
		AB-C	1991	7°	1.9″	4.8	8.4			I 424 is a "bright pale yellow star with a white companion
13h 12.9m	−59° 49′	CorO 152	2000	146°	25.0″	6.3	9.4	*G*0V		very close," and CorO 152 is a "wide unequal orange and reddish pair."
13h 21.4m	−64° 03′	h 4579	2000	97°	5.0″	7.9	8.6	*G*0	binary	Barker, 100-mm: "A tight yellow-white pair in a nice black background full of faint stars."
13h 22.6m	−60° 59′	J Cen								Showcase pair. Jaworski, 100-mm: An "easily split wide
		AB-C	2000	346°	60.7″	4.5	6.2	*B*2.5V		pair of white suns, well seen at 50×."
13h 31.4m	−42° 28′	See 180	1991	234°	3.6″	6.8	9.2	*K*0III	binary	Gould, 175-mm, 100×: "Good pair. This bright, deep yellow star has a tiny companion, close. Delicate at 100×, easy at 180×."
13h 32.1m	−63° 03′	Δ 137	2000	358°	16.0″	7.5	8.5	*B*0.5III		Barker, 100-mm: A "beautiful pair in a rich star field, a pretty view. White and yellow."
13h 37.8m	−35° 04′	See 184	1999	302°	2.8″	7.5	9.6	*G*3V	binary	Gould, 175-mm, 100×: "Quite good pair. This yellow star has a less bright, fairly close companion. Not difficult."
13h 38.1m	−58° 25′	R 223	1991	13°	2.6″	6.6	9.9	*K*1III	binary	Hartung: This "fine orange star dominates the field sown profusely with stars; it has a faint white companion close [to it], which 75-mm will just show with attention."
13h 39.2m	−49° 00′	h 4600	1999	118°	16.6″	7.8	9.3	*K*2/3III	optical	Gould, 175-mm, 100×: An "easy fairly wide pair, in a fairly thin field. Both stars are yellow."
13h 41.7m	−54° 34′	Q Cen	2000	163°	5.5″	5.2	6.5	*B*8V	binary	Showcase pair. Jaworski, 100-mm: A "lovely close pair, well separated at 167× . . . yellow white and bluish. It forms a pretty sight: a right angle triangle with two other nearby stars in the field."

RA	Dec.	Name	Year	P.A.	Sep.	m_1	m_2	Spec.	Status	Comments
$13^h42.3^m$	−33° 59′	h 4608	2000	188°	4.2″	7.5	7.4	*F*5	p. binary	Gould, 175-mm, 100×: A "fine bright yellow pair, even and neat . . . 15′ northwest is a 9th-magnitude star, which has three fairly wide 12th-magnitude companions."
$13^h43.8^m$	−40° 11′	Howe 95	1991	186°	1.1″	7.5	7.9	*F*2V	binary	Gould, 175-mm: A "fairly bright, very close pair. At 180×, it is a neat, nearly even, pale yellow double in a thin field."
$13^h44.0^m$	−59° 14′	Δ 142	2000	90°	33.2″	6.5	7.6	*B*8V		The data suggest a bright star with an easy, conspicuous companion. Gould, 175-mm, 100×: A "wide and uneven white pair in a quite starry field."
$13^h47.2^m$	−62° 35′	CorO 157	1991	321°	7.1″	7.2	9.9	*K*0I		Hartung: "Here is a beautiful field [that is] accented by an orange star with a faint ashy companion well separated [from it] . . . easy with 105-mm."
$13^h48.9^m$	−35° 42′	Howe 94	1998	358°	11.6″	6.6	10.2	*G*3IV-V	binary	Hartung: A "well separated unequal pair . . . deep yellow and reddish, these colors clear with 105-mm."
$13^h49.2^m$	−62° 06′	Δ 143	2000	37°	13.1″	7.7	8.0	*K*2/*K*3II/III		Gould, 175-mm: An "easy pair of stars, deep yellow and white — a nice contrast. The field is medium starry and cluster NGC 5316 is 1/2° northeast."
$13^h49.3^m$	−40° 31′	Δ 146	1991	86°	66.1″	7.0	7.5	*F*3V		Gould, 175-mm, 100×: A "very wide and quite bright pair, light and deep yellow in color. The field is thin."
$13^h51.5^m$	−48° 18′	CapO 61	1999	131°	30.5″	7.4	7.4	*G*6III		Gould, 175-mm, 100×: A "wide pair, both stars yellow and bright, which dominate a thin field."
$13^h51.8^m$	−33° 00′	3 Cen	2000	106°	7.9″	4.5	6.0	*B*5III *B*8V	binary	Showcase pair. 325-mm: A pair of beautifully bright stars, golden yellow in color, that are just kissing at 62×. Gould: "Pale yellow and white." Hartung: An "unequal white pair."
$13^h52.0^m$	−47° 52′	h 4619	2000	199°	23.3″	7.0	8.4	*K*1III	optical	Gould, 175-mm, 100×: A "deep yellow and dull yellow wide pair in a not very starry field."
$13^h52.1^m$	−52° 49′	N Cen	1999	289°	18.1″	5.2	7.5	*B*8V	optical	Showcase pair. Hartung, referring to the pair as Rmk 18: "This bright unequal pale yellow pair in a well sprinkled star field is a fine object for small apertures."
$13^h53.2^m$	−31° 56′	4 Cen	1998	185°	14.8″	4.7	8.5	*B*4IV		Showcase pair. Hartung: An "attractive pair for small apertures; the stars are pale yellow and ashy."
$13^h53.5^m$	−35° 40′	Y Cen AB	1997	308°	1.0″	6.3	6.4	*F*4V	p= 292 yr	Gould, 175-mm: "This bright yellow star has a wide, obvious companion . . . [which] 180× just splits into a nearly even pair. The close AB pair is now probably just beyond the reach of 100-mm."
		H V 124 AE	1998	6°	67.5″	6.3	8.7	*F*4V		
$13^h56.3^m$	−54° 08′	R 227	1995	8°	1.9″	6.5	7.5	*A*1V	p. binary	Gould, 175-mm: "Good pair. A close and slightly uneven pair of pale yellow stars, in a fairly starry field."
$14^h03.0^m$	−31° 41′	β 1197	1994	221°	2.5″	6.5	7.8	*F*8V	binary	Gould, 175-mm, 100×: "Rather nice. A close, golden yellow pair in a field scattered with fairly bright stars."

Centaurus *(continued)*

RA	Dec.	Name	Year	P.A.	Sep.	m_1	m_2	Spec.	Status	Comments
$14^h 03.8^m$	−60° 22′	β Cen	1991	234°	0.9″	0.6	4.0	*B*1III	binary	*(Hadar)* Hartung: "This brilliant bluish white star . . . [has a] close companion. Good definition is essential to see it and I have found a neutral filter useful; 200-mm will then show it clearly."
$14^h 07.7^m$	−49° 52′	Slr 19	1998	316°	1.3″	7.1	7.4	*G*3V	p=203 yr	Gould, 325-mm, 120×: A "fine golden yellow pair, very close and nearly even. It is in a scattered field, and 240× makes it neat and easy."
$14^h 20.8^m$	−42° 25′	Pol 9								Gould, 175-mm: This is broad trapezium of four stars
		AB-CD	1999	212°	79.2″	7.2	8.9	*A*1V		made by "two fairly bright stars [AB-CD], that have two
		CorO 168								wide fainter companions [possibly referring to close field
		CD	1991	202°	1.7″	8.9	9.0	*F*2/3		stars]."
$14^h 22.6^m$	−58° 28′	Δ 159	1994	159°	8.9″	5.0	7.6	*G*8III+*F*5V		Showcase pair. Gould, 175-mm, 100×: A "bright wide pair, bright yellow and dull yellow, with two wide faint companions . . . there are some faint pairs and star groups in the field."
$14^h 37.2^m$	−37° 32′	Howe 75	1999	214°	4.1″	8.0	8.6	*A*1V	binary	Gould, 175-mm, 100×: A "neat and easy white pair in a plain field."
$14^h 39.6^m$	−60° 50′	α Cen	2000	224°	13.3″	0.1	1.2	*G*2V *K*1V	p=79.9 yr	Showcase pair *(Rigil Kent)*. Gould: "One of the great doubles, at present easy in any telescope. The colors are mid-yellow and deep yellow and are seen best with small telescopes."
$14^h 48.5^m$	−35° 51′	h 4702								Gould, 175-mm, 100×: A "quite attractive easy pair. It is
		AB-C	1998	216°	9.8″	6.9	9.4	*K*0III	binary	an orange star with a much less bright companion; the field close to the pair is thin."
$14^h 57.6^m$	−35° 23′	h 4718	1991	63°	1.8″	7.4	8.7	*M*1III	binary	Gould, 175-mm, 100×: "Good effect — a close, rather uneven pair in a modestly starry field; 180× makes it easy."
$14^h 59.5^m$	−30° 43′	h 4722	1998	338°	8.5″	7.1	9.3	*A*3/5-*F*0	binary	Gould, 175-mm, 100×: This binary is a "pleasing and easy uneven pair in a sparse field."

Cepheus

RA	Dec.	Name	Year	P.A.	Sep.	m_1	m_2	Spec.	Status	Comments
$20^h 08.9^m$	+77° 43′	κ Cep	2003	120°	7.2″	4.4	8.3	*B*9III	binary	Showcase pair. 125-mm, 83×: A close pair of stars with stunning contrast — brilliant peach white and a small powder blue, both of these colors seen vividly.
$21^h 02.1^m$	+56° 40′	Σ2751	2003	355°	1.6″	6.2	6.9	*B*8III	binary	125-mm, 200×: A tight but not difficult binary. It's a pair of bright white stars, nearly alike, that are split by just a tiny sliver of space.
$21^h 11.8^m$	+59° 59′	Σ2780	2003	214°	1.0″	6.1	6.8	*B*0II	binary	125-mm, 200×: A bright pair of stars, strikingly close together. It's a gloss-white star with a slice of another star sticking out of it. A figure-8 looks wide compared to this.

RA	Dec.	Name	Year	P.A.	Sep.	m_1	m_2	Spec.	Status	Comments
21ʰ 15.6ᵐ	+78° 36′	Σ2796	1999	42°	26.4″	7.4	9.6	*A3V*		125-mm, 50×: Interesting field — it shows three bright stars in a neat triangle, and one of these (the one listed here) has a small gray companion. Webb calls the pair "white, ashy."
21ʰ 18.5ᵐ	+80° 21′	Σ2801	2000	272°	2.0″	7.9	8.6	*F6V*	p. binary	125-mm, 200×: A difficult but attractively close double. It's a lovely grapefruit-orange star that becomes a pair of touching stars for moments at a time. It's mildly unequal.
21ʰ 19.1ᵐ	+61° 52′	Es 137	2003	74°	45.0″	6.7	10.3	*B2V*		125-mm, 50×: Grand! A bright white star with a tiny, misty dot next to it. The contrast is profound, but the companion is seen easily. The separation looks smaller than what's listed.
21ʰ 19.3ᵐ	+58° 37′	Σ2790	2002	45°	4.6″	5.8	9.3			This pair is in the low-power field with Arg 107. Webb: First seen as "red, blue," then years later the primary was "orange."
21ʰ 27.4ᵐ	+59° 45′	OΣ 440	1996	175°	8.9″	6.3	10.4	*M3III*		The data suggest a bright star with a very faint companion. Webb: "Gold, no color given."
21ʰ 28.7ᵐ	+70° 34′	β Cep	1999	248°	13.2″	3.2	8.6	*B2III*		Showcase pair. A bright pair that's ideally close and easy. 60-mm, 25×: A brilliant white star touched by a little green globe, both stars seen vividly in striking contrast.
21ʰ 39.0ᵐ	+57° 29′	Σ2816								60-mm, 25×: Remarkable sight! A pair and a triple just 13′ apart, inside the giant open cluster IC 1396. Σ2816 is exotic — a bright white star between a green companion and a violet one.
		AC	2003	120°	11.7″	5.7	7.5	*O6*		
		AD	2003	339°	19.7″	5.7	7.5			
21ʰ 40.4ᵐ	+57° 35′	Σ2819	2003	59°	12.7″	7.4	8.6	*F5V*		
21ʰ 44.9ᵐ	+62° 28′	MLR 16	2003	33°	17.6″	6.0	9.5	*O9II*		125-mm, 50×: The bright white primary of this pair is just inside the field of 9 Cephei. The data suggest its companion should be easy, but it was not seen.
21ʰ 49.1ᵐ	+66° 48′	Σ2836	2003	154°	11.8″	6.5	10.4	*F4V*		Gould, 200-mm, 80×: "This fairly bright yellow star has a much fainter companion, moderately separated; various 8th- to 10th-magnitude stars are scattered about the field."
21ʰ 51.0ᵐ	+61° 37′	OΣ 451	2002	218°	4.1″	7.7	8.6	*A2*		125-mm: OΣ 451 is attractively close — it's a pair of pearly white stars, mildly unequal, that are split by hairs at 83×.
21ʰ 51.6ᵐ	+65° 45′	Σ2843	2004	149°	1.4″	7.0	7.3	*A1*	binary	125-mm, 200×: Grand! A sharp pair of kissing twins, whitish amber yellow in color. Not brilliantly bright, but very comfortably bright and well defined.
21ʰ 52.0ᵐ	+55° 48′	Σ2840	2004	197°	18.0″	5.6	6.4	*B6IV-V*	optical	60-mm: Very nice! A bright, pretty pair that's attractively close while wide enough to be easy. It's a pair of pure white stars, mildly unequal, that are split by only a modest gap.
21ʰ 53.8ᵐ	+62° 37′	S 800	1999	145°	62.6″	7.1	7.9			60-mm, 25×: Very nice view! A bright wide pair, pure white and green-white, that's right on the edge of the little star cluster NGC 7160.

Cepheus *(continued)*

RA	Dec.	Name	Year	P.A.	Sep.	m_1	m_2	Spec.	Status	Comments
$21^h 58.2^m$	+82° 52′	Σ2873	2003	67°	13.5″	7.0	7.5	*F6IV-V*		125-mm, 50×: A bright, pretty pair of amber-yellow stars, very nearly equal, that are wide enough to be very easy without looking overly wide.
$22^h 03.3^m$	+60° 51′	Σ2860	2003	257°	12.4″	7.9	9.2	*K0*		This pair is in a broad gathering of 7th-magnitude stars. Webb calls it "very yellow, blue."
$22^h 03.8^m$	+64° 38′	ξ Cep	2004	275°	7.9″	4.4	6.4	*A3*	p=3,800 yr	Showcase pair. 60-mm, 40×: Beautiful combination — a lemon-white star with a royal-blue companion. These stars are bright, boldly seen, and split by hairs.
$22^h 03.9^m$	+59° 49′	OΣ 461								60-mm, 25×: This multiple star looks like a broad little cluster. The shape is pretty — three white stars in a straight line, all exactly alike, with a smaller star off to one side.
		AB	2000	297°	11.1″	6.7	11.4	*B1V*		
		AC	2003	41°	88.9″	6.7	10.0	*B1V K0II*		
		AD	2001	73°	184.7″	6.7	7.8	*B9V*		
		AE	2001	38°	237.0″	6.7	7.0	*B9III*		
		DE	2000	346°	136.2″	7.8	7.0	*A*		
$22^h 08.6^m$	+59° 17′	Σ2872								60-mm, 25×: An easy wide pair of twins, boldly white, like the color of virgin snow. The close BC pair was not resolved.
		A-BC	1999	316°	21.6″	7.1	8.0	*B9.5V*		
		BC	2003	299°	0.8″	8.0	8.0	*A1V*		
$22^h 10.6^m$	+70° 08′	Σ2883	2003	252°	14.4″	5.6	8.6	*F2V*	binary	60-mm, 25×: Very nice contrast for the separation. A bright white star with a close little gray companion in a field packed with stars.
$22^h 11.8^m$	+59° 44′	Σ2880	1999	352°	4.1″	7.8	9.7	*G8II*		125-mm, 83×: Fantastic contrast for the separation — a grapefruit-orange star touching a ghostly wisp of light! Bright white Lambda (λ) Cephei is wide in the field.
$22^h 12.9^m$	+73° 18′	Σ2893	2003	347°	28.8″	6.2	7.9		optical	125-mm, 83×: A very wide unequal pair with beautiful colors. It's a bright yellowish peach star with a cobalt-blue companion — this lovely contrast is seen vividly.
$22^h 21.8^m$	+66° 42′	Σ2903	1999	97°	3.8″	7.1	7.8	*A7V+G0III*	binary	60-mm, 45×: An easy and attractively close binary. A tight pair of amber-yellow stars, nearly equal, that seem to pop out at you like a pair of headlights. Webb: "Yellow, blue."
$22^h 29.2^m$	+58° 25′	δ Cep								Showcase double star and famous variable. 60-mm, 25×: A bright, wide pair of stars with striking colors — they look vivid citrus orange and deep royal blue. Webb: "Very yellow, blue."
		AC	2004	191°	40.6″	4.2	6.1	*B7V*		
$22^h 33.2^m$	+70° 22′	Σ2923	2003	47°	9.5″	6.3	9.2	*A0V*	binary	125-mm, 83×: Interesting binary. It's a bright white star with a tiny speck next to it; these stars are quite wide apart for a binary so unequal.
$22^h 46.1^m$	+58° 04′	OΣ 480	2000	116°	30.7″	7.6	8.6	*F8*		125-mm, 50×: Splendid view! A bright, conspicuous wide pair that sits on the edge of cluster NGC 7380 — both the pair and the cluster are seen in the view. Lemon white, azure white.

Cepheus *(continued)*

RA	Dec.	Name	Year	P.A.	Sep.	m_1	m_2	Spec.	Status	Comments
$22^h 47.5^m$	+83° 09′	OΣ 482	1992	38°	3.5″	5.0	9.7	*K*3III		The data suggest a bright star with a companion. Webb: "Very yellow, no color given."
$22^h 49.0^m$	+68° 34′	Σ2947	2002	56°	4.5″	6.9	7.0	*F*4V	p. binary	60-mm, 45×: Grand! A pretty pair of kissing twins, yellowish peach in color, next to the brilliant gold star Iota (ι) Cephei that's wide in the field.
$22^h 51.4^m$	+61° 42′	Σ2950	2003	282°	1.3″	6.0	7.1		p. binary	The data suggest a close pair of bright stars. Webb: "Yellow, ash."
$22^h 52.7^m$	+67° 59′	OΣΣ 238	1999	280°	69.2″	7.0	7.6	*F*2		60-mm, 25×. A bright and conspicuous pair in the field with Σ2947 (above). It's a pair of white stars, nearly alike, that are very wide apart.
$22^h 52.7^m$	+60° 55′	Σ2953	1997	137°	8.4″	7.6	9.5	*A*0V+*G*0III	binary	125-mm, 83×: A dim but striking binary. It's a straw-yellow star that almost swallows a little ghostly dot, which is just beyond its glow.
$22^h 54.2^m$	+76° 20′	Σ2963	2000	2°	1.9″	7.9	8.5	*A*3	binary	125-mm, 200×: A difficult binary but attractive for how tight. It's a gloss-white star, seen for moments as a touching pair of stars. It looks very slightly unequal.
$22^h 56.7^m$	+78° 30′	Σ2971	2000	4°	5.5″	8.0	8.9	*G*5	binary	125-mm: A dim but attractively close binary. It's a lemon-white star and a pearly white star, mildly unequal, that are split by a very small gap.
$23^h 07.9^m$	+75° 23′	π Cep	2002	350°	1.1″	4.6	6.8	*G*0IV	p=160 yr	The data suggest a very close pair of bright stars. Webb: "Very yellow, purple." Smyth: "Deep yellow; purple."
$23^h 18.6^m$	+68° 07′	ο Cep	2004	221°	3.2″	5.0	7.3	*G*8III	p=1,505 yr	The data suggest a splendid sight — a bright star and a small star very close together. Webb: "Yellow, yellowish green." Smyth: "Orange yellow; deep blue; the colors in fine contrast."
$00^h 09.3^m$	+79° 43′	Σ2	2003	19°	0.8″	6.7	6.9	*A*4IV	p=540 yr	The separation is slightly widening for this slow-moving binary. Webb: "Yellow."
$00^h 16.2^m$	+76° 57′	Σ13	2003	52°	0.9″	7.0	7.1	*B*8V	p=971 yr	The data make this binary sound like a slower-moving clone of Σ2 (above). It should change very little between now and 2025. Webb: "Yellow white."
$00^h 21.4^m$	+67° 00′	OΣ 6	2002	159°	0.6″	7.5	8.8	*B*8.5V	p=335 yr	This binary is probably resolvable with 300-mm until at least 2025.
$01^h 49.9^m$	+80° 53′	OΣ 34	2003	107°	0.5″	7.6	8.1	*A*0V	p=173 yr	This binary begs study since its period is poorly known. Only the largest amateur scopes can resolve it — at least 350-mm is suggested.
$03^h 06.1^m$	+79° 25′	Σ320	2002	231°	4.7″	5.7	9.1			The data suggest a bright star with a dim companion. Webb: "Gold, blue. Test for moderate apertures." Smyth: "Orange, smalt blue . . . charming object."

Cepheus *(continued)*

RA	Dec.	Name	Year	P.A.	Sep.	m_1	m_2	Spec.	Status	Comments
$04^h 10.0^m$	+80° 42′	Σ460	2004	138°	0.8″	5.6	6.3	*G8III A4V*	p=372 yr	125-mm, 200×: Profound tightness! A brilliant straw-yellow star with the outer edge of another star coming out of it. They look like two stars that are almost, but not quite, one.

Cetus

RA	Dec.	Name	Year	P.A.	Sep.	m_1	m_2	Spec.	Status	Comments
$00^h 21.9^m$	−23° 00′	h 1957 A-BC	1999	26°	6.0″	7.7	9.1	*G0*	binary	Gould, 175-mm: An "attractive uneven pair, golden yellow and dull red, in a field of scattered stars."
$00^h 27.7^m$	−16° 25′	h 1968	2003	233°	33.1″	7.3	8.0	*F6V*		Gould, 175-mm, 65×: An "easy wide pair in a thin field, deep yellow and dull yellow."
$00^h 30.0^m$	−03° 57′	12 Cet AB	1998	200°	11.6″	6.0	10.8	*M1III*		Smyth: A "topaz yellow; bright blue; . . . beautiful, but most difficult test object."
$00^h 34.5^m$	−04° 33′	Σ39 AB-C	2003	45°	19.8″	7.1	8.6	*G1IV*		Gould, 175-mm, 65×: "Quite good. This a bright star in a thin field — a mid-yellow star with a wide and easy, dull yellow companion. Webb: "Yellow white, bluish."
$00^h 52.2^m$	−22° 37′	Stone 3	2000	249°	1.9″	7.6	8.4	*F8V*	p. binary	Gould, 175-mm, 65×: A "tight yellow pair, with a faint companion to the south; the field is sparse."
$00^h 52.7^m$	−24° 00′	β 734	1998	347°	10.8″	5.6	9.6	*K1III*		Gould, 175-mm, 100×: "Nice effect. This bright golden star has a small companion near, and moderate stars are scattered in the field." Ferguson, 150-mm: "Yellow white, very faint blue."
$00^h 53.2^m$	−24° 47′	WNO 1	2002	7°	5.4″	6.6	9.2	*F6IV-V*	binary	Gould, 175-mm, 100×: An "elegant uneven pair, pale yellow and dull brown; some moderate stars in the field [about 1° north is a 5th-magnitude star]."
$00^h 58.2^m$	−15° 41′	S 390	2002	216°	6.4″	7.8	7.9	*F6+F7*	binary	Gould, 175-mm, 65×: An "easy, bright, and attractive even pair, pale yellow in color; the field is thin, with a scatter of moderate stars."
$00^h 59.4^m$	+00° 47′	Σ80	2003	337°	28.3″	7.8	9.0	*K5II-III*		The data suggest an easily spotted pair, about 1¼° west-southwest of 26 Ceti (below). Gould, 175-mm, 65×: An "easy wide pair, deep yellow and ashy, in a broad scatter of stars."
$01^h 03.8^m$	+01° 22′	26 Cet	2003	254°	15.9″	6.1	9.5	*F1V*		Gould, 175-mm, 65×: An "easy and attractive wide pair; it is a bright light yellow star with a dull, faintish (possibly yellow) companion." Smyth: "Pale topaz; lilac tint." Webb: "Pale yellow, blue."
$01^h 07.2^m$	−01° 44′	Σ91	2002	316°	4.6″	7.4	8.6	*F9V*	binary	Gould, 175-mm, 65×: "Good pair. A fairly close pair, light yellow in color, in a moderately starry field."
$01^h 08.6^m$	−10° 11′	η Cet	2003	305°	257.7″	3.6	10.5	*K1.5III*		Smyth: "A bright star with a companion in a barren field . . . yellow; livid."

Cetus *(continued)*

RA	Dec.	Name	Year	P.A.	Sep.	m_1	m_2	Spec.	Status	Comments
$01^h 13.9^m$	-07° 37′	Σ101	1999	346°	20.9″	7.6	10.3	*G5*		Σ101 is in the low-power field with 37 Ceti (below). Webb: "Yellow, no color given." Espin calls the companion "blue."
$01^h 14.4^m$	-07° 55′	37 Cet	2003	331°	48.4″	5.2	7.8	*F6V G9V*		Data suggest a bright star with a wide companion. Webb: "Yellow [and] lilac or violet." Smyth: "White; light blue."
$01^h 19.8^m$	-00° 31′	42 Cet A-BC	2003	17°	1.6″	6.5	7.0	*A2V+G4III*	p. binary	125-mm, 200×: An attractively tight double. It's a pair of white stars, mildly unequal, that are separated by only a tiny sliver of space.
$01^h 20.0^m$	−15° 49′	h 2036	1999	342°	2.2″	7.4	7.6	*G0IV*	p. binary	Gould, 175-mm: A "bright and close golden pair in a sparse field, easy and effective at 100×."
$01^h 22.5^m$	−19° 05′	h 2043	2003	73°	5.2″	6.5	8.6	*F6IV*	binary	Gould, 175-mm: A "fine close uneven pair . . . pale yellow and dull darker yellow, nicely separated and easy at 100×. It's at the south end of a large pattern of middling stars."
$01^h 28.1^m$	−17° 16′	h 3437	2002	247°	11.9″	7.4	9.4	*F2III*	optical	Gould, 175-mm, 65×: An "easy and well-separated yellowish pair, uneven in brightness. There is a circlet of stars around the pair, east and north."
$01^h 31.6^m$	−19° 01′	h 2052	2001	115°	80.0″	6.9	7.5	*A7III K0*		The data suggest a good pair for low power with stars that probably contrast in color. Gould, 175-mm, 65×: An "extremely wide, rather bright pair in a thin field."
$01^h 51.5^m$	−10° 20′	ζ Cet	2002	41°	188.6″	3.8	10.1	*K0III*		Smyth: "A bright star with a distant companion, in a poor field . . . topaz yellow; white, with a small star to the north."
$01^h 55.9^m$	+01° 51′	Σ186	2004	63°	1.0″	6.8	6.8	*F9V*	p=162 yr	Gould, 175-mm: "Fine object. This is a bright golden pair in a thin field; a close pair of even stars, just split with 180×."
$01^h 59.0^m$	−22° 55′	H II 58	1998	303°	8.5″	7.3	7.6	*F2*		This pair is about 40′ southeast of 56 Ceti. Gould, 175-mm, 65×: A "bright yellow pair, nearly even, in a thin field."
$02^h 03.8^m$	−00° 20′	61 Cet	1998	194°	43.0″	6.0	10.8	*G5II-III*		150-mm, 60×: Profound contrast! This bright, pale lemon star has a tiny ghostly dot next to it, just beyond the edge of its glow. Webb: The primary star is "pale orange."
$02^h 12.8^m$	−02° 24′	66 Cet	2003	234°	16.4″	5.7	7.7	*F8V G1V*	optical	Showcase pair. 60-mm, 25×: Splendid combination. A very yellowish white star with a small gray companion. Both stars are comfortably bright while strongly contrasting.
$02^h 26.0^m$	−15° 20′	H III 80	2003	296°	11.9″	5.9	9.1	*A6V*	binary	Gould, 175-mm, 65×: This "bright cream/pale yellow star has a little companion; it is an easy and attractive pair in a thin field [about 1½° west of Sigma (σ) Ceti]."
$02^h 31.5^m$	+01° 06′	Σ274	2002	220°	13.8″	7.5	7.6	*A*	optical	Gould, 175-mm, 65×: A "neat, even, and easy pale yellow pair in a thin field."
$02^h 34.1^m$	−05° 38′	Σ280	1998	345°	3.0″	8.0	8.0	*K1III*	binary	Gould, 175-mm, 65×: "An easy and fairly bright pair of stars, neatly separated [in the pretty combination of] deep gold yellow and yellow orange."

Cetus *(continued)*

RA	Dec.	Name	Year	P.A.	Sep.	m_1	m_2	Spec.	Status	Comments
02ʰ 35.9ᵐ	+05° 36′	ν Cet	2000	79°	7.9″	5.0	9.1	G3III		The data suggest a bright star with a companion. Smyth: "Pale yellow; blue . . . a very delicate object . . . for the companion can only be seen in glimpses."
02ʰ 36.0ᵐ	−21° 24′	h 3511	1998	98°	14.7″	7.2	8.7	G2III	optical	Gould, 175-mm, 65×: "Attractive. An easy uneven pair, gold and bluish, at the end of a little line of stars."
02ʰ 41.2ᵐ	−00° 42′	84 Cet	1993	308°	3.6″	5.8	9.7	F6V		Gould, 175-mm: This "bright pale yellow star has a tiny companion very close [which is] obvious at 100×." Smyth calls the companion "lilac," and Webb calls it "ash."
02ʰ 43.0ᵐ	−20° 17′	h 3524	2002	161°	19.8″	7.7	9.2	G0	optical	Gould, 175-mm, 65×: An "easy wide pair in a field of scattered stars; yellow and dark yellow."
02ʰ 43.3ᵐ	+03° 14′	γ Cet	2002	299°	2.3″	3.6	6.2	A1V	binary	125-mm, 200×: Author's favorite pair. It's a brilliant star with a sharp little globe on its edge, like a star touching a planet, in the beautiful combination of whitish banana yellow and pale opal green.
02ʰ 52.7ᵐ	+06° 28′	Σ323	1997	279°	2.7″	7.8	7.9	B9	binary	The data suggest a modestly bright pair of twins. Gould, 175-mm, 65×: A "close pair of white stars in a scattered field."
02ʰ 57.2ᵐ	−00° 34′	Σ330	1998	192°	8.9″	7.2	9.1	G8III	binary	Gould, 175-mm: "Fair double star. It is a moderately bright golden star, with an easy [but faint?] dull yellow companion; the field is thin."
03ʰ 18.4ᵐ	−00° 56′	95 Cet	2004	258°	1.2″	5.6	8.0	G9IV	p=282 yr	A 160-mm should be able to split this binary until at least 2025.

Chamaeleon

RA	Dec.	Name	Year	P.A.	Sep.	m_1	m_2	Spec.	Status	Comments
08ʰ 22.8ᵐ	−76° 26′	h 4109	1999	130°	26.0″	7.2	8.2	A0V	optical	The data and location suggest a fine view at very low power. A 1¼° field will show a bold curved line of bright stars including Alpha (α) and Theta (θ) Chamaeleontis.
10ʰ 31.9ᵐ	−81° 55′	h 5444	2000	236°	41.4″	7.0	9.1	B5III-IV		The data and location suggest an ordinary but easily noticed wide pair. It's part of a triangle of three prominent stars, easily spotted between Mu (μ) and Delta¹ – Delta² (δ¹ – δ²) Chamaeleontis.
10ʰ 45.3ᵐ	−80° 28′	δ¹ Cha	1991	84°	0.8″	6.2	6.5	K0III	p. binary	Gould, 175-mm: Delta¹ – Delta² (δ¹ – δ²) Chamaeleontis form a very wide and bright pair, "pale yellow (δ²) and deep yellow (δ¹) in color. δ¹ is an extremely close pair, nicely split but still very tight with 330×."
11ʰ 41.0ᵐ	−83° 06′	h 4468	2000	140°	25.3″	6.3	11.4	K0III		The data and location suggest a bright primary star (that probably has a deep color) in a thin field.
11ʰ 59.6ᵐ	−78° 13′	ε Cha	1997	211°	0.4″	5.3	6.1	B9V	p. binary	A 300-mm should be able to split this pair. Smaller apertures can probably show an elongated star.

Circinus

RA	Dec.	Name	Year	P.A.	Sep.	m_1	m_2	Spec.	Status	Comments
13ʰ 54.6ᵐ	−66° 54′	Δ 145	2000	48°	24.0″	7.8	8.9	*B*9V	optical	The data and location suggest a wide and conspicuous pair in a rich starry field — probably a fine sight for low power.
13ʰ 58.5ᵐ	−65° 48′	h 4632	1991	14°	6.5″	6.4	9.5	*K*0III	binary	Gould, 175-mm, 100×: "A good object. A bright orange star and a fainter white one, moderately separated. The field is quite rich in fainter stars, with an asterism in the SW."
14ʰ 37.8ᵐ	−67° 56′	WFC 153	1947	246°	34.1″	6.1	9.9			The data suggest a bright star with a dim companion; probably an easy pair with 150-mm.
14ʰ 42.5ᵐ	−64° 59′	α Cir	2000	226°	15.7″	3.2	8.5	*A*		Showcase pair. Hartung: A "very bright yellow star" with a "fainter red companion . . . easy with 75-mm." Gould, 175-mm, 100×: "A little arc of fairly bright stars is next to it, just 10′ to the SE."
14ʰ 45.2ᵐ	−55° 36′	Δ 169	1999	105°	69.2″	6.1	7.6	*B*2III		Gould, 175-mm, 100×: "Good color difference. A very wide bright pair, white and dull orange, in a moderately starry field."
14ʰ 48.7ᵐ	−66° 36′	I 369	2000	34°	46.6″	5.9	9.0	*B*2.5V		This pair is ½° east-southeast of the large open cluster vdB-Ha 164. Gould, 175-mm, 100×: "Though listed as a double [in the WDS], it is really a broad, easy triple — a star with two companions in a bright triangle."
14ʰ 54.2ᵐ	−66° 25′	h 4707	1997	289°	0.8″	7.5	8.1	*G*0V	p=346 yr	A 300-mm should be able to split this binary; its separation is increasing.
15ʰ 01.3ᵐ	−67° 59′	Gli 213	2000	333°	5.2″	7.1	9.3	*B*4V	binary	Gould, 175-mm, 100×: "Quite good. An easy fairly close pair, whose companion is moderately faint. The field is faint but starry."
15ʰ 14.0ᵐ	−61° 21′	I 329	1991	339°	0.9″	6.7	7.7	*B*5V	binary	This pair is about ½° southwest of Delta (δ) Circini. Gould, 175-mm: This is a "fairly bright, dull yellowish star in a thin field . . . at 330× it becomes a very close, uneven pair that is just separated."
15ʰ 23.4ᵐ	−59° 19′	γ Cir	1996	196°	0.8″	5.0	5.7	*B*5IV+*F*8	p=270 yr	Gould, 175-mm: A "bright pale yellow star in a fairly starry field — elongated at 180×, and barely split at 330×." Hartung: "A red star lies about 3′ to the N."
15ʰ 29.5ᵐ	−58° 21′	CapO 16	1991	32°	2.4″	7.0	8.0	*A*8V	binary	Gould, 175-mm, 100×: A "fairly bright whitish pair that is tight and moderately uneven. The field is modestly starry in patches."

Columba

RA	Dec.	Name	Year	P.A.	Sep.	m_1	m_2	Spec.	Status	Comments
05ʰ 13.6ᵐ	−31° 54′	h 3735	1998	149°	7.2″	8.2	8.4	*F*2	p. binary	Gould, 175-mm, 100×: "An easy, moderately bright pair in a broad patch of mostly faint stars."
05ʰ 15.1ᵐ	−36° 39′	h 3740	1999	286°	23.7″	6.9	8.3	*K*1II		Gould, 175-mm: A "wide, fairly bright deep yellow pair in a sparse field."

Columba *(continued)*

RA	Dec.	Name	Year	P.A.	Sep.	m_1	m_2	Spec.	Status	Comments
05ʰ25.9ᵐ	−35° 21′	h 3760								Gould, 175-mm, 100×: "Good triple in a plain field. It's an easy, fairly bright yellow pair and a wider third companion. All three should be easy with 100-mm."
		AB	1999	222°	7.4″	7.8	8.4	*K*0III	binary	
		AC	1999	287°	29.1″	7.7	10.6			
05ʰ31.2ᵐ	−42° 19′	Δ 22	2000	169°	7.3″	7.2	7.8	*A*8V	binary	Gould, 175-mm, 100×: An "easy and modestly bright white pair in a scatter of stars."
05ʰ38.6ᵐ	−41° 18′	h 3781	1999	136°	16.0″	8.0	9.4	*F*3V	optical	Gould, 175-mm: An "easy and fairly bright wide pair in a quite starry field; another pair, wide and fainter is northwest."
06ʰ19.8ᵐ	−39° 30′	h 3849	2000	54°	39.5″	6.7	8.1	*K*2/3III		Gould, 350-mm, 110×: This "bright orange star has a quite bright wide companion, a fine effect. The field is moderately starry."
06ʰ24.0ᵐ	−36° 42′	h 3857								Gould, 175-mm, 100×: "Good colors." This is a bright orange star between two companions — one is "very wide and bluish," the other "faintish and much closer."
		AB	1999	255°	13.1″	5.7	9.8	*G*6III		
		Δ 28								
		AC	1999	74°	63.6″	5.7	6.9			
06ʰ25.5ᵐ	−35° 04′	h 3858								Gould, 175-mm,100×: "Nice effect. A bright deep yellow star with a very wide companion, which in turn is a neat close pair. The AB pair is easy and obvious with an 8 × 50 finderscope."
		AB	1991	47°	131.7″	6.4	7.6	*K*3III		
		BC	1999	309°	3.8″	7.6	8.2	*A*4V	binary	
06ʰ25.8ᵐ	−40° 59′	h 3860	1999	228°	8.6″	7.3	8.8	*A*5V	binary	Gould, 175-mm, 100×: An "easy white pair, about medium bright, that stands out in a rather sparse field."
06ʰ29.1ᵐ	−40° 22′	Δ 29	1999	118°	64.6″	7.7	8.0	*K*2IV/V		Gould, 175-mm: "These two pairs are about 6′ apart. Δ 29 is a very wide and nearly even pair, orange and dull yellow in color; I 4 is a nearly even pair that is just split at 180×."
06ʰ30.7ᵐ	−40° 27′	I 4	1991	306°	0.8″	7.3	7.5	*B*3IV	binary	
06ʰ35.4ᵐ	−36° 47′	β 755								Gould, 175-mm: A "pale yellow star and a cream colored, and a very close split at 100×. John Herschel's faint companion (C) is also visible."
		AB	1999	260°	1.5″	5.9	6.9	*A*0V	p. binary	
		h 3875								
		AC	1999	302°	21.0″	6.0	11.5	*A*0V		

Coma Berenices

RA	Dec.	Name	Year	P.A.	Sep.	m_1	m_2	Spec.	Status	Comments
12ʰ04.3ᵐ	+21° 28′	2 Com	2003	236°	3.7″	6.2	7.5	*F*01V-V	binary	125-mm, 83×: Grand! A bright pair of stars split by hairs, yellow-white and green-white in color. The companion is also called "rosey" (Webb), "lilac tinted" (Smyth), and "deep yellow" (Hartung).
12ʰ14.1ᵐ	+32° 47′	Σ1615	2003	87°	26.6″	7.0	8.6	*K*1III *F*8V		Webb, listed under Canes Venatici: "Yellow, ashy."
12ʰ17.5ᵐ	+28° 56′	OΣ 245	1999	281°	8.3″	5.7	10.2	*A*3IV		Gould, 175-mm, 100×: "Easy. A bright white star in a scattered field, which has a small companion fairly close to it."
12ʰ20.7ᵐ	+27° 03′	Σ1633	2004	245°	8.9″	7.0	7.1	*F*3V+*F*3V	binary	60-mm: A splendid and easy binary. It's a fairly bright pair of peach-white stars, exactly alike, that are split by hairs at 25×.

Coma Berenices *(continued)*

RA	Dec.	Name	Year	P.A.	Sep.	m_1	m_2	Spec.	Status	Comments
$12^h 22.5^m$	+25° 51′	12 Coma								110-mm, 32×: This super-wide pair has a pretty primary star. It's a bright Sun-yellow star with a small nebulous companion. Webb: "Yellow, pale blue." Franks calls the companion "lilac."
		AC	2002	168°	65.2″	4.9	8.9	*G*5III+*A*5		
$12^h 24.4^m$	+25° 35′	Σ1639	2004	325°	1.7″	6.7	7.8	*A*7V+*F*4V	p=575 yr	It should now be possible to split this binary with a 110-mm; the separation is slowly widening. Webb: "White, green white."
$12^h 28.9^m$	+25° 55′	17 Com	2002	251°	146.2″	5.2	6.6	*A*0		60-mm, 25×: Good object for very low power. It's a very bright pair of stars, gloss white and very bluish white, that are super-wide apart. Mildly unequal.
$12^h 35.1^m$	+18° 23′	24 Com	2004	270°	20.1″	5.1	6.3	*K*2III	optical	Showcase pair. 60-mm, 25×: A bright, wide and easy pair with pretty colors. The stars are citrus orange and fainter royal blue, and they're attractively close while wide enough to be easy.
$12^h 45.4^m$	+14° 22′	Σ1678	2003	172°	36.7″	7.2	7.7	*B*8V		125-mm, 50×: An ordinary but conspicuous pair. It's a bright pair of stars, very wide apart, that are easily noticed about 1° west-northwest of 29 Comae. They're blue-white and yellow-white.
$12^h 51.9^m$	+19° 10′	Σ1685	2002	202°	16.0″	7.3	7.8	*A*+*F*8III		125-mm, 50×: An attractively close pair that's fairly bright. It's a blue-white star and a green-white, slightly unequal, that are split by only a modest gap.
$12^h 52.2^m$	+17° 04′	32-33 Com	2000	51°	195.6″	6.5	7.0	*M*0III		60-mm, 25×: Grand sight! The colors of this pair are bright grapefruit orange and fainter pearly white; these colors are vivid. The stars are super-wide apart but look like a couple.
$12^h 53.3^m$	+21° 15′	35 Com								Gould, 175-mm, 180×: "Good triple. A close pair of bright orange-yellow stars, and a wide [third] companion that is possibly blue." Smyth: "[A] pale yellow, [B] indistinct, [C] cobalt blue."
		AB	2004	186°	1.2″	5.2	7.1	*G*7III	p=359 yr	
		AC	2004	127°	27.9″	5.0	9.8			
$13^h 10.0^m$	+17° 32′	α Com	2004	12°	0.4″	4.9	5.5	*F*5V + *F*6V	p=25.8 yr	A super-fast-moving binary. Smyth notes that it went from "quite round" to elongated during the course of a decade. The separation should reach 0.6″ by 2011 and then rapidly diminish.
$13^h 16.9^m$	+17° 01′	β 800	2004	107°	7.5″	6.7	9.5	*K*1V *M*1V	p=770 yr	125-mm, 62×: This striking binary is about 40′ south of the globular NGC 5053. It's a bright yellowish crimson star, almost touching a little dot of whitish opal. Hartung: "Orange and red."
$13^h 21.8^m$	+17° 46′	Σ1737	1998	219°	14.9″	7.9	10.3	*F*0	binary	Smyth: "A delicate double star, [just past] the right foot of Boötes . . . white; blue."
$13^h 28.4^m$	+15° 43′	OΣ 266	2004	356°	2.0″	8.0	8.4	*F*5	p. binary	This double star should be resolvable with 125-mm. Webb: Both stars are "green white."

Corona Australis

RA	Dec.	Name	Year	P.A.	Sep.	m_1	m_2	Spec.	Status	Comments
$18^h 06.4^m$	−41° 45′	h 5011	1998	345°	28.5″	7.6	8.5	*A0V*		Gould, 175-mm, 100×: A "wide pair of moderate brightness, just south of a gathering of stars. Some 5′ northeast is a less bright and closer pair, in pleasing contrast."
$18^h 06.8^m$	−43° 25′	h 5014	2002	8°	1.7″	5.7	5.7	*A5V A5V*	p=450 yr	Gould, 175-mm, 180×: "Fine object. A beautiful even pair, nicely apart, with the partially resolved globular NGC 6541 in the field. The stars look yellow."
$18^h 33.4^m$	−38° 44′	κ CrA	1998	358°	21.3″	5.6	6.2	*B9V A0III*	optical	Showcase pair. Gould, 175-mm, 100×: An "easy bright wide pair, in a field lacking other bright stars. It is a tiny pair of touching stars with an 8 × 50 finder! Both stars look white."
$18^h 43.8^m$	−38° 19′	λ CrA	1999	214°	29.2″	5.1	10.0	*A2V*		Gould, 175-mm, 100×: This "bright cream-white star has an obvious wide companion, dull tone yellow in color. The field is fairly starry."
$18^h 51.0^m$	−41° 04′	h 5066	1999	84°	10.1″	6.5	9.5	*B5/7III*	optical	Gould, 175-mm, 100×: An "effective uneven pair, of easy separation . . . the primary star is white, and the field is moderately rich in fainter stars west and south."
$19^h 01.1^m$	−37° 04′	BrsO 14	1999	280°	12.8″	6.3	6.6	*B8V B9V*		Gould, 175-mm, 100×: A bright and well separated even pair, both stars white. A few minutes north are the nebulae NGC 6726-7-9, and the globular cluster NGC 6723; about 1/2° west is Epsilon (ε) CrA."
$19^h 06.4^m$	−37° 04′	γ CrA	1999	62°	1.3″	4.5	6.4	*F8V F8V*	p=122 yr	Gould, 175-mm, 180×: This is a "bright pale yellow star in a sparse field; at 180×, it is a close separated pair . . . a fine pair for 100-mm or more."

Corona Borealis

RA	Dec.	Name	Year	P.A.	Sep.	m_1	m_2	Spec.	Status	Comments
$15^h 18.3^m$	+26° 50′	Σ1932	2004	261°	1.6″	7.3	7.4	*F6V+F6V*	p=203 yr	125-mm, 200×: Grand! A bright and perfect figure-8 — a pair of identical stars, yellow-white in color, that are almost but not quite fully separated. Webb: "Green-white [both stars]."
$15^h 23.2^m$	+30° 17′	η CrB	2004	107°	0.5″	5.6	6.0	*F8V+G0V*	p=41.6 yr	This is one of the fastest-moving binaries that amateurs can now still resolve. When the separation was 1.0″, a 125-mm at 350× showed an amber-yellow figure-8.
$15^h 38.2^m$	+36° 15′	Σ1964 AC	2004	85°	15.3″	7.9	8.1	*F5*		60-mm: A fine easy pair, easily noticed wide in the field with Zeta (ζ) Coronae Borealis. It's a pair of identical stars, gloss white in color, that are split by only a small gap at 25×.
$15^h 39.4^m$	+36° 38′	ζ CrB	2003	306°	6.3″	5.0	5.9	*B7V B9V*	binary	Showcase pair. 60-mm, 40×: A lovely close pair of vivid stars — a very bright white and a modestly bright white. Smyth: "Bluish white; smalt blue." Webb: "Flushed white, bluish green."

Corona Borealis *(continued)*

RA	Dec.	Name	Year	P.A.	Sep.	m_1	m_2	Spec.	Status	Comments
$15^h42.7^m$	+26° 18′	γ CrB	2004	114°	0.7″	4.0	5.6	*B9IV+A3V*	p=92.9 yr	This fast-moving binary should be split by 200-mm; the separation is still increasing. Smyth: "Flushed white; [companion] uncertain." Webb: "Greenish, white."
$15^h46.4^m$	+36° 27′	Σ1973	2002	321°	30.5″	7.6	8.8	*F5*		125-mm, 50×: An easy wide pair in an interesting field. It's a fairly bright white star with a little gray companion; they are centered in a cross-shaped asterism.
$16^h11.7^m$	+33° 21′	OΣ 305	1991	263°	5.6″	6.4	10.2	*K2III*	binary	Gould, 175-mm, 100×: "Not difficult" despite great contrast. "A bright orange/deep yellow star with a tiny close companion, in a moderately starry field."
$16^h13.8^m$	+28° 44′	Σ2029	1999	187°	6.1″	8.0	9.6	*F4IV*	binary	125-mm, 83×: Striking contrast for the separation! A gloss-white star with a shadowy spot next to it split by only a small gap.
$16^h14.7^m$	+33° 52′	σ CrB	2004	236°	7.0″	5.6	6.5	*G0V G1V*	p=889 yr	Showcase pair. 60-mm: A bright, vivid pair of lemon-white stars, mildly unequal, that are split by a hair at 25×. Smyth: "Creamy white; smalt blue." Hartung: "Bright deep yellow pair."
$16^h22.4^m$	+33° 48′	ν^1- ν^2 CrB	2001	164°	360.8″	5.4	5.6	*M2III*		60-mm, 25×: Grand sight at low power! A brilliant pair of grapefruit-orange stars, super-wide apart, that are identical even in color. Webb: Both stars are "yellow."
$16^h22.9^m$	+32° 20′	H V 38	1998	17°	31.7″	6.4	9.8	*A4V*		125-mm, 50×: Pretty pair. A bright star with a wide little companion in the pretty combination of lemon yellow and whitish Atlantic blue. Smyth, referring to the pair as 23 Herculis: "White; violet."

Corvus

RA	Dec.	Name	Year	P.A.	Sep.	m_1	m_2	Spec.	Status	Comments
$12^h09.5^m$	−11° 51′	Σ1604								125-mm, 50×: Grand! A broad triple star shaped as a triangle, made of three different colors — grapefruit orange, sky blue, and silvery nebulous. The bright primary is especially pretty.
		AB	2003	88°	9.2″	6.6	9.4	*G3V*		
		AC	2003	28°	10.4″	6.9	8.1			
$12^h11.4^m$	−16° 47′	S 634	1998	298°	4.9″	7.2	8.8	*G6V*	p. binary	125-mm, 83×: This pair has a pretty hue. It's a fairly wide pair of stars quite different in brightness but alike in color — they're an exactly equal mix of yellow and orange.
$12^h15.8^m$	−23° 21′	β 920	2000	304°	1.8″	6.9	8.2	*F7V*	binary	Hartung: A "close yellow and white pair . . . 105-mm will now resolve the stars quite easily."
$12^h29.9^m$	−16° 31′	δ Crv	2003	217°	24.9″	3.0	8.5	*B9.5V*		Showcase pair. 60-mm, 25×: A beautiful straw-yellow star that is brilliantly bright with a sharp little gray dot beside it — it looks like a star with a planet! Smyth: "Pale yellow; purple."
$12^h35.7^m$	−12° 01′	Σ1659								Webb listed under Virgo: "Dembowski notices that this triangularly shaped triple is within a large triangle [of field stars]."
		AB	2003	352°	27.5″	8.0	8.3	*G0*		
		AC	2003	69°	42.2″	8.0	11.0			

Corvus *(continued)*

RA	Dec.	Name	Year	P.A.	Sep.	m_1	m_2	Spec.	Status	Comments
$12^h 41.3^m$	−13° 01′	Σ1669	2003	312°	5.3″	5.9	5.9	*F5V F5V*		60-mm, 35×: Grand sight! A close pair of bright twins, peach white in color, that completely dominate their field. Webb: Both stars are "yellow white."
$12^h 54.0^m$	−18° 02′	S 643	2002	295°	23.7″	7.1	8.2	*A5III A0*		125-mm, 50×: A very conspicuous wide pair, brighter than average, that sits in a barren field. It's a whitish gold star with a small gray companion.

Crater

RA	Dec.	Name	Year	P.A.	Sep.	m_1	m_2	Spec.	Status	Comments
$11^h 17.0^m$	−07° 08′	β 600								125-mm, 50×: This pair is included because of its pretty primary star. It's a bright Sun-yellow star and a small aquamarine, super-wide apart.
		AC	2002	99°	54.3″	6.2	8.2	*A8IV*		
$11^h 24.9^m$	−17° 41′	γ Crt	1968	93°	5.3″	4.1	7.9	*A5V*		Showcase pair. Hartung: "A fine object for small apertures, 75-mm showing both the stars clearly." Smyth: "Bright white; gray" . . . [an] 8th-magnitude star . . . is at a distance."
$11^h 29.6^m$	−24° 28′	Jc 16	2000	82°	8.2″	5.8	8.6	*A0V*		125-mm, 83×: Grand sight! A very bright star almost touching a very tiny one, in the pretty combination of straw yellow and silvery azure.
$11^h 48.4^m$	−10° 19′	H VI 115	2002	66°	89.9″	6.3	9.2	*F7V*		80-mm, 22×: This wide pair is included because of its pretty primary star. It's a very bright star, pale peach in color, that has a little nebulous companion. It's a fine, easy pair at low power.

Crux

RA	Dec.	Name	Year	P.A.	Sep.	m_1	m_2	Spec.	Status	Comments
$12^h 04.8^m$	−62° 00′	Δ 117								Jaworski, 100-mm: "A grouping of three stars [in a boomerang shape], set in a nice field of faint stars. It is two bluish white stars, roughly equal . . . and a fainter component."
		AB	2000	149°	22.7″	7.4	7.8	*B8/9I*		
		AC	2000	18°	25.1″	7.4	10.0			
$12^h 24.9^m$	−58° 07′	BrsO 8	2000	335°	5.2″	7.8	8.0	*G0IV/V*	p. binary	Jaworski, 100-mm: This double star is "an average close pair of nearly equal brighness . . . well separated at 167×, but could be seen as double at 50×. Both stars look white."
$12^h 26.6^m$	−63° 06′	α Cru								Showcase system *(Acrux)*. Jaworski, 100-mm, 167×: This "magnificient object is three stars in a cone shape — a close pair of blue-white stars that sparkle like gems, and a contrasting yellow-green."
		AB	2000	114°	3.9″	1.3	1.6	*B1V*	p. binary	
		AC	1991	202°	90.0″	1.3	4.8	*B4IV*		
$12^h 31.2^m$	−57° 07′	γ Cru								Jaworski, 100-mm, 80×: "The stars of this triple form a nearly perfect right triangle; a rich orange star and a bluish white star form the base, and a bluish star forms the apex."
		AB	2000	26°	127.3″	1.6	6.5	*M3.5III*		
		AC	1879	82°	155.1″	1.6	9.5			
$12^h 43.5^m$	−58° 54′	h 4543								The data suggest a bright star with a pretty color and a tiny companion.
		AC	2000	95°	36.5″	6.5	9.8	*K0-1III*		

Crux *(continued)*

RA	Dec.	Name	Year	P.A.	Sep.	m_1	m_2	Spec.	Status	Comments
$12^h 46.4^m$	−56° 29′	h 4548	2000	166°	51.7″	5.0	8.9	*B*3V		The data suggest a very bright star with a dim companion. Jaworski, 100-mm, 80×: "Yellowish and light blue."
$12^h 47.7^m$	−59° 41′	β Cru AC	1991	23°	373.0″	1.3	7.2	*B*7II		*(Mimosa)* Hartung: A "crimson red star in the same field as this brilliant white star [forms this pair] . . . an excellent object for small apertures."
$12^h 54.6^m$	−57° 11′	μ Cru	2001	17°	35.0″	3.9	5.0	*B*2IV-V		Showcase pair. Jaworski, 100-mm: A "wide, easy pair, well seen at 80×. The stars are white and yellowish, with a moderate degree of contrast." Hartung: It "lies in a starry field."

Cygnus

RA	Dec.	Name	Year	P.A.	Sep.	m_1	m_2	Spec.	Status	Comments
$19^h 12.1^m$	+49° 51′	Σ2486	2004	206°	7.3″	6.5	6.7	*G*3V *G*3V	p=3,100 yr	150-mm, 36×: Grand! A bright pair of kissing twins, vivid gold in color. Webb: In a "singular and beautiful field."
$19^h 30.7^m$	+27° 58′	β Cyg	2003	55°	34.7″	3.4	4.7	*K*3II+*B*0V		Showcase pair *(Albireo)* — the favorite double star of many. 60-mm, 25×: This wide couple is a stunning pair of deeply colored stars, brilliant citrus orange and vivid royal blue, inside a dark cloud within a field packed with stars. Smyth: "Topaz yellow; sapphire blue; the colours in brilliant contrast."
$19^h 41.8^m$	+50° 32′	16 Cyg	2003	134°	39.1″	6.0	6.2	*G*1.5V		Showcase pair. 60-mm, 25×: A bright wide pair of whitish gold stars, exactly alike, in a wonderful starry field.
$19^h 45.0^m$	+45° 08′	δ Cyg	2004	225°	2.5″	2.9	6.3	*B*9.5IV	p=780 yr	125-mm, 200×: Grand! A brilliant star with a crisp little globe on its edge in the lovely combination of lucid white and greenish silver. Smyth: "Pale yellow; sea green."
$19^h 45.7^m$	+36° 05′	Σ2578	2003	125°	14.7″	6.4	7.0	*B*9.5V		125-mm, 50×: A wide pair of bright stars quite similar in brightness but different in color — they look pale lemon and greenish white.
$19^h 46.4^m$	+33° 44′	17 Cyg	2001	67°	26.3″	5.1	9.3	*F*5V		110-mm, 100×: Wide pair — a bright grapefruit-orange star with a misty dot next to it. Webb: "Very yellow, blue . . . beautiful field." Smyth: "Golden yellow; pale blue."
$19^h 46.6^m$	+32° 53′	S 726 AD	2003	192°	29.2″	6.2	9.2			Castle, 150-mm: "An extremely wide and very nice pair at 171×; it is a clone of 17 Cyg [above], and these two pairs lie in a straight N-S line with each other."
$19^h 49.0^m$	+44° 23′	Σ2588 A-BC	2002	159°	9.9″	7.7	8.1	*B*8III	binary	125-mm, 50×: A bright easy binary. It's a pair of nearly equal white stars that are fairly wide apart.
$19^h 50.6^m$	+38° 43′	19 Cyg	2003	114°	56.3″	5.4	10.5	*M*2III		An extremely wide pair with a pretty primary star. Castle, 150-mm, 171×: It is a "reddish star that looks like a naked eye view of Mars, with a very dim blue companion."
$19^h 55.1^m$	+30° 12′	OΣ 390	2003	22°	9.4″	6.6	9.5	*B*6V+*A*5V	binary	Castle, 150-mm, 171×: "A nice triple." It's a white star with two wide companions — one is bluish and the other "dim and difficult."

RA	Dec.	Name	Year	P.A.	Sep.	m_1	m_2	Spec.	Status	Comments
$19^h 55.6^m$	+52° 26′	ψ Cyg	2003	178°	2.9″	5.0	7.5	*A4V*	binary	125-mm, 83×: Grand! A bright star with a little globe on its edge in the pretty combination of straw yellow and ocean blue.
$19^h 57.9^m$	+42° 16′	Σ2607								The data suggest a bright star with a small companion. Castle, 150-mm: "A close pair that I could split at 273× . . . both stars appear white."
		AB-C	2003	289°	3.0″	6.6	9.1	*A3V*	binary	
$19^h 58.6^m$	+38° 06′	Σ2609	2003	22°	1.9″	6.7	7.6	*B5IV*	binary	The data suggest a fairly bright binary. Castle, 150-mm, 171×: A "very close pair that appears bluish and white but the color is subtle."
$20^h 01.4^m$	+50° 06′	26 Cyg	2003	150°	41.4″	5.2	8.9	*K1II-III*		125-mm, 50×: A very wide pair with pretty colors. It's a brilliant citrus-orange star and a whitish lilac, and the companion is strongly contrasting but easy to see. Webb: "Gold or yellow, blue."
$20^h 03.5^m$	+36° 01′	Σ2624								The data suggest a fairly bright binary. Castle, 150-mm: "A very close and difficult challenge . . . 366× made a clean split."
		AB	2002	175°	1.9″	7.1	7.7	*O9.5I*	binary	
$20^h 06.0^m$	+35° 47′	SHJ 314								Heckman, 250-mm, 92×: This is a "nice triangular asterism" formed by a pair of whitish twins (AF) and a dimmer third star (D).
		AD	2003	301°	11.0″	6.8	9.5	*B1V*	optical	
		AF	2003	28°	34.7″	6.8	7.3			
$20^h 09.3^m$	+35° 29′	Σ2639	2002	302°	5.7″	7.8	8.7	*B0.5IV*	binary	The data suggest an easy fairly close binary. Webb: "Yellow white, ashy white. Beautiful."
$20^h 10.6^m$	+33° 38′	S 738								Castle, 150-mm, 171×: AB is "a nicely matched, very wide pair . . . in a nice group of 8th-magnitude stars."
		AB	1998	106°	42.1″	7.8	8.4	*B9V*		
		BC	1999	183°	1.8″	8.4	9.7		binary	
$20^h 12.3^m$	+32° 05′	Σ2649	2003	150°	21.4″	7.9	9.9	*A2*	optical	Castle, 150-mm: A "very nice, very wide pair at 171× . . . white and bluish." Webb: "Yellow white, ashy."
$20^h 12.5^m$	+51° 28′	AC 17	1991	82°	4.3″	6.2	10.6	*K2.5III*		The data suggest a bright star with a close dim companion. A 150-mm should be able to split it.
$20^h 13.6^m$	+53° 07′	Σ2658	2000	108°	5.4″	7.2	9.4	*F5V*		Castle, 150-mm: An "attractive wide pair at 171×, but no color noted."
$20^h 14.5^m$	+36° 48′	29 Cyg	2002	156°	215.8″	5.0	6.7	*A2V*		Heckman, 250-mm, 92×: A "nice wide pair with color contrast. Blue-white with yellow-orange."
$20^h 18.1^m$	+40° 44′	Σ2666	2003	246°	2.8″	6.0	8.2	*O9V*	binary	Castle, 150-mm, 171×: "A very close little pretty pair, the stars blue and gold. It's in a nice field of 8th-magnitude and dimmer stars."
$20^h 18.4^m$	+55° 24′	Σ2671	2004	338°	3.6″	6.0	7.5	*A2V*	binary	Heckman, 250-mm, 125×: A "close pair featuring magnitude and color contrast. White with yellow orange."
$20^h 20.3^m$	+39° 24′	Σ2668								125-mm, 83×: Almost a miniature Delta (δ) Cygni (see opposite page). It's a bright white star almost touching a tiny silvery companion. Webb: "Yellow white, ash."
		AB-C	2003	282°	3.4″	6.3	8.5		binary	

RA	Dec.	Name	Year	P.A.	Sep.	m_1	m_2	Spec.	Status	Comments
20ʰ 22.2ᵐ	+40° 15′	γ Cyg								BC is only 141″ from Gamma (γ) Cygni and is probably resolvable with 300-mm. According to the listed separation, it is tight like a typical binary star — and probably is one.
		A-BC	1980	196°	141.0″	2.2	5.4	*F*8I		
		BC	1929	302°	1.9″	10.0	11.0			
20ʰ 22.9ᵐ	+42° 59′	ΟΣΣ 207								Heckman, 250-mm, 92×: A "beautiful pair featuring exceptional color contrast. Deep yellow and deep blue."
		AC	2003	64°	86.9″	6.4	8.0	*G*8III		
20ʰ 23.0ᵐ	+39° 13′	ΟΣΣ 206	2003	255°	42.8″	6.7	8.6	*B*6III		Heckman, 250-mm, 92×: This pair has "contrasting magnitudes with very nice color contrast. Blue and yellow-orange."
20ʰ 26.4ᵐ	+56° 38′	Σ2687	2003	117°	25.8″	6.4	8.3	*B*9V		125-mm, 50×: A bright star with an obvious wide companion in the pretty colors of whitish banana yellow and pale blue-green. Webb: "White, ash." The pair is about 2° east of 33 Cygni.
20ʰ 37.5ᵐ	+31° 34′	48 Cyg	2003	177°	182.7″	6.3	6.5	*B*8III		Webb: "White, yellow white . . . in a splendid region. 50′ north are two similiar but smaller pairs."
20ʰ 37.7ᵐ	+33° 22′	Σ2705	1998	262°	3.0″	7.5	8.5		binary	Webb: "Yellow, blue." A "curious rhombus" star formation is in the field.
20ʰ 39.6ᵐ	+40° 35′	ΟΣ 410								Webb: "Two very yellow, gold."
		AB	2003	5°	0.9″	6.7	6.8	*B*8III	p. binary	
		AB-C	2004	69°	68.1″	6.8	8.7			
20ʰ 41.0ᵐ	+32° 18′	49 Cyg	2003	46°	3.0″	5.8	8.1	*G*2III	binary	125-mm, 200×: A striking but difficult pair. It's a bright gold star that shows, with close attention, a little gray companion at its edge.
20ʰ 45.6ᵐ	+30° 43′	52 Cyg	1997	69°	6.4″	4.2	8.7	*G*9III		125-mm, 83×: Striking! A brilliant star almost touching a glimpse star, in the pretty colors of whitish pumpkin orange and grayish sapphire.
20ʰ 47.2ᵐ	+42° 25′	ΟΣ 414	2002	94°	9.9″	7.4	8.9	*B*7V	binary	Heckman, 250-mm, 92×: "A very nice color contrast. Yellow and blue."
20ʰ 47.2ᵐ	+34° 22′	T Cyg	1941	115°	10.0″	4.9	10.0	*K*3III		Hartung (referring to the pair as β 677): "This bright orange star shines like a jewel in a fine starry field. . . [and has] a "very faint point well clear [of it]."
20ʰ 47.4ᵐ	+36° 29′	λ Cyg	2004	7°	0.9″	4.7	6.3		p=391 yr	A 200-mm should be able to split this binary (now moving fairly rapidly) until at least 2025. Smyth: Both stars are "bluish."
20ʰ 48.7ᵐ	+51° 55′	Σ2732	2003	73°	4.3″	6.4	8.6	*B*8	binary	Heckman, 250-mm, 92×: This binary is a "challenging pair with a large magnitude contrast. The bright white primary overwhelms the dim companion."
20ʰ 58.5ᵐ	+50° 28′	Σ2741	2003	25°	2.0″	5.9	6.8	*B*5V	binary	The data suggest a close pair of fairly bright stars. Dekanich, 200-mm: Both stars look "blue white . . . a difficult pair at low power."

Cygnus *(continued)*

RA	Dec.	Name	Year	P.A.	Sep.	m_1	m_2	Spec.	Status	Comments
20h 59.8m	+47° 31′	59 Cyg								The data suggest a bright star with faint companions. Smyth: "A delicate triple star in the Swan's tail . . . orange tint; [others] both blue."
		Aa-B	2000	353°	20.4″	4.6	9.4	*B*1		
		Aa-C	2000	140°	26.5″	4.6	11.6			
21h 02.3m	+39° 31′	H IV 113	2003	299°	18.7″	6.6	9.5	*K*3II-III	binary	125-mm, 50×: Lovely pair of stars – bright yellowish apricot and dim sky blue. They're attractively close while wide enough to be easy.
21h 06.8m	+34° 08′	Σ2760	2003	31°	4.0″	7.9	8.7	*A*4III	p. binary	The data suggest a close but not too difficult double star. Webb: "Yellowish white, ashy."
21h 06.9m	+38° 45′	61 Cyg	2004	151°	31.1″	5.3	6.1	*K*5V *K*7V	p=659 yr	Showcase pair and remarkable object. It's two bright stars that are wide apart but show orbital motion. (It's just 11 light-years away.) 60-mm, 25×: Both stars are amber yellow.
21h 08.6m	+30° 12′	Σ2762								The data suggest a bright star with a close dim companion. Webb: "Yellow white, ruddy." Smyth: "Dull white; pale lilac; and there is a third star [in the field]."
		AB	2004	304°	3.2″	5.7	8.1	*B*9V	binary	
21h 19.0m	+39° 45′	OΣ 434	2003	122°	24.3″	6. 7	9.9	*B*9V		This pair is about 1/2° northeast of Sigma (σ) Cygni; the data suggest a fairly bright star with a wide companion. Heckman, 250-mm, 96×: An "easy blue pair with a dim companion."
21h 19.7m	+53° 03′	S 786	2003	300°	47.1″	6.9	9.2	*K*2		These pairs are about 10′ apart. Heckman: S 786 is a "wide pair of contrasting magnitudes . . . yellow and blue." 125-mm, 50×: Σ2789 is a pair of twins, like headlights in a sea of stars.
21h 20.0m	+52° 59′	Σ2789	1998	115°	6.8″	7.7	7.9	*F*8V		
21h 20.8m	+32° 27′	OΣ 437	2003	22°	2.3″	7.2	7.4	*G*4V	p. binary	Hartung: "75-mm easily resolves this close orange-yellow pair." Heckman, 250-mm, 185×: "Superb yellowish pair of equal magnitude."
21h 25.8m	+36° 40′	69 Cyg								The data suggest a bright star with a wide companion. It shares a low-power field with the cluster NGC 7063 and 70 Cygni. Heckman, 250-mm, 96×: An "easy blue pair with a large magnitude contrast."
		AC	2003	99°	53.3″	5.9	10.2	*B*0I		
21h 39.5m	+41° 44′	OΣ 447								Heckman, 250-mm, 96×: An "easy pair with beautiful yellow and blue color contrast."
		AE	2002	45°	29.6″	7.7	8.5	*K*0		
21h 44.1m	+28° 45′	μ Cyg	2004	312°	1.9″	4.7	6.2	*F*6V *G*2V	p=789 yr	Heckman, 250-mm: "A nice whitish pair, cleanly but just barely separated at 125×." Smyth: "White; blue." Webb: "Yellow, [with] tawny or blue or lilac."
21h 49.2m	+50° 31′	Σ2832	2004	213°	13.3″	7.8	8.3	*B*9IV	optical	Heckman, 250-mm, 71×: "Among my favorite pairs in Cygnus. A very nice bluish pair of nearly equal magnitude."

Delphinus

RA	Dec.	Name	Year	P.A.	Sep.	m_1	m_2	Spec.	Status	Comments
20h 19.4m	+14° 22′	Σ2665								Webb apparently resolved this pair and gave the same separation as listed. The data make it seem very difficult, but the largest aperture he used was 200-mm.
		A-BC	2003	13°	3.4″	6.9	9.6	*A*0+*G*	binary	

Delphinus *(continued)*

RA	Dec.	Name	Year	P.A.	Sep.	m_1	m_2	Spec.	Status	Comments
20h 24.4m	+19° 35′	Σ2679	2002	77°	24.2″	7.9	9.7	*A*2V		125-mm, 50×: This wide pair shows lovely contrast, though the companion is hard to see. It's a gloss-white star with a ghostly speck next to it, visible only with averted vision.
20h 30.3m	+10° 54′	1 Del	2003	349°	0.9″	6.2	8.0	*A*1	binary	The data suggest a bright star with a close faint companion. Hartung: A "close pair . . . in 1961 [when the separation was 1.1″], 150-mm separated them."
20h 31.0m	+20° 36′	β 363	1962	70°	11.8″	6.2	10.0	*A*1		The bright white primary is in a splashy group of bright stars, a fine sight with an 8 × 50 finderscope. Its companion was not seen with 125-mm, but it's probably easy with more aperture.
20h 31.2m	+11° 16′	Σ2690								125-mm, 50×: An wide and pretty pair, nicely placed. It's a bright pair of peach-white stars, very nearly identical, that are wide in the field with brilliant white Epsilon (ε) Delphini.
		Aa-BC	2003	255°	17.3″	7.1	7.4	*B*8V		
20h 37.5m	+14° 36′	β Del								This super-fast binary is now (2006) at its widest but is rapidly closing. Hartung: A "close bright yellow binary."
		AB	2004	359°	0.6″	4.1	5.0	*F*5IV+*F*2V	p=26.6 yr	
20h 39.1m	+15° 50′	β 288	1934	159°	6.8″	6.0	12.4	*B*3V		The data suggest a bright star with a very faint companion; Webb notes that it is just 9′ west-southwest of Alpha (α) Delphini — did he resolve it?
20h 39.1m	+10° 05′	κ Del	1970	282°	34.7″	5.0	11.8	*G*1IV+*K*2IV		Smyth: "A delicate but wide double star . . . white; pale lilac."
20h 40.3m	+03° 26′	ΟΣ 409	2000	84°	16.8″	7.1	10.2	*K*0		125-mm, 50×: A lovely amber-yellow star with a tiny nebulous companion split by a wide gap.
20h 44.9m	+12° 19′	Σ2723	2001	135°	1.0″	7.0	8.3	*A*3IV	binary	Hartung listed this as a binary for amateurs. He says: "In 1961 [when the separation was 1.2″] the stars were clearly apart with 200-mm."
20h 46.2m	+15° 54′	Σ2725	2004	10°	6.1″	7.5	8.2	*K*0	p=2,851 yr	60-mm, 35×: Remakable sight! This pair is in the same field with Gamma (γ) Delphini (below) and looks like a smaller duplicate of it!
20h 46.7m	+16° 07′	γ Del	2004	266°	9.1″	4.4	5.0	*K*1IV *F*7V	p=3,249 yr	Showcase pair. 60-mm, 25×: A stunning pair of stars, close together and brilliant, that are grapefruit orange in color. Smyth: "Yellow; light emerald." Gore: "Reddish yellow, greyish lilac."
20h 47.8m	+06° 00′	13 Del	1991	199°	1.5″	5. 6	8.2	*A*0V	binary	The data suggest a bright star with a close, faint companion. Webb: "White, olive."
20h 55.7m	+04° 32′	Σ2735	2003	281°	1.9″	6.5	7.5	*G*6III-IV	binary	125-mm, 200×: Grand sight! This is a bright Sun-yellow star almost touched by a bright pearly white, and both stars shine vividly though they're mildly unequal.
20h 58.5m	+16° 26′	Σ2738								125-mm, 50×: Interesting triple. It's a pair of white stars that are super-wide apart; one of them has a small blue companion.
		AB	2002	255°	14.9″	7.5	8.6	*F*5V *A*0	optical	
		AC	2002	104°	209.8″	7.5	8.1			

Dorado

RA	Dec.	Name	Year	P.A.	Sep.	m_1	m_2	Spec.	Status	Comments
$04^h 24.2^m$	−57° 04′	Rmk 4	2001	246°	5.5″	6.9	7.2	*G4V*	p. binary	Gould, 175-mm, 100×: A "nicely separated, nearly even yellow pair . . . galaxy NGC 1574 is visible near the western edge of the field."
$04^h 40.3^m$	−58° 57′	h 3683	2002	89°	3.4″	7.3	7.5	*G5V*	p=240 yr	Gould, 17-mm, 100×: "Fine effect — a bright, fairly close yellow pair; the field is thin but has two fairly bright stars in it."
$05^h 27.0^m$	−68° 37′	I 276	1999	163°	1.4″	6.7	7.0	*F0IV-V*	p. binary	Gould, 175-mm: "Good double in a fine area. A nearly even yellow pair, barely split at 100×. Various clusters and nebulae are in the field, which is part of the Large Magellanic Cloud."
$06^h 12.2^m$	−65° 32′	Δ 26	1998	120°	20.6″	6.9	8.1	*F6V+F7IV*	optical	Gould, 175-mm, 100×: A "wide easy bright pair" that sits in a row of three stars — this star, a less bright star, and bright yellow Eta² (η^2) Doradus.

Draco

RA	Dec.	Name	Year	P.A.	Sep.	m_1	m_2	Spec.	Status	Comments
$09^h 37.9^m$	+73° 05′	Σ1362	2001	126°	4.8″	7.0	7.2	*F1V+F2V*	binary	The data suggest a fairly bright pair of stars. Geertsen, 275-mm, 70×: "Ivory and cool gray. In a 9 × 50 finderscope, it appears slightly elongated."
$11^h 49.2^m$	+67° 20′	Σ1573	1999	178°	11.1″	7.5	8.3	*G8III*	optical	110-mm, 32×: A modestly dim, but attractively close pair. It's an amber-yellow star and a smaller silvery blue split by just a tiny gap.
$12^h 32.1^m$	+74° 49′	Σ1654	2000	24°	3.7″	7.7	9.4	*G9III*	binary	Webb: "Yellow, blue." Geertsen, 100-mm, 62×: "There is a third bright star in the same field of view."
$13^h 13.5^m$	+67° 17′	Σ25	1999	296°	179.0″	6.6	7.1	*K2III*		60-mm, 25×: This super-wide pair forms a nice little asterism. The view shows a neat right triangle of three stars — a peach-white star, a plain white star, and a peach-white field star.
$13^h 27.1^m$	+64° 44′	ΟΣΣ 123	2000	147°	68.9″	6.6	7.0	*F0*		60-mm, 25×: A bright, very wide pair — *both* stars look hauntingly solid blue. Webb calls this "a striking object" but says nothing more; did he also see two blue stars?
$14^h 33.9^m$	+55° 14′	Σ1860	1998	109°	0.9″	8.0	9.0	*A5*	binary	Webb calls this binary "very white, ashy white."
$14^h 41.0^m$	+57° 57′	Σ1872	2004	48°	7.9″	7.5	8.3	*K0*	p. binary	125-mm: A pretty pair of pure lemon-yellow stars that are just kissing at 83×. They look more unequal than their listed magnitudes.
$14^h 41.6^m$	+51° 24′	Σ1871	2004	310°	1.8″	8.0	8.1	*F3V*	binary	Geertsen, 100-mm, 233×: "This [binary] is reasonably bright. Both orange stars appear equal, just touching, and surrounded by shimmering diffraction rings."
$14^h 42.1^m$	+61° 16′	Σ1878	2004	317°	4.2″	6.3	9.2	*F4V*	binary	125-mm, 83×: A difficult but striking pair. It's a bright white star almost touching an elusive companion — a shadowy presence just barely glimpsed.

Draco *(continued)*

RA	Dec.	Name	Year	P.A.	Sep.	m_1	m_2	Spec.	Status	Comments
$14^h 44.1^m$	+61° 06′	Σ1882	2003	0°	11.5″	6.9	9.2	*F3V*		125-mm, 83×: This pair is in the same field with Σ1878 (previous page). The data make it sound similar to that pair, but it looks very different. It's an easy wide pair with an easily seen companion.
$15^h 20.1^m$	+60° 23′	ΟΣΣ 138								60-mm, 25×: Nice little asterism! It's a broad triple shaped as a neat triangle — a pair of lemon-white twins at the base and a small nebulous star at the apex. It's about 1½° north-northwest of Iota (ι) Draconis (below).
		AB	2002	196°	152.1″	7.6	7.8	*F0*		
		AC	2002	163°	88.9″	7.6	9.3			
$15^h 24.9^m$	+58°58′	ι Dra	2002	50°	254.7″	3.4	8.9	*K2III*		Smyth: "A bright star with a distant companion . . . orange tint, pale yellow; several other stars in the field."
$15^h 47.6^m$	+55° 23′	β 946	1991	132°	2.2″	5.9	9.5	*A3*	binary	The data suggest a bright star with a close little companion; another 6th-magnitude star is about 15′ away. Ferguson, 150-mm: "Bluish white and very faint blue."
$15^h 51.1^m$	+52° 54′	Σ1984	2003	279°	6.3″	6.9	8.9	*A1V*	binary	125-mm, 50×: An easily split binary with nice contrast for the separation. It's a gloss-white star almost touched by a sharp little nebulous globe.
$16^h 09.0^m$	+57° 56′	Es 2651	1999	140°	12.3″	6.3	12.1	*A1V*		125-mm, 50×: Surprisingly good pair — the pale little companion of this bright white star looks much brighter than its listed magnitude. It's easily spotted, about 1° east-southeast of Theta (θ) Draconis.
$16^h 23.8^m$	+61° 42′	Σ2054	2004	353°	1.0″	6.2	7.1	*G8III*	binary	125-mm, 350×: A striking but difficult pair. It's a figure-8 formed by a bright yellow star and a ghostly shadow; the ethereal companion is barely possible to see.
$16^h 24.0^m$	+61° 31′	η Dra	1996	139°	4.8″	2.8	8.2	*G8III*		125-mm, 350×: Fantastic contrast for the separation! A brilliant lemon-yellow star almost touched by an elusive wisp of light.
$16^h 36.2^m$	+52° 55′	16-17 Dra	2003	196°	90.0″	5.4	5.5	*B9.5V B9V*		125-mm, 350×: Striking triple! It's a pair of kissing stars and a wide third companion — all three stars are very bright and nearly alike. They're gloss white in color.
		17 Dra	2003	107°	3.0″	5.4	6.4	*B9.5V*	p. binary	
$16^h 48.9^m$	+59°30′	ΟΣ 316	2003	348°	47.4″	7.8	8.9	*G5*		Geertsen, 8 × 50: An "easy finderscope pair, with a magnitude difference that's obvious. Both members are white."
$16^h 56.4^m$	+65° 02′	20 Dra	2004	68°	1.0″	7.1	7.3	*F2IV*	p=422 yr	125-mm, 200×: Very attractively close binary. It's a lemon-white figure-8, mildly unequal, surrounded by beautiful diffraction rings.
$17^h 05.3^m$	+54° 28′	μ Dra	2004	14°	2.3″	5.7	5.7	*F7V*	p=672 yr	Showcase pair. 60-mm: A sharp pair of identical stars, goldish white in color, that are almost fully split at 120×. They're just bright enough to be easy with 60-mm.
$17^h 29.0^m$	+50° 52′	Σ2180	2000	260°	3.0″	7.8	8.1	*A7IV*	binary	The data suggest an attractively close binary. Geertsen, 100-mm: The stars are "bluish white and ivory . . . easily separated with 175×."

Draco *(continued)*

RA	Dec.	Name	Year	P.A.	Sep.	m_1	m_2	Spec.	Status	Comments
$17^h 32.2^m$	+55° 11′	ν Dra	2003	311°	63.4″	4.9	4.9	*A4 A6V*		The Dragon's Eyes. 60-mm, 25×: A pair of brilliant white twins, very wide apart, that shine like sparkling diamonds. Smyth: "Both [stars] pale grey." Webb: "Yellow white. A grand object."
$17^h 35.0^m$	+61° 53′	26 Dra								AB-C is a showcase pair. Gould, 200-mm: "A bright yellow star with a wide faint companion; at 200× the primary star shows a close companion that's seen with difficulty."
		AB	2000	331°	1.6″	5.3	8.5	*G0V*	p=76.1 yr	
		AB-C	1996	241°	25.1″	5.2	8.1	*F8*		
$17^h 38.6^m$	+55° 46′	Σ2199	2004	57°	2.0″	8.0	8.6	*F8V*	p=1,298 yr	Geertsen, 80-mm, 233×: "A very close yellow pair that challenges a small scope . . . I was able to see both members just touching, surrounded by shimmering diffraction rings."
$17^h 41.9^m$	+72° 09′	ψ Dra	2003	15°	30.0″	4.6	5.6	*F5IV F8V*		Showcase pair. 60-mm, 25×: A wide pair of bright stars, mildly unequal, that are very lemony white in color. Smyth: "Both [stars] pearly white." Webb: "Whitish yellow, lilac."
$17^h 59.2^m$	+64° 09′	Σ2273	1999	283°	21.3″	7.6	7.7	*F4IV*	optical	150-mm, 60×: An ordinary but pretty couple. It's a pair of peach-white stars, nearly equal, that are wide and easy but close enough to be attractive.
$18^h 00.2^m$	+80° 00′	40-41 Dra	2003	232°	18.6″	5.7	6.0	*F7V*	optical	Showcase pair. Geertsen, 100-mm, 62×: A "nice, bright pair with nearly equal magnitudes." Webb: "Whitish yellow, paler yellow . . . grouped finely with a smaller lilac star."
$18^h 01.4^m$	+65° 57′	Σ2284	1999	191°	3.0″	8.0	9.4	*F7IV*	binary	The data suggest an attractively close binary. Geertsen, 100-mm, 62×: "White and blue. The separation looks wider than the listed one." Webb: "Yellowish, ash."
$18^h 02.8^m$	+75° 47′	Σ2302								150-mm, 60×: An easy wide pair with very pretty contrast — a snow-white star with a pure gray companion. Both stars look sharply defined.
		AC	2002	279°	22.8″	6.9	9.4	*A0V*		
$18^h 02.9^m$	+56° 26′	Σ2278								150-mm, 150×: Neat little triple — a small pair in a peanut shape, with a wide third companion. All the stars look white.
		AB	2004	28°	35.9″	7.8	8.1	*A9V*		
		BC	2004	146°	6.0″	8.1	8.5	*A0*	binary	
$18^h 23.9^m$	+58° 48′	39 Dra	2000	347°	3.7″	5.1	8.1	*A1V*	binary	125-mm, 200×: Profound contrast! A brilliant white star with a little smoke puff next to it split by only a small gap.
$18^h 33.9^m$	+52° 21′	Σ2348								The data suggest a bright star with a wide companion. Geertsen, 100-mm, 62×: "A challenging pair for 100-mm, with a large magnitude difference. Light yellow and greenish."
		AB-C	2002	271°	25.3″	5.5	8.7	*G9III*		
$18^h 38.4^m$	+63° 32′	Σ2377	2002	338°	16.7″	7.0	9.8	*K2III*	optical	Geertsen, 100-mm, 62×: "A surprisingly easy pair, considering the data. [It's three stars in a straight line:] a pale yellow star, a white companion, and a close field star."
$18^h 38.9^m$	+52° 21′	Σ2368	1998	322°	1.8″	7.6	7.8	*A3*	binary	125-mm, 200×: An attractively close binary with a pretty color. It's a pair of identical stars, split by just a tiny gap, that are amber yellow with flashes of blue.

Draco *(continued)*

RA	Dec.	Name	Year	P.A.	Sep.	m_1	m_2	Spec.	Status	Comments
18ʰ 51.2ᵐ	+59° 23′	ο Dra	2003	319°	36.5″	4.8	8.3	G7III-IV		Showcase pair. 60-mm, 25×: A bright star with a wide little companion in the pretty combination of yellowish peach and clear gray. Webb: "Very yellow, ash."
18ʰ 53.5ᵐ	+75° 47′	Σ2452	2002	218°	5.6″	6.7	7.4	A1V	binary	Geertsen, 100-mm, 62×: "This bright pair pops right out of an otherwise indistinguishable field . . . the stars look blue and pale yellow."
18ʰ 57.3ᵐ	+62° 24′	Σ2440	2003	122°	17.2″	6.6	9.6	G8III		Geertsen, 100-mm, 62×: "This pair and a field star form a little triangle, which sits at one corner of a large star triangle! Greenish white and blue." Franks: "Pale yellow, blue."
18ʰ 57.5ᵐ	+58° 14′	Σ2438	2003	359°	0.9″	7.0	7.4	A2IV	p=231 yr	Smyth: "A close double star in the back of Draco's neck . . . white, pale red."
19ʰ 02.1ᵐ	+52° 16′	Σ2450								Gould, 200-mm, 135×: "Good pair. This bright, deep yellow-orange star has a much fainter companion fairly close."
		A-BC	2002	299°	4.1″	6.5	9.5	G8III	binary	Ferguson, 150-mm: "Yellowish white and faint blue."
19ʰ 16.9ᵐ	+63° 12′	Σ2509	2003	329°	1.8″	7.4	8.2	F6V	binary	Smyth: "A close double star on Draco's neck . . . both white . . . a very difficult object."
19ʰ 29.5ᵐ	+78° 16′	Σ2571	2002	18°	11.3″	7.7	8.3	F0IV	optical	125-mm, 60×: A pretty pair. It's a couple of sharp disks, yellow-white and bluish white in color, that are split by only a modest gap.
19ʰ 40.2ᵐ	+60° 30′	Σ2573	2002	26°	18.4″	6.5	8.9	A5V	optical	The data suggest a bright star with a small companion. Geertsen, 100-mm, 62×: "Aquamarine and white. A 9 × 50 [finder] will split this pair under a dark sky."
19ʰ 48.2ᵐ	+70° 16′	ε Dra	2002	20°	3.2″	4.0	6.9	G7III	p. optical	125-mm, 200×: Fantastic contrast for the separation! A brilliant star with a little dot at its edge in the beautiful colors of Sun yellow and powder blue.
19ʰ 52.8ᵐ	+64° 11′	Σ2604	2003	183°	27.7″	6.9	9.0	G5		The data suggest a fairly bright star with a wide companion. Geertsen: With 100-mm, the stars are "light yellow with greenish; a 9 × 50 finder also splits them, but barely."
20ʰ 04.7ᵐ	+63° 53′	Σ2640	2002	15°	5.5″	6.3	9.5	A2II-III	binary	Gould, 200-mm, 135×: "These two pairs are in the same field [about ³/₄° south of 65 Draconis]. Σ2640 is a close, very uneven pair; Σ2642 is an even, close, but not bright pair."
20ʰ 05.6ᵐ	+63° 42′	Σ2642	1999	187°	1.9″	9.3	9.4	G0	binary	
20ʰ 28.2ᵐ	+81° 25′	75 Dra	2000	282°	196.6″	5.5	6.7	G9III		60-mm, 25×: A fine object for low power. It's a pretty pair of very bright stars, mildly unequal, that are super-wide apart. They look coppery white and pearly white.

Equuleus

RA	Dec.	Name	Year	P.A.	Sep.	m_1	m_2	Spec.	Status	Comments
20ʰ 59.1ᵐ	+04° 18′	ε Equ								AB-C is a showcase pair. 125-mm, 500×: Fantastic triple.
		AB	2003	286°	0.7″	6.0	6.3	F6IV	p=101 yr	It's a figure-8 and a wide third member in the combination
		AB-C	2004	67°	10.5″	5.3	7.1		binary	of bright straw yellow (A and B) and dim silvery blue (C).

Equuleus *(continued)*

RA	Dec.	Name	Year	P.A.	Sep.	m_1	m_2	Spec.	Status	Comments
21h 02.2m	+07° 11′	λ Equ	2004	215°	3.0″	7.4	7.6	*F8*	binary	125-mm: A smaller clone of lovely Gamma (γ) Virginis as that binary was seen in 1997. It's a bright pair of identical stars, azure white in color, that are split by a hair at 83×.
21h 10.3m	+10° 08′	γ Equ	1994	264°	1.5″	4.7	8.7	*A9V*		The data suggest a bright star with a small close companion. Webb: "Yellow, white. A striking pair."
21h 13.5m	+07° 13′	S 781								125-mm, 50×: A fine object at low power. It's a pair of vivid stars, very nearly equal, that are super-wide apart. They're blue-white and green-white, and they seem to jump out of their field.
		AB-D	2002	172°	184.0″	7.4	7.2	*A7V*		
21h 18.6m	+11° 34′	β 163	2000	260°	0.6″	7.4	8.9	*G0V + G6V*	p=78.5 yr	A 300-mm should be able to split this very fast-moving binary, and the gap continues to widen. Hartung split it with 200-mm when the separation was 1.0″.
21h 19.7m	+09° 31′	Σ2786	1997	188°	2.8″	7.5	8.2	*A3IV*	binary	125-mm: A dim but easy and attractively close binary. It's a pair of gloss-white stars, very nearly equal, in the shape of a figure-8 at 83×.
21h 22.9m	+06° 49′	β Equ								Smyth: "Lucid white; dusky; blue." [John Herschel] called it: " A pretty test object . . . a coarse triple, and one of the small stars itself is a pretty first class double star."
		AC	2002	302°	72.0″	5.2	11.6			
		AE	2002	273°	93.6″	5.2	12.1			

Eridanus

RA	Dec.	Name	Year	P.A.	Sep.	m_1	m_2	Spec.	Status	Comments
01h 38.8m	−53° 27′	Δ 4	1999	105°	10.3″	7.2	8.5	*F5IV-V*	optical	This pair forms a wide arc with q^1 and q^2 Eridanus. Gould, 175-mm, 100×: "An easy yellow pair in a thin field. Two 12th-magnitude companions are offset from the pair southwest."
01h 39.8m	−56° 12′	ρ Eri	1999	190°	11.2″	5.8	5.9	*K0V K5V*	p=484 yr	Gould, 175-mm, 100×: An "easy bright yellow pair, the stars even in brightness; a 7th-magnitude star is nearby, otherwise the field is faint and sparse."
01h 56.0m	−51° 37′	χ Eri	1956	204°	4.9″	3.7	10.7	*G8III*		The data suggest a brilliant star with a tiny companion. Hartung: "The brightness of this golden yellow star makes the faint companion difficult, and 250-mm is needed to show it to me."
02h 43.3m	−40° 32′	h 3527	1993	43°	2.1″	7.0	7.2	*B9.5V*	p. binary	Gould, 400-mm, 64×: A "close and fairly bright even pair. Both members are white, and there is a modest scatter of stars about."
02h 49.8m	−20° 15′	h 3533	2002	268°	38.5″	7.6	8.2	*K5*		The data and location suggest a conspicuous pair about 3/4° north-northwest of Tau² (τ²) Eridani. Gould, 175-mm, 65×: A "moderately bright, very wide, deep yellow pair."
02h 58.3m	−40° 18′	θ Eri	2002	90°	8.4″	3.2	4.1	*A4III A1V*	p. binary	Showcase pair. Hartung: "This brilliant white pair is one of the gems of the southern sky and completely dominates the field of a few scattered stars." Gould: "One of the best double stars."

Eridanus *(continued)*

RA	Dec.	Name	Year	P.A.	Sep.	m_1	m_2	Spec.	Status	Comments
$03^h02.7^m$	−07° 41′	ρ^2 Eri	1990	66°	1.5″	5.4	8.9	*K*0II-III		This bright sun sits in a line of three bright stars — ρ^1, ρ^2, and ρ^3 Eridani. Hartung: An orange-yellow star with a companion that is "rather faint and just visible with 105-mm."
$03^h03.0^m$	−02° 05′	Σ341	1998	223°	8.7″	7.6	10.0	*F*5	binary	This binary star forms a cone shape with 5 and 7 Eridani. Gould, 175-mm, 100×: An "easy, but not bright, uneven pair, with a yellow primary."
$03^h12.4^m$	−44° 25′	JC 8								Gould, 350-mm, 160×: "A good object. It is a bright yellow star . . . with a fairly close companion, that is gray by contrast. The primary itself is a very close pair, probably the limit of 175-mm."
		AB	2002	170°	0.7″	6.4	7.4	*F*7III+*A*0V	p=45.2 yr	
		h 3556								
		AB-C	2002	190°	3.7″	6.4	8.8	*F*5V		
$03^h18.7^m$	−18° 34′	h 3565	2002	121°	7.8″	5.9	8.2	*F*0IV	p. binary	The data suggest a bright star with a companion. Gould, 175-mm, 100×: "Nice effect — an easy uneven pair in a thin field, which has a pale yellow primary."
$03^h19.5^m$	−21° 45′	τ^4 Eri								Gould, 175-mm: "This bright mustard-yellow star has one easy companion (C), and several very wide. 180× shows the small, close companion B."
		AB	1987	289°	5.9″	4.0	9.5	*M*3III		
		AC	1964	114°	39.3″	4.0	10.8	*M*3III		
$03^h32.3^m$	−07° 05′	S 411	2002	88°	18.8″	7.4	9.2	*F*5III	optical	Gould, 175-mm, 100×: "Easy pair. A yellow star, which has a fainter companion, moderately wide. There are some bright stars wide in the field."
$03^h39.8^m$	−40° 22′	Δ 15	1999	328°	7.6″	6.9	7.7	*A*3V	binary	This binary star is about 30′ east of y Eridani. Gould, 350-mm, 64×: "An easy cream white pair, bright and moderately uneven. The field shows a few bright stars well out."
$03^h44.1^m$	−40° 40′	h 3589	1993	350°	5.0″	6.7	9.3	*K*1III	binary	Gould, 400-mm, 64×: "Good effect — this bright, deep yellow star in a thin field has a much less bright companion, close."
$03^h48.6^m$	−37° 37′	f Eri	2002	217°	8.4″	4.7	5.3	*B*9V *A*1V	p. binary	Showcase pair. Hartung: "This beautiful pale yellow pair dominates a field of scattered stars and is a fine sight with 75-mm." Gould, 400-mm, 64×: This pair is "not unlike [Theta] θ Eri [on the previous page]."
$03^h52.7^m$	−05° 22′	30 Eri	2000	134°	8.3″	5.5	10.4	*B*8V		The magnitudes suggest a bright star with a faint companion. Smyth: "A delicate double star . . . yellow; pale blue."
$03^h54.3^m$	−02° 57′	32 Eri	2004	348°	6.8″	4.8	5.9	*G*8III *A*2V	binary	Showcase pair. 60-mm, 40×: A bright pair of pretty stars — grapefruit orange and silvery blue. They're attractively close but wide enough to be easy. Smyth: "Topaz yellow, sea green."
$04^h02.1^m$	−34° 29′	β 1004	2002	70°	1.2″	7.3	7.9	*G*1/2V	p=282 yr	A 150-mm should be able to split this binary, but the separation is closing.
$04^h14.4^m$	−10° 15′	39 Eri	1996	141°	6.3″	5.0	8.5	*K*2III	binary	Gould, 175-mm: This "bright orange star has a delicate white companion in contrast. A good pair."

Eridanus *(continued)*

RA	Dec.	Name	Year	P.A.	Sep.	m_1	m_2	Spec.	Status	Comments
04h 15.1m	−30° 04′	h 3632	1998	163°	11.0″	7.8	9.7	*A0V*	optical	This pair is in a neat line of four stars in a field without bright stars. Gould, 175-mm, 100×: An "easy but not bright white pair, in a mostly faint field."
04h 15.3m	−07° 39′	o² Eri								This bright star has a famous double companion — a white dwarf paired with a red dwarf. Hartung calls the primary (A) "orange-yellow" and the other stars (B and C) "indigo-blue."
		A-BC	2002	104°	83.0″	4.4	9.7	*G9V*		
		BC	1995	337°	8.9″	9.5	11.2	*A*	p=252 yr	
04h 19.0m	−33° 54′	h 3642	1998	159°	6.0″	6.5	8.7	*A3V*	binary	The data suggest a bright star with a companion. Gould, 175-mm, 100×: A "rather nice, uneven pair with an easy separation, white . . . 41 Eri is in the field [16′ to the west-northwest]."
04h 21.5m	−25° 44′	h 3644								Gould, 175-mm, 100×: A "bright yellow star with an easy wide companion in a sparse field; the primary is itself a close binary but is now near minimum separation."
		AB-D	1999	41°	44.6″	6.2	8.2	*K0*		
04h 23.3m	−05° 00′	h 342	1998	235°	17.3″	7.8	9.6	*K3III*		This pair is in a group of three stars that form a neat arc. Gould, 175-mm, 100×: This "middling bright star has an easy less bright companion . . . the colors are deep gold yellow and dull yellow."
04h 26.9m	−24° 05′	β 311	1999	319°	0.5″	6.7	7.1	*A3V*	p=596 yr	This binary is probably resolvable with 300-mm until at least 2025.
04h 27.9m	−21° 30′	β 184	1994	251°	1.7″	7.4	7.7	*F6V*	p. binary	Gould, 175-mm, 100×: "A neat and moderately bright close pair . . . in a patchy field of stars. Both stars look yellow."
04h 28.9m	−25° 12′	Stone 8								Gould, 175-mm, 100×: "A good object in a nondescript field. It's a neat and easy yellow pair, uneven in brightness."
		AC	1998	352°	7.0″	7.9	9.5	*K0IV*	binary	
04h 31.4m	−13° 39′	Σ560	2002	45°	29.6″	6.3	9.3	*A2V*		The data suggest a bright star with a small companion. Gould, 175-mm, 100×: "A wide, easy, very uneven pair with a white primary." [It's next to another 6th-magnitude star 19′ west.]
04h 35.2m	−09° 44′	Σ570	2004	260°	12.6″	6.7	7.6	*A1*	optical	Gould, 175-mm, 100×: "Good object. A bright, attractive white pair that's slightly uneven; the field is faint [with a 5th-magnitude star 3/4° to the north-northwest]." Webb: "White, bluish."
04h 38.0m	−13° 02′	Σ576	2002	172°	12.3″	7.3	7.9	*B9.5IV*	optical	Gould, 175-mm, 100×: "A bright and easy pair of white stars, in a moderately starry field. A 12th-magnitude companion is 30″ southwest."
04h 39.6m	−21° 15′	β 1236								Gould, 175-mm, 100×: A "bright yellow star with a wide, dull tone companion that's perhaps yellowish brown; it sits at the end of a line of dim stars in a thin field."
		AC	1998	314°	40.5″	7.3	9.0	*K0/1III*		
04h 43.6m	−08° 48′	55 Eri	2004	318°	9.2″	6.7	6.8	*G5III*		Gould, 175-mm, 100×: "An easy, even, and fairly bright pale yellow pair." Smyth: "Both [stars] yellowish white." Hartung: "Deep and pale yellow."

Eridanus *(continued)*

RA	Dec.	Name	Year	P.A.	Sep.	m_1	m_2	Spec.	Status	Comments
$04^h 56.4^m$	−05° 10′	62 Eri	2003	75°	66.2″	5.5	8.9	*B6V*		Smyth: "A wide double star . . . white; lilac; a third star of the 10th mag [forms] a scalene triangle." Webb, Gould: The primary star is "pale yellow."
$05^h 00.7^m$	−13° 30′	Σ631	2002	107°	5.9″	7.5	8.8	*A0*	binary	Gould, 175-mm, 100×: "An attractive, fairly bright pair in a thin field [about 1° south of 64 Eridani]. The stars are uneven and moderately close."
$05^h 03.0^m$	−08° 40′	Σ636	1998	105°	3.6″	7.1	8.5	*A0V*	binary	Gould, 175-mm, 100×: "Fairly good. This uneven pair is fairly bright and fairly close, and lies in a modestly starry field [about 20′ south of the Witch Head Nebula, IC 2118]."
$05^h 08.3^m$	−08° 40′	Σ649	2004	68°	21.6″	5.8	9.0	*B8V*		This pair is just 13′ west-northwest of Lambda (λ) Eridani. Gould, 175-mm, 100×: "Easy pair with the color contrast of cream white and dull red-brown." J. Herschel: the companion is "very ruddy."

Fornax

RA	Dec.	Name	Year	P.A.	Sep.	m_1	m_2	Spec.	Status	Comments
$02^h 23.2^m$	−29° 52′	β 738	1999	213°	1.6″	7.6	8.0	*G1V*	p=560 yr	This binary is gradually widening, and it should now be possible to split it with 130-mm.
$02^h 33.8^m$	−28° 14′	ω For	2002	245°	10.6″	5.0	7.7	*B9.5IV*	binary	Showcase pair. 60-mm, 45×: A bright star and a small star, goldish white and smoke gray in color, just wide enough apart to be easy. Gould, 175-mm, 100×: "A bright white star with an easy less bright companion."
$02^h 43.8^m$	−27° 54′	β 261	1991	101°	3.1″	7.9	9.2	*G5III*	binary	Gould, 175-mm, 100×: "Attractive effect — a neat, rather delicate yellow pair . . . in the field with a little group of three wide pairs to the west."
$02^h 44.2^m$	−25° 30′	BrsO 1	1998	192°	12.5″	7.0	8.5	*G1V*	optical	Gould, 175-mm, 100×: "An easy uneven pair with a yellow primary, amid a stream of scattered stars."
$02^h 48.6^m$	−37° 24′	h 3532	2000	145°	5.3″	7.0	8.1	*F3/5IV*	binary	Gould, 175-mm, 100×: "An attractive, fairly bright pair in a blank patch, with some stars wide in the field." Hayes, 114-mm, in *Revue des Constellations*: "Yellow with pale blue."
$02^h 50.2^m$	−35° 51′	η^2 For	1991	18°	4.9″	6.0	10.0	*K0III*		Ferguson, 150-mm: A "bright yellowish white star with an extremely faint companion." Gould: $\eta^2 - \eta^3$ form a wide pair for a finderscope in a "field otherwise thin."
$03^h 12.1^m$	−28° 59′	α For	2002	299°	4.8″	4.0	7.2	*F8V*	p=269 yr	Showcase pair. Gould, 175-mm: "This very bright pale yellow star has a much less bright companion, fairly close. A very good object in an attractive field."

Gemini

RA	Dec.	Name	Year	P.A.	Sep.	m_1	m_2	Spec.	Status	Comments
$06^h 14.9^m$	+22° 30′	η Gem	2004	259°	1.8″	3.5	6.2	*M*3.5I-II	p=474 yr	Hartung: "This brilliant orange star dominates a field sown with scattered stars . . . in 1961 [when the separation was 1.3″], 105-mm showed the faint [companion] star pretty steadily."
$06^h 20.6^m$	+18° 03′	ΟΣΣ 75	2000	129°	47.1″	7.7	8.9	*G*7III		60-mm, 25×: This pair is much prettier than the data suggest. It's a grapefruit-orange star with a small gray companion, and they look no wider than the showcase pair Albireo, Beta (β) Cygni.
$06^h 22.8^m$	+17° 34′	Σ899	2001	18°	2.2″	7.4	8.0	*A*0V	binary	125-mm, 200×: A difficult but striking pair. It's a pearly white star almost touching a very ethereal presence; this close companion is like a patch of mist that barely seems to exist.
$06^h 27.8^m$	+20° 47′	15 Gem	2004	203°	25.2″	6.7	8.2	*K*2III		60-mm, 25×: Grand sight! It's a bright star with a small wide companion in the pretty combination of tangerine orange and bluish red. Amber-yellow 16 Geminorum is wide in the field.
$06^h 29.0^m$	+20° 13′	18 Gem								60-mm, 25×: Grand sight! A brilliant Sun-yellow star with a small blue companion super-wide apart. They're a fine couple at low power. Dembowski: "Bluish white, ashy yellow."
		A-BC	2002	331°	111.6″	4.1	8.0	*B*6III		
$06^h 32.3^m$	+17° 47′	20 Gem	2002	211°	19.7″	6.3	6.9	*F*8III	optical	60-mm, 25×: A fine combination. It's a bright pair of mildly unequal gloss-white stars that are wide enough to be very easy but close enough to make an attractive couple.
$06^h 37.7^m$	+16° 24′	γ Gem								Webb: "Brilliant white [primary]. With low power, minute stars radiate from it every way. Pretty field."
		AB	1997	296°	140.5″	1.9	11.2	*A*0IV		
		AC	1907	335°	143.5″	1.9	10.9			
$06^h 43.9^m$	+25° 08′	ε Gem	2002	95°	110.6″	3.1	9.6	*G*8I		60-mm, 25×: Beautiful pair at low power! It's a brilliant Sun-yellow star with a tiny speck of light next to it. They're super-wide apart but look like a couple. Smyth: "Brilliant white; cerulean blue."
$06^h 44.0^m$	+13° 14′	30 Gem	1904	184°	27.2″	4.5	11.1	*K*0III		125-mm, 200×: Difficult but striking. A very bright citrus-orange star that shows a ghostly speck of light at the edge of its glow. The little companion is probably at the limit of 125-mm.
$06^h 45.2^m$	+30° 50′	Σ957	2003	90°	3.6″	7.5	9.4	*A*0	binary	The data suggest a close binary star that's probably resolvable with 100-mm. Webb: "White, ash."
$06^h 54.6^m$	+13° 11′	38 Gem								Showcase pair. 60-mm, 25×: A bright lemon-white star nearly touching a little grayish smokeball. Both stars are vivid with striking contrast. Smyth: "Light yellow, purple."
		AB	2004	326°	7.3″	4.7	7.8	*F*0V	p=1,944 yr	
		AC	1999	328°	118.4″	4.7	10.3			
$06^h 57.0^m$	+24° 57′	Σ991	2001	164°	3.4″	8.0	9.2	*A*0	binary	125-mm: Very nice contrast for the separation. This is a peach-white star with a small gray companion split by a hair at 83×.

Gemini *(continued)*

RA	Dec.	Name	Year	P.A.	Sep.	m1	m2	Spec.	Status	Comments
06h 58.1m	+14° 14′	ΟΣΣ 80								60-mm, 25×: This triple is a fine little asterism. It's a neat broad triangle of three stars — a pair of yellow-white stars and a dim silvery star.
		AB	2002	54°	124.4″	7.3	7.4	B9		
		AC	2002	112°	80.0″	7.3	8.4			
07h 00.6m	+12° 43′	Σ1007								125-mm, 50×: Splendid view! There are two very wide, easy pairs in the field — a pair of white twins (Σ1007) and a modestly unequal pair that's peach white and pink red (h 3288).
		AD	2000	28°	67.9″	7.4	7.7	A3		
07h 02.3m	+12° 35′	h 3288	2002	247°	38.5″	7.3	8.7	A2		
07h 08.4m	+15° 56′	45 Gem	1982	10°	12.1″	5.4	11.3	G8III		The data suggest a bright star with a very dim companion. Webb: "Gold, no color given."
07h 08.5m	+24° 59′	Σ1023	2004	105°	24.2″	8.9	9.4	K5		60-mm: These wide pairs are in the same field at 25×, and an 8 × 50 finderscope shows a nice asterism — a broad rectangle of stars that includes a little white pair (ΟΣ 83).
07h 09.6m	+25° 44′	ΟΣ 83								
		AC	2002	80°	120.5″	7.2	7.8	G5V		
07h 12.8m	+27° 13′	Σ1037	2004	312°	1.1″	7.2	7.3	F8V	p=119 yr	This fast-moving binary should currently be splittable by 100-mm, but the gap is closing. Webb: "Yellowish."
07h 12.8m	+15° 11′	Weisse 14	1997	160°	2.1″	7.8	8.9	B9.5IV	binary	125-mm, 125×: A difficult pair. It appeared for moments as a dim white peanut — a gratifying sight for a binary star as dim and close as this one.
07h 18.1m	+16° 32′	λ Gem	1997	33°	9.7″	3.6	10.7	A3V		Smyth: "Brilliant white, yellowish . . . observed under the most favorable circumstances . . . but the companion was seen best under an averted eye." Webb: "Greenish blue, no color given."
07h 20.1m	+21° 59	δ Gem	2003	227°	5.8″	3.6	8.2	A9III K3V	p=2,239 yr	125-mm, 83×: Fantastic contrast! This is a brilliant amber-yellow star with a ghostly, barely seen speck of light next to it. Webb and Smyth call the companion purple, Hartung calls it red.
07h 24.1m	+21° 27′	Σ1081	1999	236°	1.8″	7.7	8.5	B9	binary	125-mm, 200×: A difficult but gratifying pair. It appeared as a ragged white rod pinched in the center. This much resolution is wonderful for a binary as dim and close as this one.
07h 25.6m	+20° 30′	Σ1083	2003	45°	6.7″	7.3	8.1	A5	binary	60-mm, 45×: Nice view! A milky yellow star and a small silvery white split by a hair; the bright white star 61 Geminorum is also in the field at 25×.
07h 26.0m	+14° 06′	Σ1088								Webb saw this multiple as two pairs in the same field, both somewhat alike.
		AB	2002	196°	10.9″	7.4	9.4	A0V	optical	
		AC	2002	238°	112.9″	7.4	8.7			
		Σ1087								
		CD	1997	38°	23.0″	8.7	11.7	A2		
07h 27.4m	+15° 19′	Σ1094	1999	102°	2.2″	7.6	8.5	A0V	binary	125-mm, 200×: A dim but attractively close binary star. It's a gloss-white star and an ivory white, slightly unequal in brightness, that are just kissing each other.

Gemini *(continued)*

RA	Dec.	Name	Year	P.A.	Sep.	m_1	m_2	Spec.	Status	Comments
$07^{h}27.7^{m}$	+21° 27′	63 Gem	1997	324°	43.0″	5.3	10.9	*F*5V+*F*5V		125-mm, 200×: Fantastic contrast! The companion is a ghostly presence, seen intermittently, inside the halo of a brilliant yellow star. Smyth: "Pale white; purple."
$07^{h}27.7^{m}$	+22° 08′	S 548	2004	277°	35.6″	7.0	8.9	*K*5		Webb, referring to the pair as Hh 264: Knott calls the pair "orange, blue, [with] colors very fine."
$07^{h}32.8^{m}$	+22° 53′	Σ1108	2003	179°	11.6″	6.6	8.2		optical	60-mm, 25×: Striking contrast for the separation! A bright white star with a tiny nebulous companion split by only a tiny gap. Webb: "Yellowish, bluish. Beautiful."
$07^{h}34.6^{m}$	+31° 53′	α Gem								Showcase pair *(Castor)*. 60-mm: A stunning pair of brilliant stars, lemon white in color, that are just kissing at 65×. They're mildly unequal. C looks like a field star, but it's worth noticing.
		AB	2004	62°	4.2″	1.9	3.0	*A*1V *A*2V	p=445 yr	
		AC	2001	164°	71.0″	1.9	9.8			
$07^{h}44.4^{m}$	+24° 24′	κ Gem	2002	241°	7.2″	3.7	8.2	*G*8III		The data suggest a brilliant star with a small companion. It's harder to resolve than the data suggest — 125-mm has repeatedly failed to do so. Smyth: "Orange; pale blue."
$07^{h}45.3^{m}$	+28° 02′	β Gem	1900	280°	29.7″	1.1	13.7	*K*0III		*(Pollux)* A stunningly brilliant star with an extremely faint companion. Smyth: "Orange tinge; ash colored."
$07^{h}47.5^{m}$	+33° 25′	π Gem								The data suggest a bright star with dim companions. Smyth: "A most delicate triple star . . . topaz yellow; bluish; dusky."
		AB	1998	215°	19.3″	5.1	11.4	*M*1III		
		AC	1998	343°	92.0″	5.3	10.4			
$07^{h}48.4^{m}$	+18° 20′	Σ1140	2003	274°	6.4″	7.0	8.7	*K*0III	binary	125-mm, 83×: An easy close pair with pretty colors. It's a grapefruit-orange star with a small gray companion just a small space apart. At 50×, the bright orange star 81 Geminorum is wide in the field.

Grus

RA	Dec.	Name	Year	P.A.	Sep.	m_1	m_2	Spec.	Status	Comments
$21^{h}42.5^{m}$	−37° 56′	h 5288	1999	60°	19.5″	7.6	9.1	*A*5V	optical	The data and location suggest a conspicuous primary star in a desolate field. Gould, 175-mm, 100×: "This moderately bright, pale yellow star has a wide and easy, fainter companion."
$22^{h}12.0^{m}$	−38° 18′	h 5319	1992	132°	2.1″	7.7	7.7	*F*3V	binary	Gould, 175-mm, 100×: An "elegant close pair, even in brightness, that is in a wide scatter of moderate stars. Pale yellow."
$22^{h}18.3^{m}$	−53° 38′	Hdo 298	1988	39°	5.0″	5.4	9.7	*G*3V		Gould, 175-mm, 180×: "A bright orange star in a thin field; at 180×, it has a small companion that's testing."
$22^{h}22.7^{m}$	−45° 57′	π^1 Gru	1975	201°	2.8″	6.5	10.7			Hartung: "The two components of this star . . . make a fine sight. The first is a bright orange red star . . . the second star is brighter and yellow." 150-mm should be able to show their companions.
$22^{h}23.1^{m}$	−45° 56′	π^2 Gru	1985	220°	5.1″	5.6	11.4	*F*3III-IV		
$22^{h}24.6^{m}$	−41° 27′	Jc 19	1998	65°	17.8″	6.7	8.2	*F*5V		Gould, 175-mm, 100×: "A fine pale yellow pair in a thin field; uneven but easy and attractive."

Grus *(continued)*

RA	Dec.	Name	Year	P.A.	Sep.	m_1	m_2	Spec.	Status	Comments
22ʰ 25.9ᵐ	−45° 07′	I 136	1991	278°	2.1″	7.8	8.9	G8IV	p. binary	Gould, 175-mm, 100×: "Very nice. A close yellow pair, unequal in brightness, in a straggle of moderate stars across the field."
22ʰ 42.6ᵐ	−47° 13′	CorO 252	1998	125°	7.5″	6.0	11.1	G0V		This pair is just 20′ south of Beta (β) Gruis. Gould, 350-mm, 110×: This "bright deep yellow star has a much fainter companion quite close; a wide bluish companion is 60″ south."
22ʰ46.7ᵐ	−46° 56′	h 5362	1998	141°	10.4″	6.6	9.9	A9III	optical	Gould, 350-mm, 110×: An "easy uneven pair with a cream white primary; in a thin field with a few brighter stars wide in the field."
23ʰ 06.9ᵐ	−38° 54′	β 773	1999	209°	1.0″	5.7	8.2	A1V	binary	Hartung: "This bright pale yellow star has a companion very close, which 150-mm shows with care."
23ʰ 06.9ᵐ	−43° 31′	θ Gru								Gould, 175-mm: This "bright yellow star has a bright
		AB	1999	107°	1.4″	4.5	6.6	A1V	p. binary	companion extremely wide apart; it is best seen with an
		AC	1999	292°	159.5″	4.5	7.8	G2V		8 × 50 finder. At 180×, member B is seen as a small point close to the primary."
23ʰ 07.2ᵐ	−50° 41′	Δ 246	1999	254°	8.8″	6.3	7.1	F6-7IV-V	p. binary	The data suggest a bright pair of stars. Gould, 350-mm, 110×: A "fine and easy, pale yellow pair in a thin field."
23ʰ 20.8ᵐ	−50° 18′	Rst 5560								Gould, 175-mm: An "easy and rather nice pale yellow pair
		AB	1999	233°	1.3″	6.2	8.9	A8		[referring to AB-C] that is without many stars in the field.
		Δ 248								With 180×, the difficult B star was seen very close to the
		AB-C	2000	212°	16.8″	6.1	6.6	A8		primary."
23ʰ 23.9ᵐ	−53° 49′	Δ 249	2000	212°	26.5″	6.1	7.1	A4III		Gould, 175-mm, 50×: A "wide easy pair in a moderately starry field; both stars look cream white."

Hercules

RA	Dec.	Name	Year	P.A.	Sep.	m_1	m_2	Spec.	Status	Comments
16ʰ 08.1ᵐ	+17° 03′	κ Her	2003	13°	27.4″	5.1	6.2	G7III		Showcase pair. 60-mm, 25×: A wide, bright pair with striking colors — a grapefruit-orange star and a smaller whitish scarlet. Smyth: "Light yellow; pale garnet."
16ʰ 13.3ᵐ	+13° 32′	49 Ser	2003	355°	4.1″	7.4	7.5	G9V	p=1,354 yr	125-mm, 83×: A close pair of bright pearly white stars, nearly alike. Smyth: "A close double . . . absurdly placed on the left arm of Hercules . . . pale white, yellowish."
16ʰ 21.9ᵐ	+19° 09′	γ Her	1999	228°	43.0″	3.8	10.1	A9III		The data suggest a very bright star with a faint companion. Smyth: "An open double star in a dark field on the Hero's left arm . . . silvery white, lilac."
16ʰ 28.9ᵐ	+18° 25′	Σ2052	2003	123°	2.1″	7.7	7.9	K1V	p=224 yr	325-mm, 250×: A difficult but attractively close binary star. It's a bright citrus-orange star, which was seen intermittently as a pair of kissing stars. They look nearly equal.
16ʰ 29.4ᵐ	+10° 36′	Σ2051	2002	19°	13.8″	7.7	9.4	G5III	optical	125-mm, 50×: A fine combination. A peach-white star with a small blue companion, attractively close while wide enough to be easy.

Hercules *(continued)*

RA	Dec.	Name	Year	P.A.	Sep.	m_1	m_2	Spec.	Status	Comments
16ʰ31.6ᵐ	+05° 26′	Σ2056	2000	313°	6.8″	7.8	9.2	*A*3	binary	60-mm, 62×: An attractively close binary star. It's an amber-yellow star and a smaller bluish silver just a small space apart.
16ʰ31.8ᵐ	+45° 36′	Σ2063	2003	196°	16.3″	5.7	8.7	*A*2V	binary	150-mm, 60×: A pretty pair. It's a very bright star with a small companion in the lovely combination of Sun yellow and pale scarlet. They're modestly wide apart.
16ʰ38.7ᵐ	+48° 56′	42 Her	1996	92°	27.2″	4.9	11.8	*M*3III		The data suggest a bright star with a very faint companion. Smyth: "A very delicate triple . . . orange; blue; the third star, which is still more minute, makes a neat triangle."
16ʰ39.6ᵐ	+23° 00′	Σ2079	2000	91°	17.2″	7.6	8.1	*F*0		350-mm, 54×: Nice effect — this yellow-white pair and a red-tinged field star form a perfect triangle!
16ʰ40.6ᵐ	+04° 13′	36-37 Her	2004	229°	69.1″	5.8	6.9	*B*9.5V		60-mm, 25×: Nice couple for low power. It's a wide pair of bright stars in a barren field, yellowish white and yellowish peach in color. Slightly unequal. Smyth: "Pale blue; blue."
16ʰ41.3ᵐ	+31° 36′	ζ Her	2004	229°	0.9″	3.0	5.4	*G*1IV	p=34.4 yr	125-mm, 500×: This very fast-moving binary was resolved in 1992, when the separation was 1.0″. It's a brilliant Sun-yellow star with a little nebulous globe on its edge.
16ʰ42.4ᵐ	+21° 36′	Σ2085	2000	312°	6.1″	7.4	9.2	*A*0IV	binary	125-mm: Splendid colors and closeness. An amber-yellow star and a small ocean blue split by a hair at 125×.
16ʰ44.2ᵐ	+23° 31′	Σ2094	2001	72°	1.2″	7.5	7.9	*F*5III	binary	Harshaw resolved this tight binary star with 200-mm; he calls both stars "white."
16ʰ45.1ᵐ	+28° 21′	46 Her	2001	160°	5.3″	7.4	9.2	*F*7III	binary	150-mm, 100×: Beautiful but difficult pair. It's a straw-yellow star almost touched by a wispy, ghostly smudge of light — much harder than the data suggest. Smyth: "Pale white, sky blue."
16ʰ48.7ᵐ	+35° 56′	Σ2104	2001	19°	5.6″	7.5	8.8	*F*2	binary	150-mm, 60×: An attractively close binary with pretty colors. It's a straw-yellow star and a smaller silvery blue split by a hair at 60×.
16ʰ51.8ᵐ	+28° 40′	Σ2107	2003	98°	1.4″	6.9	8.5	*F*5IV	p=268 yr	Webb resolved this binary star when the separation was only 1.1″. He calls it "yellowish, bluish."
16ʰ55.0ᵐ	+25° 44′	56 Her	2000	91°	18.0″	6.1	10.8	*G*8III		150-mm, 100×: A beautiful pair for its contrast — the companion is a ghostly smudge of light inside the halo of a bright orange star. Smyth: "Light yellow; pale red."
16ʰ56.7ᵐ	+14° 08′	OΣ 318	2003	241°	2.9″	7.0	9.6	*G*9III	binary	Ferguson saw the close little companion of this binary star with 150-mm; he calls the pair yellowish white and blue. There is a wide pair in the in the field about 30′ to the south.
17ʰ03.7ᵐ	+13° 36′	Σ I 33	2002	118°	305.1″	5.9	6.2	*A*1V *K*1III		A most interesting super-wide pair — two bright stars alike in magnitude but very different in temperature. 60-mm, 25×: Pure white and peach white.

Hercules *(continued)*

RA	Dec.	Name	Year	P.A.	Sep.	m_1	m_2	Spec.	Status	Comments
17ʰ 04.8ᵐ	+28° 05′	Σ2120	2003	231°	22.6″	7.4	9.3	*K*0III		Webb: "Yellow, very blue in '29 [1829] . . . red, blue, colors remarkable in '36." [1836]
17ʰ 11.7ᵐ	+49° 45′	Σ2142	2000	113°	4.7″	6.2	9.4	*A*5III	binary	Ferguson saw the dim companion of this bright star with 150-mm; he calls the pair "bluish white and blue."
17ʰ 12.1ᵐ	+21° 14′	Σ2135	2004	192°	8.4″	7.6	8.9	*K*0	p. binary	125-mm: A dim but pretty pair. It's an amber-yellow star and a small silvery lime with just a modest gap between them.
17ʰ 14.6ᵐ	+14° 23′	α Her	2003	104°	4.8″	3.5	5.4	*M*5I-II	p=3,600 yr	Showcase pair — perhaps the loveliest of all combinations. 60-mm, 120×: It's a great star almost touched by a sharp little globe; a brilliant orange-red and a vivid bluish turquoise.
17ʰ 15.0ᵐ	+24° 50′	δ Her	2001	282°	11.0″	3.1	8.3	*A*3IV		Showcase pair. 125-mm, 83×: Grand sight! A brilliant star with a tiny companion, just a small space apart, in the pretty combination of Sun yellow and whitish powder blue.
17ʰ 17.3ᵐ	+33° 06′	68 Her	2001	59°	4.2″	4.8	10.2	*B*1.5V		The data suggest a difficult pair — a bright star with a close, very dim companion. Webb gives the same separation as listed and calls it "white, no color given." Did he really resolve it?
17ʰ 23.7ᵐ	+37° 09′	ρ Her	2004	319°	4.1″	4.5	5.4	*B*9.5III	binary	Showcase pair. 60-mm, 45×: A bright pair of kissing stars, mildly unequal, that are azure white in color. Smyth: "Bluish white; pale emerald." Webb: "Greenish white, greenish."
17ʰ 24.5ᵐ	+36° 57′	OΣ 329	2003	12°	33.2″	6.3	9.9	*G*5III+*F*0V		The data and location suggest a bright star with a very dim companion just a 15′ southeast of Rho (ρ) Herculis (above). Webb included it in his catalog.
17ʰ 24.6ᵐ	+15° 36′	Σ2160	2003	65°	3.7″	6.4	9.3	*B*9V	binary	Smyth, listed under Ophiuchus: "Brilliant white; violet tint . . . [primary] is followed by a ruddy star which, with another . . . forms a neat isosceles triangle."
17ʰ 26.2ᵐ	+29° 27′	Σ2165	2003	61°	9.9″	7.7	9.6	*A*	p. binary	150-mm: An easy pair with pleasing separation. It's a pearly white star with a small nebulous companion split by only a modest gap at 60×.
17ʰ 29.5ᵐ	+34° 56′	Σ2178	2001	228°	10.5″	7.3	9.1	*K*0	optical	125-mm, 50×: A miniature Albireo (showcase pair Beta [β] Cygni). It's a citrus-orange star and a small silvery sapphire fairly wide apart. Webb: "Yellowish, bluish."
17ʰ 36.0ᵐ	+21° 00′	Σ2190	2003	22°	10.3″	6.1	9.5	*A*7IV	binary	125-mm, 50×: Striking contrast for the separation! A bright Sun-yellow star with a little speck of light close beside it.
17ʰ 41.1ᵐ	+24° 31′	Σ2194	2003	7°	16.2″	6.5	9.3		optical	125-mm, 50×: Splendid view! The field shows a bright yellow star (84 Herculis), a bright orange star (83 Herculis), and this wide pair — a dusty gold star with a little sapphire companion.

RA	Dec.	Name	Year	P.A.	Sep.	m_1	m_2	Spec.	Status	Comments
17ʰ46.0ᵐ	+39° 19′	Σ2224	1998	349°	7.6″	6.7	10.0		binary	325-mm, 100×: This binary begs attention. The data suggest a fairly easy pair, but the author could not see the small companion. Harshaw, with 200-mm, didn't see it either.
17ʰ46.5ᵐ	+27° 43′	μ Her								110-mm, 100×: A nice pair for its striking contrast. The companion is a ghostly wisp of light, seen only intermittently, inside the halo of a brilliant orange star. Smyth: "Pale white; purple."
		A-BC	2000	249°	34.9″	3.4	9.8	*G*5IV		
17ʰ50.3ᵐ	+25° 17′	Σ2232	2003	138°	6.3″	6.7	8.9	*A*1V	binary	125-mm, 83×: Splendid view — a bright banana-yellow star with an ashy green companion, split by just a modest gap, inside a large black cloud with milky edges.
17ʰ52.0ᵐ	+15° 20′	OΣ 338	2001	171°	0.8″	7.2	7.4	*G*8III	binary	Webb, listed under Ophiuchus: "Gold, green white . . . three other pairs in the field [all three are dim pairs]."
17ʰ53.3ᵐ	+40° 00′	90 Her	1999	116°	1.5″	5.3	8.8	*K*3III	binary	The data suggest a bright star with a close little companion. Webb: Dembowski calls it "gold, azure."
17ʰ56.4ᵐ	+18° 20′	Σ2245	2002	292°	2.6″	7.4	7.6	*A*0III	binary	125-mm: An easy close binary, quite pretty. It's a fairly bright pair of identical stars, reddish white in color despite the spectral type, that are just touching at 140×.
17ʰ59.0ᵐ	+30° 03′	Σ2259	2002	278°	19.7″	7.3	8.4	*G*5II	optical	125-mm, 50×: A wide, easy pair in a beautiful field. It's a pumpkin-orange star with a small blue companion. They're next to a bright white star (Nu [ν] Herculis) in a rich starry field.
18ʰ01.5ᵐ	+21° 36′	95 Her	2003	257°	6.3″	4.9	5.2	*A*5III	binary	Showcase pair. 60-mm: A bright pair of identical stars split by a hair at 45×. To the author they look pure gold, but Smyth calls them "light apple green; cherry red."
18ʰ03.1ᵐ	+48° 28′	Σ2277	2003	128°	26.5″	6.3	8.9	*A*1V		125-mm, 50×: This pair is included because of its pretty primary star. It's a bright lemon-yellow star with a dim azure-white companion quite wide apart.
18ʰ05.8ᵐ	+21° 27′	OΣ 341								125-mm, 50×: A neat little multiple star. It's a peach-white star inside a wide arc of three dim stars. Very wide member G looks more like a field star.
		AB-C	2002	175°	27.8″	7.6	9.8	*M*1V		
		AE	2002	38°	66.1″	7.4	10.3			
		AF	1999	356°	112.3″	7.4	11.0			
		AG	2002	239°	132.8″	7.4	7.6			
18ʰ06.5ᵐ	+40° 22′	Σ2282	2004	83°	2.5″	7.9	8.7	*A*1V	binary	125-mm, 200×: A dim but attractively close binary. It's a gloss-white star and an ash white, mildly unequal, that are split by hairs.
18ʰ07.8ᵐ	+26° 06′	100 Her	2003	183°	14.2″	5.8	5.8	*A*3V		60-mm, 25×: Interesting object. It's a bright pair of identical stars, fairly close together, that are possibly a chance alignment — a remarkable coincidence if they are. They're white.

Hercules *(continued)*

RA	Dec.	Name	Year	P.A.	Sep.	m_1	m_2	Spec.	Status	Comments
18ʰ 10.1ᵐ	+16° 29′	Σ2289	2003	220°	1.2″	6.7	7.2	A0V+G0III	p=3,040 yr	125-mm, 200×: Difficult but gratifyingly close. A bright banana-yellow star with part of an ash-gray star sticking out of it — but seen only intermittently. Hartung: "Deep and pale yellow."
18ʰ 25.0ᵐ	+27° 24′	Σ2315	2004	122°	0.6″	6.6	7.8	A0V+A4V	p=2,094 yr	250-mm should be able to split this binary; the gap is slowly widening.
18ʰ 27.8ᵐ	+24° 42′	Σ2320	2004	1°	1.2″	7.1	8.9	B9V	binary	Webb called this binary "white, ash" and measured the separation as 1.8″ at the time. Does it move fast enough for amateurs to measure changes? It begs further study.
18ʰ 33.8ᵐ	+17° 44′	Σ2339								The data and location suggest an attractively close pair inside a giant triangle (about 2° in size) of 6th-magnitude stars. Webb: "White, blue."
		AB-CD	2000	275°	2.0″	7.5	8.7			
18ʰ 35.5ᵐ	+23° 36′	OΣ 359	2004	6°	0.7″	6.4	6.6	G9III-IV	p=211 yr	A 175-mm should be able to split this binary until at least 2025.
18ʰ 35.9ᵐ	+16° 59′	OΣ 358	2004	154°	1.8″	6.9	7.1	F8V	p=380 yr	125-mm, 200×: Ideal brightness! A pair of hair-split twins, whitish banana yellow in color. They're just dim enough to show no glare without being hard to see.
18ʰ 39.3ᵐ	+20° 56′	Σ2360	2000	358°	2.4″	8.0	9.2	B5IV	binary	Webb calls this binary "white, ash."
18ʰ 45.7ᵐ	+20° 33′	110 Her								Smyth: "A most delicate but wide double star . . . pale yellow; dusky . . . [the companion] is certainly a *minimum visible* of my telescope."
		AC	2000	58°	73.0″	4.2	11.0			
18ʰ 49.0ᵐ	+21° 10′	Σ2401	1999	38°	4.0″	7.3	9.3	B3V	binary	125-mm, 83×: Striking binary! It's a bright white star almost touched by a little silver dot, and these stars are seen with good definiton and strong contrast.
18ʰ 52.3ᵐ	+14° 32′	Σ2411	2003	95°	13.4″	6.6	9.6	G9III	binary	150-mm, 150×: Striking but difficult binary. The companion is a just a dim vapor of light, seen only with averted vision, that's almost touching a lovely tangerine-orange star.
18ʰ 54.7ᵐ	+22° 39′	113 Her								Webb: "Double companion." If Webb split the faint BC pair (his largest aperture was 200-mm), can you?
		AB	1999	36°	35.0″	4.6	11.1	G4III+A6V		
		BC	1999	317°	8.7″	11.1	11.1			

Horologium

RA	Dec.	Name	Year	P.A.	Sep.	m_1	m_2	Spec.	Status	Comments
02ʰ 19.9ᵐ	−55° 57′	h 3497	2000	83°	36.5″	6.0	10.6	K5III		Gould, 175-mm, 100×: This "bright orange star has an easy wide companion in a field mostly faint."
02ʰ 28.0ᵐ	−58° 08′	DAW 1								Gould, 400-mm, 64×: A "bright yellow star with an easy fainter companion"; at 140×, the primary itself is a pair — an "even, very close pair at right angles to the wide companion."
		AB	1995	32°	1.2″	8.0	8.5	F8	binary	
		h 3503								
		AC	1999	301°	17.7″	8.1	9.6		binary	

Horologium *(continued)*

RA	Dec.	Name	Year	P.A.	Sep.	m_1	m_2	Spec.	Status	Comments
02h 38.9m	−54° 50′	h 3520	1999	205°	20.6″	7.7	8.6	F0IV/V	optical	Gould, 400-mm, 64×: A "wide, easy and bright yellow pair in a faint field [some 23′ southwest of Zeta (ζ) Horologii]."
02h 39.7m	−59° 34′	Δ 7 A-BC	1999	97°	36.5″	7.7	7.7	G8/K0III		Gould, 175-mm, 100×: A "wide and fairly bright pair of deep yellow stars in a sparse field." [He makes it sound like the pair almost jumps out at you!]
03h 04.6m	−51° 19′	Δ 10	1999	70°	38.3″	7.6	8.5	G1V		Gould, 400-mm, 64×: A "very wide, deep yellow pair, in a scattered field of fairly bright stars. Probably a good pair for small scopes at very low power."
03h 24.6m	−45° 40′	h 3576	1993	341°	2.9″	7.3	8.8	A2V	binary	Gould, 400-mm, 64×: A "creamy, pale yellow star just split; at 140× it is an easy, uneven and fairly close pair. There is a little gathering of stars 10′ to 15′ to the south."
03h 24.6m	−51° 04′	h 3575	1999	48°	35.8″	6.7	10.2	A0V		Gould, 400-mm, 64×: "A bright cream white star with an easy wide companion, with a field quite starry north and east."
03h 56.6m	−39° 55′	h 3611	1999	139°	4.1″	8.0	8.7	A3	binary	Gould, 175-mm: "Quite good." There are two conspicuous stars in the field — one white and the other yellow; the white star becomes an "easy, moderately close pair with 100×."
04h 19.3m	−44° 16′	h 3643	1999	115°	70.4″	5.5	8.6	K2III		Gould, 175-mm, 100×: This "bright, deep yellow star has a moderately wide companion, perhaps bluish."

Hydra

RA	Dec.	Name	Year	P.A.	Sep.	m_1	m_2	Spec.	Status	Comments
08h 15.8m	+02° 48′	Σ1210	2002	113°	15.8″	7.3	9.5	B9.5IV	optical	125-mm, 50×: Nice effect. This pair and a field star form a neat straight line — a plain white field star beside a bluish white star with a green-white companion.
08h 28.5m	−02° 31′	Σ1233 AB-C	2002	330°	18.0″	6.4	10.5	A5		The data suggest a bright primary star with a very faint companion. Webb, listed under Monoceros: "Yellowish, no color given."
08h 37.9m	−06° 48′	h 99	2002	181°	60.2″	6.8	8.3	G0		125-mm, 50×: Very nice view! The field shows three bright stars in a neat triangle, and one of these has a very wide little companion. All these stars look white.
08h 39.7m	+05° 46′	Σ1255	2004	30°	26.2″	7.3	8.6	G1V		125-mm, 50×: An easy, wide pair in the field with bright white Delta (δ) Hydrae. It's an azure-white star and a pearly white star, mildly unequal, that are split by a wide gap.
08h 43.7m	−07° 14′	F Hya	2002	311°	79.0″	4.7	8.2	G1I		Showcase pair. 125-mm, 50×: A vivid super-wide pair with lovely colors — a brilliant Sun-yellow star and a small aquamarine. Webb, listed as S 579 under Monoceros: "Fine yellow, beautiful blue."

Hydra *(continued)*

RA	Dec.	Name	Year	P.A.	Sep.	m_1	m_2	Spec.	Status	Comments
08ʰ 45.3ᵐ	−02° 36′	Σ1270	2003	264°	4.7″	6.9	7.5	*F2IV*	binary	60-mm: Very pretty view — a pair of peach-white stars, split by a hair at 45×, with a bright yellow star and a bright white star also in the field.
08ʰ 46.8ᵐ	+06° 25′	ε Hya AB-C	2004	299°	3.0″	3.5	6.7	*F8V*	p=990 yr	*Revue des Constellations:* "Brilliant pair with great difference in magnitudes . . . the companion touching the first diffraction ring with 75-mm and 95-mm." Smyth: "Pale yellow; purple."
08ʰ 51.6ᵐ	−07° 11′	15 Hya AB AB-C AB-D	 2003 1998 1998	 121° 5° 56°	 1.1″ 45.7″ 54.7″	 5.8 5.5 5.5	 7.4 9.7 10.8	 *A4*	 binary	Smyth: "Most delicate triple . . . pearl white; [C and D] both purplish. It is located in a region utterly destitute of large stars." Hartung: "The close companion [B] is quite clear with 150-mm."
08ʰ 52.1ᵐ	+04° 28′	Σ1290	1998	326°	2.8″	7.4	9.2	*A2*	binary	Gould, 175-mm, 100×: "Good pair. An elegant close uneven pair in a thin field; easy and obvious at 180×."
08ʰ 55.5ᵐ	−07° 58′	17 Hya	2003	4°	4.0″	6.7	6.9	*A2 A7*	binary	60-mm, 25×: An attractively close binary star. It's a fairly bright pair of gloss-white stars, very nearly equal, that are almost but not quite apart.
08ʰ 56.8ᵐ	−17° 26′	Arg 72	2002	182°	4.0″	7.2	7.4	*F3/5V*	p. binary	125-mm: Another attractively close double star. It's a pair of identical stars, azure white in color, that are split by a hair at 83×.
09ʰ 00.8ᵐ	−09° 11′	β 409	1999	187°	9.7″	7.3	10.1	*A0*	binary	Gould, 175-mm, 100×: This pair is "in a group of fairly bright stars, and it has an easy less bright companion to the south."
09ʰ 14.4ᵐ	+02° 19′	θ Hya	2000	234°	20.7″	3.9	9.8	*B9.5V*		Gould, 175-mm, 100×: "This bright, nearly white star dominates a thin field; the much less bright companion is moderately wide and easy." Smyth: "Pale yellow, ash colored."
09ʰ 17.5ᵐ	+00° 33′	Σ1336	2003	181°	40.0″	7.0	10.2	*A0*		125-mm, 50×: An attractive pair for its great contrast, which looks about as wide as Albireo (Beta [β] Cygni). It's a fairly bright white star with a very faint dot of light next to it.
09ʰ 20.5ᵐ	−09° 33′	27 Hya	2002	212°	229.1″	4.9	7.0	*G8III F4V*		60-mm, 25×: Ideal object for low power — a super-wide pair of beautiful stars. It's a bright grapefruit-orange star and a small plum red, and the combination of these colors is striking.
09ʰ 23.3ᵐ	+03° 30′	Σ1347	2002	312°	21.2″	7.3	8.3	*F0*	optical	60-mm, 25×: This pair is well suited to a small aperture. It's a gloss-white star and a smaller bluish white, wide apart and easily seen without being overwide.
09ʰ 24.5ᵐ 09ʰ 27.3ᵐ	+06° 21′ +06° 14′	Σ1348 Σ1355	2004 2004	317° 352°	1.9″ 1.9″	7.5 7.7	7.6 7.8	*F7V* *F7V*	 p. binary	125-mm, 200×: How remarkable! These two pairs are less than 1° apart. Each pair is a duplicate of the other, and all four stars look identical! They are pairs of hair-split white stars.

RA	Dec.	Name	Year	P.A.	Sep.	m_1	m_2	Spec.	Status	Comments
09ʰ 29.1ᵐ	−02° 46′	τ¹ Hya	2002	5°	66.2″	4.6	7.3	F6V		Showcase pair. 60-mm, 25×: Fantastic combination — a bright Sun-yellow star beside a small plum red, and both these colors are seen boldly. They're very wide apart.
09ʰ 31.5ᵐ	+01° 28′	Σ1365	2002	157°	3.4″	7.4	8.0	F9III	binary	125-mm: Nice view — a close binary in a long line of three stars. It's a pair of amber-yellow stars, split by hairs at 83×, between two other yellowish stars.
09ʰ 32.2ᵐ	−10° 51′	Σ1367	1999	184°	5.2″	7.9	9.0	G0		125-mm: An attractively close pair with a bright yellow star and a bright white star in the field. It's a gloss-white star and a small pure silver, split by a hair at 83×.
09ʰ 32.9ᵐ	−14° 00′	β 910								Gould, 175-mm, 100×: This "fairly bright yellow star
		AB	1998	305°	6.6″	7.3	10.2	K0IV	binary	has an easy fainter companion." There is a star about 3′
		AC	2002	280°	169.9″	7.3	10.0	G5		to the west, which looks like a distant third companion.
09ʰ 35.4ᵐ	+03° 54′	Σ1371	2000	275°	8.1″	7.9	10.2	G0	binary	Webb, listed under Sextans: "Yellowish, no color." John Herschel called the companion "dusky red; very remarkable color."
09ʰ 35.6ᵐ	−19° 35′	S 604	2002	92°	51.4″	6.3	9.4	A2V		125-mm, 50×: A bright star with a small companion. It's a lemony white star with a small white companion, super-wide apart but they look like a couple.
09ʰ 53.6ᵐ	−19° 29′	h 4261	1999	82°	8.3″	7.8	9.4	G8 K0III	binary	125-mm, 50×: A binary with very nice contrast for the separation. It's a straw-yellow star almost touched by a tiny speck of arctic blue.
10ʰ 04.0ᵐ	−18° 06′	SHJ 110								125-mm, 50×: Grand! A wide pair of bright stars, obviously
		AC	2003	274°	20.9″	6.2	7.0	B8	optical	double at a glance, that jump out of their fairly thin field. Both stars look pure white.
10ʰ 06.8ᵐ	−24° 43′	β 217	1998	132°	2.1″	7.9	8.0	F8IV-V	p. binary	Gould, 175-mm, 100×: "A neat, tight and even pair of yellow stars in a modestly starry field."
10ʰ 36.1ᵐ	−26° 41′	β 411	1998	310°	1.3″	6.7	7.8	F6V	p=170 yr	Gould, 175-mm: Good pair. It's a "bright pale yellow star near the northwest end of a line of fairly bright stars . . . at 180× it becomes a slightly uneven close pair."
10ʰ 47.6ᵐ	−15° 16′	Σ1474								125-mm, 50×: Very nice view. This triple is a neat straight
		AB	2002	28°	66.2″	6.7	7.1	B9IV		line of stars that are nearly alike. The bright orange star
		BC	1998	17°	6.7″	7.8	7.5	F5		Nu (ν) Hydrae is in the field.
11ʰ 32.3ᵐ	−29° 16′	N Hya	2003	210°	9.4″	5.6	5.7	F8V F8V	p. binary	Showcase pair. 60-mm, 25×: A bright pair of identical stars, grapefruit orange in color, that are split by a hair at 25×. Smyth, listed under Crater: "Lucid white; violet tint."
11ʰ 36.6ᵐ	−33° 34′	h 4455	1991	241°	3.4″	6.0	7.8	K0III	binary	Hartung: "This beautiful pair is a fine object for 75-mm, with clear colors [that he does not name] in the black field."
11ʰ 41.7ᵐ	−32° 30′	h 4465								Gould, 175-mm: "The colour effect is good. This very
		AC	1999	44°	66.1″	5.4	8.3	K5III		bright red-gold star has a moderately bright, wide companion that is vaguely yellow."

Hydra *(continued)*

RA	Dec.	Name	Year	P.A.	Sep.	m_1	m_2	Spec.	Status	Comments
$11^h 52.9^m$	−33° 54′	β Hya	1998	36°	0.7″	4.7	5.5	*B*9III	p. binary	This bright pair should be resolvable with 175-mm; Hartung calls it a "bright pale yellow pair."
$11^h 56.7^m$	−32° 16′	Δ 116	1999	82°	18.9″	7.7	7.8	*G*3V		This pair forms a neat triangle with two close field stars. Gould, 175-mm, 100×: A "wide and easy yellow pair, the stars even in brightness."
$12^h 06.1^m$	−32° 58′	h 4495	1999	319°	5.8″	6.7	8.8	*G*0V	p. binary	Gould, 175-mm: A "good pair, moderately uneven and easy, in a wide-scattered star field. The primary star is yellow, and at 180× there is a wide 13th-magnitude companion."
$12^h 10.0^m$	−34° 42′	Jc 17	2000	17°	3.2″	6.4	8.0	*A*0V	binary	Gould, 175-mm, 100×: A "fine easy pair, off white and a little uneven; a 9th-magnitude companion is wide northeast, and the field has a random sprinkle of stars."
$12^h 54.3^m$	−27° 58′	h 4556	1998	80°	5.9″	7.7	8.8	*F*9V	binary	Gould, 175-mm, 100×: "Good object. An easy uneven pair, yellow in color, in a widely scattered star field."
$13^h 36.8^m$	−26° 30′	H N 69	1999	191°	10.2″	5.7	6.6	*A*7III	optical	Showcase pair. Gould, 175-mm: "Good small scope object . A bright and easy white pair that has very wide, faint companions." Hartung: A "bright pale yellow pair."
$13^h 41.5^m$	−23° 27′	h 4606 AB-C	1998	354°	31.1″	6.6	10.2	*A*1V		Gould, 175-mm, 65×: This "bright, very pale yellow star has an easy but faint, quite wide companion. The field is modestly starry."
$14^h 28.2^m$	−29° 29′	52 Hya AB-C	1999	279°	3.7″	5.0	10.0	*B*8V		Ferguson, 150-mm: "A bright blue star with an extremely faint companion."
$14^h 29.8^m$	−25° 33′	CorO 172	1999	57°	12.6″	7.9	9.7	*K*1III	optical	Gould, 175-mm, 100×: "Easy pair. A moderately bright orange star with a well separated companion; a fainter companion is also in the view, in a field not very starry."
$14^h 46.0^m$	−25° 27′	54 Hya	1998	124°	8.1″	5.1	7.3	*F*2V	binary	Showcase pair. 60-mm, 45×: A bright grapefruit-orange star with a little nebulous companion only a small gap apart. Hartung: "Yellow and reddish." Smyth: "Pale orange; violet tint."
$14^h 58.7^m$	−27° 39′	59 Hya	1998	356°	0.6″	6.2	6.8	*A*8III	p=429 yr	A 200-mm can probably still split this binary, but the separation is slowly closing. Hartung: This pair of "deep yellow disks are just apart with 200-mm."

Hydrus

RA	Dec.	Name	Year	P.A.	Sep.	m_1	m_2	Spec.	Status	Comments
$01^h 25.3^m$	−59° 30′	h 3435	1999	357°	25.3″	7.1	9.4	*F*2IV		This ordinary pair offers something to see — it sits where there are no other sights for amateurs besides a few faint stars. Gould, 175-mm: An "easy wide uneven pair."
$01^h 55.3^m$	−60° 19′	h 3475	2002	75°	2.5″	7.2	7.2	*F*2V	p. binary	Hartung: A "dainty, almost equal yellow pair . . . easy object." Gould, 175-mm: "The field includes some bright stars."

Hydrus *(continued)*

RA	Dec.	Name	Year	P.A.	Sep.	m_1	m_2	Spec.	Status	Comments
$02^{h}00.5^{m}$	−62° 46′	h 3479	2000	272°	32.2″	7.7	9.8	*F*3/5IV/V		The data and location suggest an ordinary but easily noticed pair about 1° south of Alpha (α) Hydri. Gould, 175-mm, 100×: A "wide and easy uneven pair in a thin field, with a few stars wide away."
$03^{h}07.5^{m}$	−78° 59′	h 3568	2000	224°	15.3″	5.7	7.7	*F*2II-III	optical	Showcase pair. Hartung: A "fine unequal pair, yellow and bluish white, in a field with a few scattered stars."

Indus

RA	Dec.	Name	Year	P.A.	Sep.	m_1	m_2	Spec.	Status	Comments
$20^{h}36.5^{m}$	−45° 33′	I 41	1993	356°	2.2″	7.7	8.5	*A*7III	p. binary	Gould, 175-mm, 100×: "A neat little close pair in a moderately starry field."
$20^{h}45.0^{m}$	−50° 29′	I 17								Gould, 175-mm: "An extremely wide, fairly bright pair, pale yellow and orange-yellow; at 180×, the pale yellow star is a very tight even pair that is just split."
		AB	1992	35°	1.0″	8.0	8.0	*A*0IV/V	binary	
		Δ235								
		AC	1999	122°	125.2″	8.0	7.5	*K*0III		
$21^{h}10.4^{m}$	−54° 34′	h 5246	1999	130°	4.0″	7.8	8.0	*K*1/2V+*F*	p. binary	Gould, 175-mm, 100×: "Good object. An even, moderately bright, and fairly close yellow pair. The field has only a few faint stars."
$21^{h}19.9^{m}$	−53° 27′	θ Ind	2000	273°	6.7″	4.5	6.9	*A*5V	p. binary	Showcase pair. Gould, 175-mm, 100×: A "fine bright uneven pair, light yellow and reddish brown. The field has a few fairly bright stars."
$21^{h}44.0^{m}$	−57° 20′	Jc 25								Gould: "A pair for binoculars or an 8 × 50 finder. A rather broad arc of three bright stars, in varying shades of yellow."
		AB	1999	4°	152.3″	6.5	6.9	*F*5/7IV/V		
		AC	1999	214°	187.0″	6.5	7.5	*K*2III		
$21^{h}48.7^{m}$	−65° 30′	I 19	1991	309°	1.3″	7.3	8.7	*F*3V	p. binary	Gould, 175-mm: This "fairly bright yellow star has a less bright companion, just separated with 135×. The field has a mild scatter of fainter stars. A wide companion is 40″ to the W."
$22^{h}23.7^{m}$	−72° 48′	h 5325	1999	267°	18.9″	8.0	8.4	*A*1V	optical	Gould, 175-mm, 100×: "A wide, nearly even pair, with a faint third star nearly in a line."

Lacerta

RA	Dec.	Name	Year	P.A.	Sep.	m_1	m_2	Spec.	Status	Comments
$22^{h}09.3^{m}$	+44° 51′	h 1735								125-mm, 50×: Nice effect. This broad triple is three stars in a straight line — a small star between two alike bright ones: pure white (A), blue-white (B), and yellow-white (D).
		AB	2003	110°	26.7″	6.7	9.7	*B*9IV		
		AD	2001	286°	109.6″	6.7	6.8			
$22^{h}18.9^{m}$	+37° 46′	Σ2894	2003	194°	15.8″	6.2	8.9	*A*8III	optical	60-mm, 25×: A close pair with a bright orange star (1 Lacertae) in the field. It's a gloss-white star with a tiny gray companion split by a fairly small gap.

Lacerta *(continued)*

RA	Dec.	Name	Year	P.A.	Sep.	m_1	m_2	Spec.	Status	Comments
22ʰ 21.0ᵐ	+46° 32′	2 Lac	1998	6°	47.6″	4.6	11.6	*B6V*		125-mm, 200×: Striking, fantastic contrast! A brilliant Sun-yellow star (despite its spectral type) with a tiny glimpse star at the edge of its glow. Smyth: "Pale yellow; orange tinted."
22ʰ 23.6ᵐ	+45° 21′	Σ2902	2004	88°	6.4″	7.6	8.2	*G5*	binary	125-mm, 83×: An easily split binary in a field packed with faint stars. It's a peach-white star and a smaller milky silver split by a fairly small gap.
22ʰ 26.8ᵐ	+37° 27′	Σ2906	2000	1°	3.7″	6.5	9.6	*B2V*	binary	Hartung: "The bright [primary] star looks on occasion distinctly bluish, which makes the companion appear reddish by contrast . . . 105-mm just brings [the companion] into view."
22ʰ 35.7ᵐ	+56° 52′	h 1791	2002	59°	17.3″	7.6	9.7	*G0*		125-mm, 50×: Splendid view! An easy wide pair between a bright orange star and a bright red star. It's a pure white star and a small blue-white, quite wide apart.
22ʰ 35.9ᵐ	+39° 38′	8 Lac	2004	185°	22.2″	5.7	6.3	*B2V B5*		Showcase pair. 60-mm, 25×: Fantastic view — a wide pair of bright white stars, nearly alike, with several dim companions. It lies in a faint cloud of nebulousity.
22ʰ 39.3ᵐ	+39° 03′	10 Lac	2002	49°	62.2″	4.8	10.3	*O9V*		60-mm, 35×: Fantastic contrast! A bright white star with a little ember of light at the edge of its glow — much closer than the listed separation suggests. Smyth: "White; violet."
22ʰ 41.5ᵐ	+40° 14′	12 Lac	2002	15°	69.1″	5.2	10.8	*B2III*		125-mm, 50×: This pair is also pretty for its striking contrast. It's a bright white star with a tiny blue dot beside it — much closer than the listed separation suggests.
22ʰ 43.1ᵐ	+47° 10′	Σ476 A-BC	1999	301°	0.5″	7.4	7.1	*A1V+G*	binary	This super-tight binary pair should be resolvable with 300-mm.
22ʰ 44.1ᵐ	+41° 49′	13 Lac	1998	129°	14.6″	5.2	10.9	*G8III*		The data suggest a bright star with a modestly close companion. Ferguson, 150-mm: "Yellowish white, very faint blue."
22ʰ 44.1ᵐ	+39° 28′	Σ2942 AB	2003	277°	2.9″	6.2	8.9	*K5III*	binary	Hartung: "This fine orange star has a companion close [referring to AB] . . . which sometimes looks greenish by contrast. [There is] a faint third star."
		β 450 AC	1998	247°	9.0″	6.1	11.6			

Leo

RA	Dec.	Name	Year	P.A.	Sep.	m_1	m_2	Spec.	Status	Comments
09ʰ 24.7ᵐ	+26° 11′	κ Leo	1975	211°	2.4″	4.5	9.7	*K3III*		The data suggest a bright star with a close, faint companion. Webb: Dembowski calls them "yellow, blue."
09ʰ 28.5ᵐ	+09° 03′	2 Leo	2004	93°	0.6″	5.7	7.3	*F9IV*	p=118 yr	Smyth: "An exquisite close double star . . . pale yellow; greenish; at times both stars looking yellow." The separation is increasing.
09ʰ 28.5ᵐ	+08° 11′	3 Leo	2000	79°	25.2″	5.8	11.1	*G9III*		The data suggest a bright star with a faint companion. Smyth: "A delicate double star . . . pale yellow, greenish, two or three other stars in the field."

Leo *(continued)*

RA	Dec.	Name	Year	P.A.	Sep.	m_1	m_2	Spec.	Status	Comments
09ʰ 32.0ᵐ	+09° 43′	6 Leo	2000	76°	37.4″	5.2	9.3	*K3III*		60-mm, 25×: A nice object for low power. It's a lovely brick-red star with a faint dot of light wide beside it. Smyth: "Pale rose tint, purple." Webb: "Deep orange, green."
09ʰ 35.2ᵐ	+14° 05′	ΟΣΣ 102	2004	47°	45.4″	7.9	9.1	*B9*		125-mm, 50×: A nice bonus in the field with 7 Leonis (below). It's a very wide pair of stars, slightly unequal, that are azure white and pearly white.
09ʰ 35.9ᵐ	+14° 23′	7 Leo	2003	80°	41.0″	6.3	9.4	*A1V*		125-mm, 50×: This pair is included because of its pretty primary star. It's a very bright star, deeply bluish white in color, with a small blue-green companion. They're very wide apart.
09ʰ 57.0ᵐ	+19° 46′	Σ1399	2003	175°	31.0″	7.6	8.4	*G0*		125-mm, 50×: Pleasant effect. An easy wide pair, peach white and silver, in a small group of stars within a barren region of the sky.
10ʰ 08.4ᵐ	+11° 58′	α Leo	2000	308°	176.0″	1.4	8.2	*B7V*		*(Regulus)* 60-mm, 25×: Beautiful contrast — a brilliant white beacon with a tiny smokeball beside it. They're super-wide apart but look like a pair. Smyth: "Flushed white, pale purple."
10ʰ 16.3ᵐ	+17° 44′	ΟΣ 215	2004	180°	1.4″	7.2	7.5	*A9IV*	p=670 yr	125-mm, 200×: A difficult but gratifying pair. It's a fairly bright star, bluish white in color, that becomes a peanut shape for scattered moments.
10ʰ 17.2ᵐ	+23° 06′	39 Leo	1998	299°	7.8″	5.8	11.4	*F7V*		The data suggest a bright star with a very faint companion. Webb: "Yellow, no color given."
10ʰ 20.0ᵐ	+19° 50′	γ Leo	2004	127°	4.6″	2.4	3.6	*K0III*	p=619 yr	Showcase pair *(Algieba)*. 60-mm, 120×: A brilliant figure-8, slightly unequal, grapefruit orange in color. Smyth: "Bright orange; greenish yellow." Webb: "Gold, greenish red."
10ʰ 20.5ᵐ	+06° 26′	Σ1426								The data suggest a wide and easy binary with a close double star for a companion. Webb: The AB pair is "yellowish."
		AB	2004	309°	0.9″	8.0	8.3	*F5*	binary	
		AB-C	2002	8°	8.0″	8.0	9.4		binary	
10ʰ 22.7ᵐ	+15° 21′	ΟΣ 216	2001	240°	1.9″	7.4	10.3	*G5*	p=306 yr	A 150-mm should be able to split this binary; its separation continues to widen.
10ʰ 25.6ᵐ	+08° 47′	Σ1431	2003	73°	3.3″	7.8	9.1	*A8III*	optical	The data and location suggest an interesting pair — two very tight stars close beside a bright star (44 Leonis). Webb: "White, bluish white."
10ʰ 26.9ᵐ	+17° 13′	ΟΣ 217	2001	146°	0.7″	7.8	8.6	*F6V*	p=140 yr	This fast-moving binary should now be splittable with 300-mm; the separation will continue to widen.
10ʰ 33.8ᵐ	+23° 21′	Σ1447	2001	124°	4.4″	7.5	8.9	*A2*	binary	125-mm: Beautiful contrast for the separation! At 50×, it's a lemon-white star almost touched by a little smokeball. Webb: "Very white, bluish."

Leo *(continued)*

RA	Dec.	Name	Year	P.A.	Sep.	m_1	m_2	Spec.	Status	Comments
$10^h 34.4^m$	+21° 36′	Σ1448 AC	2001	260°	11.1″	7.5	9.6	*K0*	optical	125-mm: A pretty pair with nice contrast for the separation. It's a bright grapefruit-orange star with a tiny silver dot close beside it. Webb: "Yellowish, no color given."
$10^h 35.0^m$	+08° 39′	49 Leo	2004	157°	2.1″	5.8	7.9	*A2V*	binary	125-mm, 200×: Grand sight! A bright yellow-white star with a crisp little globe at its edge — it looks like a star with a planet! Smyth: "Silvery white; pale blue."
$10^h 53.4^m$	-02° 15′	S 617	2003	178°	35.2″	6.3	8.7	*G9IV*		The data suggest a bright primary star, probably yellow or orange. But with 150-mm, Ferguson calls this pair "white and blue."
$10^h 55.6^m$	+24° 45′	54 Leo	2003	111°	6.3″	4.5	6.3	*A1V A2V*	binary	Showcase pair. 60-mm: A bright, easy pair with lovely contrast — a bright banana-yellow star and a smaller sapphire blue, split by hairs at 45×. Webb: "Greenish white, blue."
$11^h 00.0^m$	-03° 28′	Σ1500	1999	303°	1.4″	7.9	8.3	*F8V*	binary	This close, dim binary is about 1° south-southwest of 61 Leonis; Webb included it in his catalog. He calls both stars "yellowish."
$11^h 06.9^m$	+01° 57′	65 Leo	1991	104°	2.7″	5.6	9.7	*G9III*		Hartung: "This bright, deep yellow star has a faint [close] companion, which 105-mm will show in good conditions."
$11^h 13.7^m$	+20° 08′	Σ1517	2004	321°	0.6″	7.5	8.0	*G5V*	binary	This binary is easily spotted 24′ south-southwest of Delta (δ) Leonis, and Webb included it in his catalog. He calls both members "yellowish." It probably now needs 300-mm to split.
$11^h 14.1^m$	+20° 31′	δ Leo	1999	342°	203.9″	2.6	8.6	*A4V*		125-mm, 50×: A pretty pair, which seems much closer than its listed separation. It's a brilliant banana-yellow star with a speck of silver beside it, just beyond its glow.
$11^h 15.4^m$	+27° 34′	Σ1521	2003	96°	3.6″	7.7	8.1	*A5*	binary	125-mm: A bright, easy, and attractively close binary. It's a pair of gloss-white stars, modestly unequal, that are split by hairs at 83×.
$11^h 19.4^m$	-01°39′	Σ1529	2002	254°	9.4″	7.1	7.9	*F6IV G3*	binary	60-mm: An attractively close binary, pleasantly bright with just 60-mm. It's a lemony white star and a smaller silvery white just a tiny gap apart at 25×. Webb: "A fine object."
$11^h 23.9^m$	+10° 32′	ι Leo	2004	108°	1.7″	4.1	6.7	*F4IV*	p=186 yr	A 150-mm should be able to resolve this fast-moving binary; its separation is increasing. Smyth: "Pale yellow; light blue . . . beautiful object." Hartung: "Deep yellow and whitish."
$11^h 25.6^m$	+16° 27′	81 Leo	1998	2°	55.1″	5.6	10.8	*F2V*		325-mm, 73×: Grand sight! A brilliant yellow star with a small red companion, and the little star is vividly seen in striking contrast. They're super-wide apart but look like a couple.

Leo *(continued)*

RA	Dec.	Name	Year	P.A.	Sep.	m_1	m_2	Spec.	Status	Comments
$11^{h}26.8^{m}$	+03° 01′	83 Leo	2004	150°	28.6″	6.6	7.5	G7V	optical	60-mm, 25×: Splendid "double double." The field shows two bright stars — a yellow-white star and a lemon-yellow star — each with a small gray companion. Both pairs are very wide.
$11^{h}27.9^{m}$	+02° 51′	τ Leo	2004	181°	88.9″	5.1	7.5	G8II-III		
$11^{h}31.7^{m}$	+14° 22′	88 Leo	2004	331°	15.4″	6.3	9.1	G0IV	binary	60-mm, 25×: A striking but difficult binary for 60-mm. It's a bright white star with a tiny speck beside it — modestly wide and quite hard to see. Smyth: "Topaz yellow; pale lilac."
$11^{h}34.7^{m}$	+16° 48′	90 Leo								60-mm, 120×: Pretty and interesting triple. It's a tight pair of yellow stars and a wide little gray star all nearly in a line. Smyth: "Silvery white [A], purplish [B], pale red [C]."
		AB	2003	208°	3.4″	6.3	7.3	B4V	binary	
		AC	1996	239°	67.0″	6.3	9.8	F5		
$11^{h}36.3^{m}$	+27° 47′	Σ1555	2004	150°	0.7″	6.5	6.8	F0V	p=820 yr	A 175-mm should resolve this binary; the gap is gradually widening.
$11^{h}39.6^{m}$	+19° 00′	Σ1565	2003	305°	21.5″	7.3	8.4	F4IV	optical	110-mm, 32×: Nice effect — an easy and conspicuous pair in a nearly black field. It's a deep white star and a smaller nebulous star modestly close together.
$11^{h}48.0^{m}$	+20° 13′	93 Leo	2002	357°	74.1″	4.6	9.0	A7		60-mm, 25×: Grand sight! A bright lemon-yellow star with a drop of nebulous light wide beside it. The contrast is quite beautiful despite the large separation. Webb: "Yellow, white."
$11^{h}49.1^{m}$	+14° 34′	β Leo	1918	359°	80.3″	2.1	13.2	A3V		(*Denebola*) Webb: β Leonis is "bluish [and] dull red [and] very wide. β 603 is 19′ away."
$11^{h}48.6^{m}$	+14° 17′	β 603	2003	339°	1.0″	6.0	8.5	A8III	binary	
$11^{h}52.8^{m}$	+15° 26′	SHJ 132	2003	14°	39.0″	6.9	10.2	A2		125-mm, 50×: This pair is wide in the field with bright-white 95 Leonis. It's a beautifully deep white star with a tiny speck of light wide beside it.

Leo Minor

RA	Dec.	Name	Year	P.A.	Sep.	m_1	m_2	Spec.	Status	Comments
$09^{h}30.7^{m}$	+33° 39′	7 LMi	2004	128°	63.2″	6.0	9.7	G8III		125-mm, 50×: A very wide pair with pretty colors. A bright Sun-yellow star and a small silvery white; the companion is quite dim but easily seen. Smyth: "Bluish white, livid."
$09^{h}35.7^{m}$	+35° 49′	11 LMi	1988	55°	4.8″	4.8	12.5	G8V	p=201 yr	A 250-mm should be able to split this binary; its separation is increasing.
$09^{h}41.4^{m}$	+38° 57′	Σ1374	2003	309°	3.0″	7.3	8.7	G3IV	p. binary	125-mm, 50×: Striking contrast for the separation! A pure lemon-yellow star with a little patch of powder blue sitting on its edge. Webb, listed under Hydra: "Yellowish, very blue."
$10^{h}24.4^{m}$	+34° 11′	ΟΣΣ 104	2002	288°	208.4″	7.2	7.3	M4/6III		8 × 50: A fine sight at very low power. It's a wide pair of twins in a vivid trapezoid with 27, 28, and 30 Leonis Minoris, shaped like the keystone of Hercules. Probably red-colored with more aperture.

Leo Minor *(continued)*

RA	Dec.	Name	Year	P.A.	Sep.	m_1	m_2	Spec.	Status	Comments
$10^{h}27.0^{m}$	+29° 41′	Σ1432	2003	122°	28.7″	7.8	10.3	*F2*	optical	125-mm, 50×: This dim pair stands out in a nearly black field. It's a lemony white star with a wide misty dot beside it. Despite the listed magnitude, the companion was easily seen.
$10^{h}45.9^{m}$	+30° 41′	42 LMi	2002	174°	196.5″	5.3	7.8	*A1V*		60-mm, 25×: This pair is included because of its bright and pretty primary star. It's a brilliant white star with a green-white companion, super-wide apart.

Lepus

RA	Dec.	Name	Year	P.A.	Sep.	m_1	m_2	Spec.	Status	Comments
$04^{h}59.0^{m}$	−16° 23′	β 314								125-mm, 50×: A very pretty primary star. It's a bright amber-yellow star with a tiny glimpse star (C) wide beside it. A 150-mm should be able to split AB.
		AB	2003	323°	0.8″	5.9	7.5	*F3V+F9V*	p=55 yr	
		AB-C	1998	53°	53.3″	5.9	10.4	*F3V*		
$05^{h}12.3^{m}$	−11° 52′	ι Lep	1998	337°	12.0″	4.5	9.9	*B8V*		125-mm, 50×: Striking contrast, grand! A brilliant dusty gold star with a tiny speck of light beside it, modestly wide apart. Smyth: "White, pale violet." Webb: "Greenish, no color given."
$05^{h}13.2^{m}$	−12° 56′	κ Lep	2004	357°	2.0″	4.4	6.8	*B9V*	binary	125-mm, 200×: Grand sight! A brilliant white star with a little smoky globe at its edge. The companion is easily seen in striking contrast.
$05^{h}17.6^{m}$	−15° 13′	S 473	2002	306°	20.7″	7.0	8.5	*B9.5V*		60-mm, 25×: A pretty pair of stars. It's a gloss-white star and a small sea green attractively close together but wide enough to be easy. It's easily spotted about 1.5° northeast of Mu (μ) Leporis.
$05^{h}19.3^{m}$	−18° 31′	S 476	2002	19°	39.5″	6.3	6.5	*B3V A*		60-mm, 25×: A wide, bright, and easy pair, with a bright yellow-white star in the field. It's a pair of identical stars, beautifully deep white in color. Gould: "A third star forms a line with this pair."
$05^{h}20.4^{m}$	−21° 14′	38 Lep	1991	280°	4.1″	4.7	8.5	*A0V*	binary	Hartung: "John Herschel records this pair as 'most beautiful' . . . the colors are pale yellow and white, easily seen with 75-mm."
$05^{h}21.8^{m}$	−24° 46′	h 3752	2002	97°	3.4″	5.4	6.6	*G0*	p. binary	125-mm: Nicely placed. This is the bright, grapefruit-orange star beside globular cluster M79. With 200×, it becomes a close pair of stars that are mildly unequal.
$05^{h}26.0^{m}$	−19° 42′	h 3759	2002	318°	26.7″	5.9	7.3	*F5IV*		60-mm: Beautiful colors! A bright amber-yellow star and a vivid cobalt blue split by a modest gap at 25×. It's easily spotted 1¼° north-northwest of Beta (β) Leporis.
$05^{h}26.4^{m}$	−20° 43′	β 319	1991	231°	3.8″	7.7	10.7	*A5III/IV*	binary	Webb included this binary in his catalog, noting that it is 26′ west and 2′ north of of Beta (β) Leporis (next page). The data suggest a moderately bright star with a close, faint companion.

Lepus *(continued)*

RA	Dec.	Name	Year	P.A.	Sep.	m_1	m_2	Spec.	Status	Comments
$05^h 28.2^m$	−20° 46′	β Lep	1993	339°	2.2″	3.0	7.5	*G5II*		Hartung: “The faint companion of this brilliant deep yellow star is difficult unless the definition is good, when 150-mm should show it.”
$05^h 32.7^m$	−17° 49′	α Lep	1999	157°	35.6″	2.6	11.2	*F0I*		Gould, 350-mm, 110×: “A brilliant star, possibly pale yellow, with two very wide companions.”
$05^h 39.3^m$	−17° 51′	β 321								80-mm, 33×: This is a broad little cluster of four bright stars — a bright white star inside a triangle of three small gray ones. With 350-mm, Gould saw the the close B and D companions.
		AB	2002	153°	0.4″	6.7	7.8	*B7V*		
		h 3780								
		AC	2002	138°	89.9″	6.7	8.9			
		AE	2002	8°	76.0″	6.7	7.9			
		AF	2002	299°	132.3″	6.7	8.3			
		CD	1991	356°	1.4″	8.9	9.6			
$05^h 39.7^m$	−20° 26′	Lal 1	2002	124°	10.9″	6.9	7.9	*B8*	optical	60-mm, 25×: An easy pair with nice contrast for the separation. It’s a pure white star almost touching a smaller pure gray.
$05^h 41.3^m$	−26° 21′	h 3788	1999	155°	25.9″	7.6	9.2	*F3/5V*		125-mm: An easy, unequal pair easily noticed in a field packed with stars. It’s a pearly white star and a silvery star split by a wide gap. The companion looks brighter than its listing.
$05^h 44.5^m$	−22° 27′	γ Lep	1999	350°	96.9″	3.6	6.3	*F6V K2V*		Showcase pair. 125-mm, 50×: Fantastic colors — this is a bright Sun-yellow star with a brick-red companion; both colors are seen vividly. Webb and Hartung saw similar hues.
$05^h 49.6^m$	−14° 29′	β 94	1991	167°	2.4″	5.7	8.2	*G8III*		This pair is 43′ east-northeast of Zeta (ζ) Leporis; the data suggest a bright star with a close companion. Hartung: “75-mm shows well this deep yellow and ashy pair in a field sprinkled with stars.”

Libra

RA	Dec.	Name	Year	P.A.	Sep.	m_1	m_2	Spec.	Status	Comments
$14^h 24.7^m$	−11° 40′	Σ1837	2003	274°	1.2″	6.9	8.0	*F5III*	p. binary	Gould, 175-mm: “Good object. A bright yellowish star . . . which 180× shows as a close, moderately uneven pair. The field is thin with a brighter star (2 Librae) to the west.”
$14^h 25.5^m$	−19° 58′	SHJ 179								60-mm, 35×: Haunting sight! A bright pair of twins in a black field, like a pair of eyes in a cave. They look reddish white. Gould, 175-mm: The “companion becomes a pair with 180×.”
		AB	2003	295°	34.2″	6.6	7.2	*A2V*		
		β 225								
		BC	1999	88°	1.2″	7.2	8.4	*A4V*	binary	
$14^h 46.0^m$	−15° 28′	5 Lib	1962	252°	3.1″	6.4	10.1	*K1III*		The data suggest a very bright star with a close, faint companion. Hartung: “105-mm shows clearly the white companion of this orange star.”
$14^h 48.5^m$	−17° 20′	β 346	1999	276°	2.5″	7.5	7.9	*G1V*	p. binary	125-mm, 200×: Attractively close pair easily spotted about $1\frac{1}{2}$° south-southwest of Alpha (α) Librae (next page). It’s two vaguely yellowish stars not quite fully apart.

Libra *(continued)*

RA	Dec.	Name	Year	P.A.	Sep.	m_1	m_2	Spec.	Status	Comments
14^{h}49.3^m	−14° 09′	μ Lib	2003	340°	1.9″	5.6	6.6	*A*1	binary	125-mm, 200×: Striking contrast for the separation — a bright straw-yellow star touching a little blue-green star. Hartung: "There is no color difference between these pale yellow stars."
14^{h}49.3^m	−24° 15′	H VI 117	1998	221°	63.9″	5.8	8.6	*K*3III		Gould, 175-mm: "Good object. A bright, wide pair with good color effect; the stars are orange and bluish in contast. B also has a tiny companion, glimpsed at 100×."
14^{h}50.9^m	−16° 03′	α Lib	2002	315°	231″	2.7	5.2	*A*3IV *F*4IV		Showcase pair. 60-mm, 25×: A super-wide pair of dazzling stars that are brilliant straw yellow and bright gloss white. They totally dominate their field. Smyth: "Pale yellow; light grey."
14^{h}57.5^m	−21° 25′	33 Lib	2002	306°	25.1″	5.9	8.2	*K*5V *M*2V	p=2,130yr	Hartung, referring to the pair as HN 28: "Orange and red is the unusual color combination of this wide pair, an easy object for small apertures." Smyth: "Straw colored, orpiment yellow."
14^{h}58.9^m	−11° 09′	18 Lib	1999	39°	19.7″	5.9	9.9	*K*2III-IV		125-mm, 200×: A pretty pair. It's a bright orange star and a small dot of silver separated by a wide gap. Gould: "There are two very wide faintish companions [besides the main companion]."
15^{h}01.6^m	−03° 10′	Σ1899	2002	67°	28.6″	6.7	10.2	*K*2IV *K*0V		125-mm, 50×: A fine view at low power. It shows a triangle of three bright stars — an orange-red, a gloss white, and a citrus-orange star with a tiny wide companion.
15^{h}12.2^m	−19° 48′	ι Lib	2002	109°	57.3″	4.5	10.9	*B*9IV		125-mm, 50×: Splendid view! It shows three bright stars in a thin field — a white star (25 Librae), a yellow-orange star, and a pure yellow star with a tiny wide companion (Iota [ι] Librae).
15^{h}14.5^m	−18° 26′	SHJ 195	2002	140°	47.4″	6.8	8.3	*F*3IV/V		60-mm, 25×: Splendid view! It shows three bright stars in a perfect triangle — a straw-yellow star, a pearly white star (26 Librae), and a whitish gold star with a wide little companion (SHJ 195).
15^{h}19.2^m	−24° 16′	β 227	2000	161°	1.8″	7.5	8.6	*K*1III		Gould, 175-mm, 100×: "A neat close pair . . . [which is] the middle star in a line of three. The bright star just east is h 4756 [below], an elongated figure-8 at 330×."
15^{h}19.7^m	−24° 16′	h 4756	1992	263°	0.6″	7.9	8.3	*F*4V	p. binary	Gould, 175-mm 100×: The view shows "two fairly bright stars in an otherwise faintish field . . . one of these is orange and has an easy, less bright companion moderately separated."
15^{h}25.4^m	−21° 56′	h 4769	1999	192°	9.7″	7.9	9.7	*K*0III	binary	
15^{h}29.0^m	−28° 52′	β 1114 AB	1998	317°	0.8″	7.0	7.8	*F*6IV	binary	Gould, 175-mm, 330×: This "moderately bright yellowish star has an easy fainter companion [h 4774]; 330× splits this companion into a very close, slightly uneven pair [β 1114]."
		h 4774 AB-C	1998	9°	9.5″	7.0	9.6		binary	

Libra *(continued)*

RA	Dec.	Name	Year	P.A.	Sep.	m_1	m_2	Spec.	Status	Comments
$15^h 31.7^m$	−20° 10′	S 672	2002	280°	11.4″	6.3	8.9	*A*8V	optical	The data suggest a bright star with a fairly close companion. Gould, 175-mm: An "easy and rather nice pair, pale yellow and dull brownish, in a wide scatter of stars."
$15^h 33.2^m$	−24° 29′	Lal 123 A-BC	1998	300°	9.0″	6.9	7.0	*A*3+*F*0V		According to the Hipparcos Catalog, this pair is a magnitude brighter than the data listed. Gould, 175-mm, 100×: An "easy, fairly bright, even pair in a thin star field."
$15^h 38.7^m$	−08° 47′	Σ1962	2003	189°	11.6″	6.4	6.5	*F*8V *F*8V	optical	60-mm: Nice — a close bright pair inside inside a neat triangle of fainter stars. It's a bright pair of gloss-white stars, very nearly alike, just a small gap apart.
$15^h 39.9^m$	−19° 46′	β 122	1996	226°	1.8″	7.7	7.7	*F*5V	binary	Gould, 175-mm, 100×: "Good effect. A fairly bright yellowish pair, attractively close, in a field with a few other bright stars . . . [Kappa] κ Lib is 1/2° east [and 41 Librae is about 1/2° northwest]."
$15^h 42.8^m$	−16° 01′	β 35	1996	111°	1.8″	7.3	8.7	*F*7IV	binary	This binary is 28′ southwest of Eta (η) Librae. Gould, 175-mm, 180×: An "easy, uneven close pair, both stars yellowish, and there are two very wide fainter companions."
$15^h 43.2^m$	−25° 25′	β 354	1999	289°	5.7″	7.3	9.3	*F*1V		Gould, 175-mm, 100×: "Good pair, which should be easy with small scopes. A bright uneven pair, with a good brightness contrast; the thin field has a few fairly bright stars in it."
$15^h 46.2^m$	−28° 04′	β 620 AB	1998	170°	0.6″	7.6	7.0	*F*2IV	binary	The data suggest a fairly bright star with companions. Gould, 175-mm, 330×: A "very wide pair [meaning AC] with a white primary. At 330×, this primary becomes elongated."
		h 4803 AC	1999	213°	50.6″	6.6	9.0			

Lupus

RA	Dec.	Name	Year	P.A.	Sep.	m_1	m_2	Spec.	Status	Comments
$14^h 20.2^m$	−43° 04′	h 4672	1991	301°	3.5″	5.8	7.9	*G*8III	binary	Gould, 175-mm, 180×: "Good double. This bright yellow star has a small point close beside it, which is obvious with 180×. It sits in a faint sparse field."
$14^h 22.6^m$	−48° 19′	R 244	1991	122°	4.5″	6.1	9.5	*B*1III		Gould, 175-mm, 100×: This "bright white star has a tiny companion, close southeast; the field is rich and fine."
$14^h 30.3^m$	−49° 31′	Hdo 232	2000	19°	24.0″	5.5	12.0	*A*1V+*B*		Hartung: The companion of this bright star "can be only glimpsed occasionally with 200-mm . . . it is probably a field star, of which there are very many in the attractive field."
$14^h 37.3^m$	−46° 08′	h 4690	1999	25°	19.1″	5.6	7.7	*G*8III+*A*1V		Showcase pair. Gould, 175-mm, 100×: A "bright, easy double, that is orange and blue-white; a fine object of contrasting brightness. A bright star 12′ southwest is also orange."
$14^h 47.0^m$	−52° 23′	h 4698	2000	260°	8.9″	5.2	13.0	*G*8III		Hartung: This "fine deep yellow star in a scattered star field has a white companion which may be seen steadily with 150-mm . . . it is thus much brighter [than the magnitude given]."

Lupus *(continued)*

RA	Dec.	Name	Year	P.A.	Sep.	m_1	m_2	Spec.	Status	Comments
$14^h 51.3^m$	−47° 24′	h 4706	1999	219°	6.7″	7.7	9.0	*K*0	binary	Gould, 175-mm, 100×: "Nice pair. This dull, deep yellow star has an easy, less bright companion. The field is only middling starry, with a little arc of three faint stars some 8′ SW."
$14^h 53.4^m$	−45° 51′	Δ 171	2000	227°	17.6″	7.1	9.6	*B*3V		Gould, 175-mm, 100×: A "wide and fairly bright uneven pair, in a rather bare field. White."
$14^h 56.5^m$	−47° 53′	h 4715	1991	278°	2.2″	6.0	6.8	*B*9V	p. binary	Hartung: "A good object for small apertures, this close unequal pair [is] in a field sown with stars." Gould, 175-mm, 100×: "White [pair]; faint field."
$14^h 58.5^m$	−47° 26′	CapO 62	1999	164°	24.5″	7.4	8.4	*K*2/3III		This pair is 33′ northeast of a bright (5.6-magnitude) star. Gould, 175-mm, 100×: An "easy wide pair, yellow and white, in a moderately starry field."
$15^h 05.1^m$	−47° 03′	π Lup	2000	71°	1.6″	4.6	4.6	*B*5V	p. binary	Hartung: "105-mm clearly resolves this beautiful pale yellow pair which dominates a well sprinkled star field."
$15^h 10.7^m$	−43° 44′	CapO 415	1999	21°	49.8″	7.1	7.7	*G*5V		The data suggest a fairly bright pair of stars with a yellow or orange primary. Gould: A "very wide easy small scope pair . . . only likely to be good at very low power."
$15^h 11.6^m$	−45° 17′	Δ 178								Jaworski, 100-mm: "Beautiful wide pair . . . with the color contrast of yellow orange and white. The pair is easy with 50× and well shown at 80×."
		AC	1999	258°	30.9″	6.5	7.3	*K*0III		
$15^h 11.9^m$	−48° 44′	Δ 177	1999	143°	26.5″	3.8	5.5	*B*9.5V+*A*5V	optical	Showcase pair. Jaworski, 100-mm: An "easy wide double at 50× . . . very bright yellowish white and [plain] white, and the two show a moderate contrast in brightness."
$15^h 12.3^m$	−52° 06′	ζ Lup	1999	249°	71.8″	3.5	6.7	*K*0III *F*8V		Gould, 175-mm, 100×: A "very wide pair of bright stars, deep yellow and dull brown-yellow, which dominate a starry field. NE are two fairly bright stars — the farther one is deep orange."
$15^h 13.0^m$	−37° 15′	CorO 179	2000	228°	6.4″	8.0	8.1	*A*3/4V	binary	Gould, 175-mm, 100×: "Good object. An easy, even, and fairly bright white pair in a middling starry field."
$15^h 14.0^m$	−43° 48′	I 228	1991	16°	1.3″	8.0	8.2	*A*4/5IV/V		Gould, 175-mm: "An even and very close pale yellow pair, in a quite starry field of mostly faint stars; 180× shows it well."
$15^h 14.5^m$	−43° 23′	Δ 179	1999	46°	10.5″	7.3	8.5	*A*1V		Jaworski, 100-mm: A "beautiful close pair set in a fine field of scattered stars of varying brightness. The stars are yellow-white and bluish, and the pair is well shown at 80×."
$15^h 15.3^m$	−44° 09′	HDO 244	1999	40°	13.9″	6.7	9.6	*B*9IV	binary	This pair is 34′ northeast of e Lupi. Gould, 175-mm, 100×: An "easy uneven pair, with arcs of fainter stars west and south. Quite good."
$15^h 18.5^m$	−47° 53′	μ Lup								AC is a showcase pair. Hartung, 105-mm: "The [AB] stars are pale yellow, the wider companion [C] looks reddish, and the field is sown with stars . . . the whole effect most attractive."
		AB	1999	323°	1.0″	5.0	4.9	*B*8V		
		BC	1992	129°	22.0″	4.9	6.3			

Lupus *(continued)*

RA	Dec.	Name	Year	P.A.	Sep.	m_1	m_2	Spec.	Status	Comments
$15^h 21.5^m$	−38° 13′	Howe 76	1999	121°	5.7″	6.6	9.3	*A*0	binary	Gould, 175-mm: "This bright white star has a delicate companion southeast, fairly close; a good effect. The field is middling starry."
$15^h 22.7^m$	−44° 41′	ε Lup								Gould, 175-mm, 100×: This "very bright white star has an easy wide companion; the field is quite starry."
		AC	1999	169°	26.3″	3.6	9.1			
$15^h 30.4^m$	−41° 55′	h 4776	1999	229°	5.6″	6.3	8.4	*B*9.5V	binary	This pair is $1^1/_4$° southwest of Gamma (γ) Lupi (below). Gould, 175-mm, 100×: "Good object. A fairly bright and attractive uneven pair, in a fainter, moderately starry field."
$15^h 35.1^m$	−41° 10′	γ Lup	1996	276°	0.8″	3.0	4.5	*B*2IV	p=190 yr	A 150-mm should be able to split this binary until at least 2025. Hartung: A "close white binary."
$15^h 35.9^m$	−44° 57′	d Lup	1991	10°	2.1″	4.7	6.5	*B*3V	p. binary	Hartung: "This bright attractive pair, pale and deep yellow, is excellent for small apertures." Gould, 175-mm, 100×: A "very close, uneven pair, off-white in color."
$15^h 38.1^m$	−42° 34′	ω Lup	1933	28°	11.8″	4.3	11.2	*K*4.5III		Gould, 175-mm: "A bright deep-mustard-yellow star in a fine field . . . 180× is needed to see its faint companion. There are many stars to the N and S, and two wide faint pairs 15′ S."
$15^h 41.9^m$	−30° 09′	Arg 28								This multiple star is 48′ east-southeast of Tau (τ) Librae. Gould, 175-mm, 100×: This is a "triangle of stars, with a fainter star within the triangle; the brightest star is dull orange."
		AB	1998	24°	35.1″	7.9	10.1	*K*3III		
		AC	1991	316°	31.6″	7.9	10.4			
		AD	1998	324°	88.5″	7.9	10.5			
$15^h 44.4^m$	−41° 49′	Howe 79	1999	338°	3.2″	6.1	7.9	*A*0V+*B*	binary	Gould, 175-mm, 100×: "A close, bright and easy yellow pair, with good brightness contrast. A good pair in a starry field." Hartung: "Pale and deep yellow . . . easy for small apertures."
$15^h 47.1^m$	−35° 31′	Δ 192								Gould, 175-mm: A "bright wide pair . . . which forms a line with a close field star. It is on the edge of a dark space to the north [dark nebula B228] and a starry field to the south."
		AB-C	1998	143°	34.7″	6.9	7.3	*B*9.5V		
$15^h 56.9^m$	−33° 58′	ξ Lup	2000	49°	10.3″	5.1	5.6	*A*3V *B*9V	optical	Showcase pair. Hartung: "This beautiful pale yellow pair dominates a field of scattered stars." Jaworski, 100-mm: A "moderately wide pair of white gems . . . easy at 50×."
$16^h 00.1^m$	−38° 24′	η Lup	1999	19°	14.8″	3.4	7.5	*B*2.5IV		Showcase pair. Hartung: "Very attractive white and ashy pair, which is in fine contrast with a starry field." Gould, 175-mm, 100×: This "very bright white star has well separated companion."
$16^h 03.8^m$	−33° 04′	Howe 82	1991	346°	2.4″	7.7	7.9	*F*3V		Gould, 175-mm, 100×: "Good effect. A bright and easy pair of equals, pale yellow in color, in a patchwork field with a bright star to the south."

Lynx

RA	Dec.	Name	Year	P.A.	Sep.	m_1	m_2	Spec.	Status	Comments
06ʰ 22.1ᵐ	+59° 22′	4 Lyn	2003	143°	0.7″	6.3	7.7	A3V	p. binary	Smyth: "A close double star . . . both [stars] white . . . elegant but difficult object."
06ʰ 24.7ᵐ	+59° 40′	ΟΣΣ 72								60-mm, 25×: Interesting sight! A super-wide pair of bright white stars, exactly alike, with a brighter white star (4 Lyncis, above) in the field.
		AC	1999	323°	134.3″	7.6	7.6	K0		
06ʰ 26.8ᵐ	+58° 25′	5 Lyn								125-mm, 50×: A brilliant grapefruit-orange star and a small pinkish white, super-wide apart; B not seen. Smyth: "A coarse triple star . . . [A] orange tint; [B] blue; [C] pale garnet."
		AB	1999	140°	32.3″	5.2	11.9	K4III		
		AC	2002	272°	95.8″	5.4	7.9			
06ʰ 39.7ᵐ	+58° 06′	Σ936	1999	283°	1.2″	7.3	9.0	G5	binary	This binary is 1¼° west of 6 Lyncis, and Webb included it in his catalog. He calls the stars "yellow, blue."
06ʰ 44.9ᵐ	+59° 27′	Σ946	2004	129°	4.0″	7.3	9.1	F5	binary	This binary is about 10′ west of 12 Lyncis (below). Smyth: "A neat double star . . . bright white, blue."
06ʰ 46.2ᵐ	+59° 27′	12 Lyn								Showcase system. 125-mm, 200×: A bright, vivid, and attractively close triple — a touching pair of yellow stars and a wider bluish silver. Webb: "[A and B] Greenish white, [C] bluish."
		AB	2004	74°	1.9″	5.4	6.0	A3V	p=706 yr	
		AC	2004	309°	8.7″	5.4	7.0			
06ʰ 48.2ᵐ	+55° 42′	Σ958	2003	257°	4.6″	6.3	6.3	F5 F6	binary	60-mm, 90×: Splendid sight! A bright pair of identical stars, khaki white in color, that are split by a tiny gap at 90×. Webb: "White, yellow."
07ʰ 03.1ᵐ	+54° 10′	Σ1001	2003	66°	8.9″	7.8	9.4	G0		125-mm, 50×: An attractively close pair with good contrast. It's a very lemony white star and a small bluish turquoise, split by tiny gap at 50×.
07ʰ 05.7ᵐ	+52° 45′	Σ1009	2003	148°	4.3″	6.9	7.0	A3V	p. binary	125-mm: Splendid sight! A bright, pretty pair of golden-yellow twins split by hairs at 83×. It seems to jump out of its field, which is without other bright stars.
07ʰ 14.8ᵐ	+52° 33′	Σ1033	1994	277°	1.5″	7.8	8.4	F0	binary	125-mm: Lovely effect. 50× shows a bright orange star with a fainter amber-yellow star in the field, and 200× splits the fainter star into a pair of touching stars very nearly alike.
07ʰ 16.4ᵐ	+47° 38′	Σ1044	2002	168°	12.8″	7.8	7.8	A4V		125-mm, 50×: The data suggest a wide and easy pair of twins, but two separate observations showed a pearly white star with a close companion that's very dim.
07ʰ 22.3ᵐ	+50° 09′	20 Lyn	2003	255°	15.0″	7.5	7.7	A6II	optical	60-mm, 25×: Splendid combination! It's a pair of identical yellow-white stars that are attractively close while wide enough to be easy. Smyth: "Both [stars] silvery white."
07ʰ 22.9ᵐ	+55° 17′	19 Lyn								Showcase object. 60-mm, 25×: A fainter clone of Mizar and Alcor (see Ursa Major). It's a bright, wide pair of stars and a super-wide third star in a crooked L shape. All three are white.
		AB	2004	315°	14.8″	5.8	6.7	B8V B9V	optical	
		AD	1991	6°	215.3″	5.8	7.6			

Lynx *(continued)*

RA	Dec.	Name	Year	P.A.	Sep.	m_1	m_2	Spec.	Status	Comments
07ʰ 35.9ᵐ	+43° 02′	ΟΣ 174	2003	89°	2.1″	6.6	8.3	*F0*	binary	The data suggest that this binary has a fairly bright primary and should be splittable with 100-mm. Ferguson, 150-mm: "A pretty pair, white and blue."
07ʰ 38.9ᵐ	+42° 29′	ΟΣΣ 87	1998	177°	62.2″	7.8	7.6	*F5*		125-mm, 50×: A surprisingly pretty pair, much brighter than the data suggest. It's a wide, bright pair of gloss-white stars, exactly alike, that seem to pop out of their thin field.
08ʰ 24.7ᵐ	+42° 00′	S 565	1998	175°	83.9″	6.2	8.6	*K5III*		The data suggest a bright star with a very wide companion. Webb, referring to the pair as ΟΣ 93: Dembowski calls them "gold, azure," Franks calls them "orange red, fine blue."
08ʰ 49.0ᵐ	+38° 21′	Σ1274	2000	41°	9.1″	7.4	9.4	*A2*	binary	125-mm: Lovely contrast for the separation! It's a gloss-white star with a little whitish blue dot at its edge split by a hair at 50×. Webb: "Very white, ash."
08ʰ 50.8ᵐ	+35° 04′	Σ1282	2000	280°	3.4″	7.6	7.8	*F8*	binary	325-mm: A pretty and attractively close binary. At 98×, it's a pair of kissing twins, peach orange in color. Hartung: An "elegant deep yellow pair, clear but not bright with 75-mm."
09ʰ 18.4ᵐ	+35° 22′	Σ1333	2004	50°	1.9″	6.6	6.7	*A8V*	p. binary	125-mm, 200×: Grand sight! A lovely bright pair of hair-split twins that are pure lemon yellow in color. At 31×, brilliant orange Alpha (α) Lyncis is wide in the field.
09ʰ 18.8ᵐ	+36° 48′	38 Lyn	2004	226°	2.6″	3.9	6.1	*A3V*	binary	Showcase pair. 125-mm, 200×: A brilliant lemon-white star touching a smaller green-white star. Both stars are vividly seen in striking contrast. Webb: "Greenish white, blue."
09ʰ 21.0ᵐ	+38° 11′	Σ1338								Hartung: This "fairly bright yellow pair [AB] is not easy to
		AB	2004	298°	1.1″	6.7	7.1	*F2V+F4V*	p= 303 yr	resolve . . . and 200-mm is needed to do it [this was said
		AB-C	1986	167°	10.0″	6.1	11.4			when the separation was 1.2″]."
09ʰ 35.4ᵐ	+39° 58′	Σ1369	2003	150°	24.5″	7.0	8.0	*F2V*	optical	125-mm: Splendid view at 50×! It shows a bright white star with a blue-white companion beside a brilliant yellow-orange star.

Lyra

RA	Dec.	Name	Year	P.A.	Sep.	m_1	m_2	Spec.	Status	Comments
18ʰ 31.1ᵐ	+32° 15′	Σ2333	1998	333°	6.5″	7.8	8.6	*B9IV*	binary	325-mm, 73×: A pretty and attractively close binary. It's a pair of amber-yellow stars (despite the spectral type) very slightly unequal and just a small gap apart.
18ʰ 32.9ᵐ	+38° 50′	ΟΣΣ 171	2002	328°	150.1″	7.0	8.1	*F8 G5*		60-mm, 25×: Splendid view for low power. It shows a super-wide pair of stars, blue-white and plain white, wide in the field with the great star Vega (Alpha [α] Lyrae).

Lyra *(continued)*

RA	Dec.	Name	Year	P.A.	Sep.	m_1	m_2	Spec.	Status	Comments
$18^h 36.2^m$	+41° 17′	Σ2351	2003	340°	5.0″	7.6	7.6	*A1V A0V*	binary	325-mm, 73×: A beautiful and attractively close binary. It's a pair of nearly identical stars, radiant pure white in color, just a small gap apart. Webb: "Minute, but pretty."
$18^h 36.6^m$	+33° 28′	Σ2349	2002	204°	7.3″	5.4	9.4	*B8II-III*		The data and location suggest a bright and easily spotted star with a companion — it's the first bright star due west of Beta (β) Lyrae. Webb: "Bluish white, no color given."
$18^h 36.9^m$	+38° 46′	α Lyr	2000	182°	78.2″	0.0	9.5	*A0V*		*(Vega)* 125-mm, 50×: Grand! This stunning, lime-white beacon is surrounded by many close field stars; the 9th-magnitude star given here looks like a companion. Smyth: "Pale sapphire; smalt blue."
$18^h 38.4^m$	+36° 03′	Σ2362	1999	188°	4.0″	7.5	8.7	*A5*	binary	125-mm: An easy and attractively close binary. It's a very lemony white star with a smaller green-white split by hairs at 83×.
$18^h 41.3^m$	+30° 18′	Σ2367 AB-C	2002	192°	14.0″	7.1	8.8	*G3IV*		125-mm, 50×: A bright and easy pair with lovely colors. It's a lovely banana-yellow star with a small blue companion — attractively close while wide enough to be easy.
$18^h 42.1^m$	+34° 45′	Σ2372	2003	82°	24.8″	6.5	7.7		optical	60-mm, 25×: A splendid wide pair for low power. It's a bright blue-white star with a little nebulous companion, easily noticed about 2° northwest of Beta (β) Lyrae.
$18^h 42.9^m$	+44° 56′	Σ2380	2002	9°	25.3″	7.3	8.7	*G8III*		125-mm, 50×: An easy wide pair with a pretty primary star. It's a bright, pale lemon star with a little gray companion. Webb: "Yellow, bluish white."
$18^h 44.3^m$	+39° 40′	ε^1 - ε^2 Lyr	1998	174°	210.5″	5.0	5.3	*A4V F0V*		This showcase system is called "the Double-Double"; it's a glorious sight in any instrument. 7× 50 binoculars: a bright wide pair of stars ($\varepsilon^1 - \varepsilon^2$) that form a triangle with brilliant Alpha (α) and Zeta (ζ) Lyrae. 60-mm, 120×: It's two bright pairs in the same view, and each is a pair of touching stars. Epsilon[1] (ε^1) is straw yellow with a smaller arctic-blue companion, and Epsilon[2] (ε^2) is a pair of amber-yellow twins.
		ε^1 Lyr	2003	352°	2.1″	5.0	6.1	*A4V F1V*	p=1,725 yr	
		ε^2 Lyr	2004	82°	2.4″	5.3	5.4	*A8V F0V*	p=724 yr	
$18^h 44.8^m$	+37° 36′	ζ Lyr AD	2003	150°	43.8″	4.3	5.6	*F0IV*		60-mm, 25×: Nice. A bright, very wide, and easy pair of vivid stars, goldish white in color, in a field packed with stars. Smyth: "Topaz; greenish." Webb: "Greenish white, yellow."
$18^h 45.2^m$	+38° 19′	Σ2393	2002	24°	17.6″	7.8	10.4	*K5III*		Webb included this pair in his catalog (because of its pretty primary?). It's about two-thirds of the way from Epsilon (ε) to Zeta (ζ) Lyrae. Webb: "Red, no color given." Espin calls the companion "blue."

Lyra *(continued)*

RA	Dec.	Name	Year	P.A.	Sep.	m_1	m_2	Spec.	Status	Comments
$18^h 45.8^m$	+34° 31′	Σ2390	2000	156°	4.2″	7.4	8.6	A7V	binary	125-mm: Nice view. This binary forms an L shape with a stream of dim stars in a field with several fairly bright ones. It's a Sun-yellow star and a bluish turquoise fairly wide apart at 83×.
$18^h 47.2^m$	+31° 25′	Σ2397	1998	267°	3.9″	7.5	9.1	G3III	binary	125-mm: A faint and difficult but attractively close binary. It's a faint grapefruit-orange star with a tiny wisp of arctic blue beside it, split by hairs at 83×. It forms a triangle with two bright stars.
$18^h 49.9^m$	+32° 33′	ν^1 Lyr	1998	174°	18.5″	5.3	12.7	A3V		325-mm, 98×: A broad triple star shaped like a triangle — a bright yellowish white star with two dim companions. Smyth calls it a quadruple with a "pale yellow" primary and three blue companions."
$18^h 50.1^m$	+33° 22′	β Lyr	2002	150°	46.0″	3.6	6.7	B7V+A8		60-mm, 25×: Grand sight for low power. It's a brilliant white star with a wide little sapphire companion in a field packed with stars. Smyth: "Very white and splendid; pale grey."
$18^h 54.9^m$	+33° 58′	OΣ 525								Webb: "Yellow, no color, blue. Beautiful miniature of β Cygni with two wide companions. D. [Dembowski?] calls [B member] . . . the minimum visible in his 6.5-inch [165-mm] achromat."
		AB	2003	128°	1.8″	6.1	9.1	A8+G5III	binary	
		SHJ 282								
		AC	2004	349°	45.1″	6.1	7.6			
$18^h 58.8^m$	+40° 41′	Σ2431	2003	235°	18.8″	6.2	9.6	B3V		150-mm, 60×: Striking contrast for the separation! This is a bright white star with a little misty dot beside it. Only a small gap apart, it looks much closer than the listed separation.
$19^h 01.4^m$	+46° 56′	16 Lyr	1998	287°	46.4″	5.0	10.6	A7V		Smyth: "White, blue . . . delicate, though open object. The companion . . . is merely estimated, but it is well seen."
$19^h 02.8^m$	+31° 23′	Σ2441	2002	264°	6.0″	7.9	9.8	A3	p. binary	This double is about 1° southeast of Lambda (λ) Lyrae — it's probably quite easy to spot. The data suggest an easy pair for amateur telescopes. Webb: "Yellow, no color given."
$19^h 07.4^m$	+32° 30′	17 Lyr	2001	290°	3.7″	5.3	9.1	F0V		Smyth: "Light yellow; cerulean blue; a third star at a distance . . . very beautiful and delicate object." Webb: "Yellowish, bluish."
$19^h 08.8^m$	+34° 46′	Σ2470	2003	268°	13.6″	7.0	8.4	B5V	optical	60-mm, 25×: A beautiful "double-double" — two pairs just arcminutes apart that look exactly alike except in color. Σ2470's primary is pure white; Σ2474's is yellowish white.
$19^h 09.1^m$	+34° 36′	Σ2474	2003	262°	15.8″	6.8	7.9	G1V	optical	
$19^h 12.4^m$	+30° 21′	Σ2483	2003	317°	9.9″	8.0	9.1	B9IV	binary	125-mm, 50×: This dim binary is easily noticed just beyond M56, which lies 55′ east of the pair. It's a gloss-white star with a nebulous companion split by a fairly small gap.
$19^h 13.8^m$	+39° 09′	η Lyr	2003	80°	28.1″	4.4	8.6	B2.5IV		Showcase pair. 60-mm, 25×: Striking! A bright white star with a tiny wide companion that is just glimpsed; the pair is in a glorious field full of stars. Smyth: "Sky blue; violet tint."

Lyra *(continued)*

RA	Dec.	Name	Year	P.A.	Sep.	m_1	m_2	Spec.	Status	Comments
19ʰ 14.5ᵐ	+34° 34′	OΣ 367								125-mm, 50×: An easy wide pair, easily noticed, in a stunning field packed with stars. It's a lemony white star with a nebulous speck beside it.
		A-BC	1999	227°	33.6″	7.3	10.3	*F*5IV		
19ʰ 15.9ᵐ	+27° 27′	Σ371								125-mm, 200×: A difficult but gratifying sight. It's a white rod (AB) with a little misty dot (C) wide beside it. It's gratifying to see even a rod for a binary as close as AB.
		AB	2000	160°	0.9″	7.0	7.6	*B*8V	binary	
		OΣ 371								
		AB-C	1997	270°	47.5″	7.1	9.8			
19ʰ 20.1ᵐ	+26° 39′	OΣΣ 181	2001	0°	61.2″	7.4	7.5	*F*5		125-mm, 50×: Interesting sight — a bright pair of stars, identical in luminosity, with a mild color contrast. They look peach white and blue-white. Webb, listed under Vulpecula: "Red yellow, azure."

Mensa

RA	Dec.	Name	Year	P.A.	Sep.	m_1	m_2	Spec.	Status	Comments
04ʰ 24.9ᵐ	−77° 41′	h 3673	1998	67°	10.5″	8.0	8.3	*F*7V	optical	Gould, 175-mm, 100×: An "easy and nearly even pair of yellow stars in a thin field."
05ʰ 35.3ᵐ	−71° 08′	I 277	1900	190°	3.9″	7.8	11.0	*K*3III		Hartung: "This orange star, shining brightly in a fine field [of the Large Magellanic Cloud] sown with stars . . . has a faint ashy companion close."
06ʰ 28.6ᵐ	−78° 54′	h 3888	1999	118°	35.6″	7.6	10.3	*F*0IV		The data and location suggest a very ordinary white pair, about 1.4° west-northwest of Theta (θ) Mensae, that is at least something to see in an otherwise uninteresting field.
06ʰ 54.1ᵐ	−77° 47′	h 3932	1998	286°	8.4″	7.7	9.8	*A*5IV/V	binary	Gould, 175-mm, 100×: An "easy pair of uneven stars, moderate in brightness, in an average field [1.6° north of Theta (θ) Mensae]."

Microscopium

RA	Dec.	Name	Year	P.A.	Sep.	m_1	m_2	Spec.	Status	Comments
20ʰ 33.8ᵐ	−40° 33′	Jc 18	1998	223°	4.4″	7.8	8.5	*A*1V	binary	Gould, 175-mm, 100×: A "fine uneven pair . . . in a field moderately starry but not bright. A 10th-magnitude star is wide southeast from the pair." [Does Gould think it looks like a companion?]
20ʰ 34.8ᵐ	−33° 55′	h 5207	1998	257°	10.4″	7.9	9.6	*F*6V	optical	Gould, 175-mm, 100×: "Nice effect. An easy uneven pair, with a wide companion to the northwest. The field has a good scatter of stars."
20ʰ 40.9ᵐ	−42° 24′	h 5211	1999	300°	20.1″	6.5	10.4	*G*6III		Gould, 175-mm: This "bright yellow star has an easy, fairly wide companion; it is in a good field of mixed stars, including a triangle of stars 10′ northeast."
20ʰ 50.0ᵐ	−33° 47′	α Mic	1999	166°	20.4″	4.9	10.0	*G*7III		Hartung: A "field of a few scattered stars is dominated by [this] bright yellow star which has an ashy companion wide south [east], shown clearly with 75-mm."

Microscopium *(continued)*

RA	Dec.	Name	Year	P.A.	Sep.	m_1	m_2	Spec.	Status	Comments
20ʰ 51.7ᵐ	−40° 54′	h 5228	2000	104°	32.2″	7.4	9.3	*K*0III		Gould, 175-mm, 100×: "A very wide and easy pair with a deep yellow primary. The field has some brighter stars to the east, but is rather bare elsewhere. Suits small scopes."
21ʰ 02.2ᵐ	−43° 00′	Δ 236	1999	73°	57.5″	6.7	7.0	*G*3IV+*K*0IV		The data suggest a fairly bright pair of colored stars. Gould: "Better in an 8 × 50 finder [than a telescope], which shows a neat, close pair of yellow stars."
21ʰ 06.4ᵐ	−41° 23′	η Mic AC	1999	88°	128.3″	5.7	7.9	*F*6/7V		Gould, 175-mm, 100×: "An extremely wide pair of pale yellow stars . . . a nice pair in an 8 × 50 finder."
21ʰ 12.1ᵐ	−30° 35′	β 251	1991	232°	2.1″	7.4	9.4	*F*5V	binary	Gould, 175-mm, 135×: This "fairly bright pale yellow star has a small companion close SW. It is an attractive and delicate pair, with a pattern of moderate stars east and north in the field."
21ʰ 27.0ᵐ	−42° 33′	Mlb 6	1999	150°	2.9″	5.6	8.2	*A*	binary	Hartung: "This close unequal deep yellow and white pair is clearly resolved with 75-mm." Gould, 175-mm, 100×: "A fine double in a not very starry field . . . the companion is clear and just split."

Monoceros

RA	Dec.	Name	Year	P.A.	Sep.	m_1	m_2	Spec.	Status	Comments
06ʰ 01.8ᵐ	−10° 36′	3 Mon	1996	355°	1.9″	5.0	8.0	*B*5III	binary	The data suggest a bright star with a close companion. Hartung: A "close pair . . . 75-mm will show the companion with care." Webb: This pair is "in a large faint nebula."
06ʰ 15.5ᵐ	−04° 55′	β 567	1998	241°	3.9″	6.0	10.0	*A*5IV		Gould, 175-mm: "Good pair. A bright pale yellow star in a patchy field [another 6th-magnitude star is about 25′ north-northwest]; at 180× the small close companion shows."
06ʰ 23.8ᵐ	+04° 36′	8 Mon	2004	29°	12.1″	4.4	6.6	*A*5IV *F*5V	binary	Showcase pair. 60-mm, 25×: Striking! A brilliant white star almost touched by a silvery smokepuff, and both stars are seen vividly. Smyth: "Golden yellow; lilac."
06ʰ 26.7ᵐ	+00° 27′	Σ910 A-BC	2002	153°	66.2″	7.0	8.1	*G*5		Webb: "Yellowish. Near 77 Orionis [now known as HD 45416], a fine 6th-magnitude yellow . . . another pair south preceding [in the field]: a noble spectacle." Smyth: "Topaz yellow, plum tinge."
06ʰ 26.7ᵐ	−07° 31′	Σ914	2002	299°	20.7″	6.3	9.3	*A*0V		125-mm, 50×: A bright white star with a small ash-gray companion that is easily noticed in the field with Beta (β) Monocerotis. The pair looks wider than the data suggest.
06ʰ 28.2ᵐ	+05° 16′	Σ915	2002	43°	5.9″	7.6	8.5	*A*1	binary	125-mm: This is an attractive dim pair in a starry field near the Rosette Nebula. It's a lemon-white star with a shadowy spot beside it, split by a hair at 50×.

Monoceros *(continued)*

RA	Dec.	Name	Year	P.A.	Sep.	m_1	m_2	Spec.	Status	Comments
06h 28.8m	−07° 02′	β Mon								Showcase system. 60-mm, 120×: A close trio of white stars — two are split by a hair — that are beautufully bright and exactly alike! Smyth: "[A] white; [B and C] pale white." W. Herschel: "One of the most beautiful sights in the heavens."
		AB	2002	133°	7.1″	4.6	5.0	*B*3V *B*3		
		BC	2002	108°	2.9″	5.0	5.3	*B*3V *B*3		
06h 31.2m	+11° 15′	Σ921	1999	3°	16.1″	6.1	9.1	*B*2V	optical	125-mm, 50×: Striking pair! A bright amber-yellow star and a little blue star, attractively close. Both stars are easily seen and strongly contrasting. Webb: "Yellow white, bluish white."
06h 31.7m	+05° 46′	Σ926	1999	287°	10.9″	7.2	8.6	*A*1		125-mm: A pretty combination. A radiantly pure white star with a small gray companion attractively close while wide enough to be easy. Webb: "Yellow white, ash."
06h 34.8m	+07° 34′	14 Mon	1999	209°	10.8″	6.5	10.6	*A*0V		Smyth: "A most delicate double star . . . yellowish white, dusky . . . this is indeed a difficult object . . . [the companion] is only seen by transient glimpses."
06h 41.0m	+09° 54′	15 Mon								This bright star is at the base of the tree in the Christmas Tree Cluster (NGC 2264). 125-mm, 50×: A lemon-white star touched by a silver star with a little speck wide beside them. Smyth: "In a magnificient star field."
		Aa-B	2002	214°	2.9″	4.6	7.8		binary	
		Aa-C	2002	14°	16.6″	4.6	9.9			
		FG	2002	263°	39.5″	9.0	10.0			
06h 41.2m	+09° 28′	Σ954	2002	153°	12.9″	7.2	10.2	*B*2III		125-mm, 50×: Good view — this pair is in the field with the Christmas Tree Cluster. It's a fairly bright white star with a speck of light wide beside it.
06h 41.2m	+08° 59′	Σ953	2003	331°	7.0″	7.1	7.7	*F*5	binary	60-mm, 35×: Splendid view — it shows a fairly bright pair of stars beside the southern half of the Christmas Tree Cluster. It's a dusty gold star and a smaller gray star split by just a small gap.
06h 42.7m	+01° 43′	Σ956								60-mm, 35×: AC is a dim but easy wide pair, gloss white and ash gray in color, in a dark patch within the rich Milky Way. Member B not seen with 125-mm.
		AB	1999	195°	6.6″	8.0	10.7	*O*8	binary	
		AC	2002	158°	36.5″	8.0	9.0			
06h 50.8m	−00° 32′	β 897	1999	23°	6.3″	5.8	11.1	*F*2V		The data and location suggest a bright star with a very faint companion in a field full of stars. A 110-mm should be able to split the pair.
06h 53.9m	−04° 24′	Σ985	2002	323°	32.6″	7.8	8.4	*K*5		60-mm, 25×: A pretty view at low power. It shows a modestly close pair of stars, scarlet white and pure white, in a scattered group of dim stars.
06h 54.1m	+06° 41′	OΣΣ 79	1991	90°	115.7″	7.2	7.5	*G*5		60-mm, 25×: A splendid pair for low power. It's a super-wide pair of stars alike in brightness but with contrasting colors — one is orange-white and one is blue-white.
06h 54.1m	−05° 51′	Σ987	1998	177°	1.3″	7.1	7.2	*A*6V	binary	125-mm. 50×: This is a bright goldish white star in a field packed with stars, which becomes a pair of kissing stars at 200×. It's very slightly unequal.

Monoceros *(continued)*

RA	Dec.	Name	Year	P.A.	Sep.	m_1	m_2	Spec.	Status	Comments
07ʰ01.5ᵐ	−03° 07′	Σ1010								125-mm: This pair forms a broad arc with two field stars — an attractive little asterism. It's a gloss-white star with a little gray companion wide apart.
		AC	2003	6°	22.7″	7.7	8.8	*B3*	optical	
07ʰ04.3ᵐ	+01° 29′	ΟΣΣ 82	2004	320°	89.9″	6.5	7.6	*A0V*		60-mm, 25×: Nice view — a bright wide pair of stars in a rich starry field with cluster NGC 2324. It's a pure white star and a smaller brownish white, super-wide apart.
07ʰ05.8ᵐ	−10° 40′	Σ1019								125-mm, 50×: AC is a pretty pair with striking contrast. It's a bright white star with a speck of blue light wide beside it. Gould: "B is visible with 175-mm at 100×."
		AB	2003	282°	6.4″	6.5	10.1			
		AC	2004	295°	38.5″	6.5	9.6			
07ʰ07.9ᵐ	−04° 41′	Σ1029	1995	27°	1.7″	7.5	8.0	*A9V*	p. binary	125-mm, 200×: This is a miniature Castor (Alpha [α] Geminorum). It's a bright pair of kissing stars, blue-white in color, that look like Castor does in a 60-mm telescope.
07ʰ12.7ᵐ	−03° 11′	Σ1045	1998	236°	5.5″	8.0	9.1	*F5*	binary	The data and location of this binary suggest a nice effect — a close pair at the end of a long curving line of 9th- and 10th-magnitude stars. Webb: "White, ash."
07ʰ15.6ᵐ	−01° 52′	Σ1056	2003	299°	4.0″	8.0	8.9	*G0*	binary	125-mm: A dim but attractively close binary in a field packed with faint stars. It's a pair of white stars, mildly unequal, that are just a tiny gap apart at 83×.
07ʰ24.0ᵐ	−03° 59′	Σ1084	1998	284°	14.9″	7.2	10.0	*K0*	optical	125-mm: A pretty pair. It's a lovely grapefruit-orange star with a small dot of light wide beside it. Both stars are easily seen in striking contrast.
07ʰ32.1ᵐ	−08° 53′	Σ1112	1998	113°	23.8″	6.0	8.7	*F5V*		125-mm, 50×: This pair is both prettier and more difficult than the data suggest. It's a very bright, very lemony white star with a ghostly speck of light wide beside it.
07ʰ45.3ᵐ	−00° 26′	h 767	1998	164°	21.5″	8.0	10.6	*A0*		125-mm, 50×: Nice effect — The field shows two white stars that look identical, and one of them (h 767) has a tiny nebulous glow near its edge.
07ʰ51.0ᵐ	−03° 07′	Σ1152	1998	313°	6.0″	8.4	10.3	*G5*	binary	Webb notes that these three pairs are all close together (they're all within a 1° field in an east-west line). He calls Σ1152 "yellow" and Σ1154 "white, purplish."
07ʰ52.1ᵐ	−03° 03′	Σ1154	2000	354°	2.8″	7.1	9.3	*A5*	binary	
07ʰ54.6ᵐ	−02° 48′	Σ1157	2003	188°	0.7″	7.9	7.9	*F0*		
08ʰ06.5ᵐ	−09° 15′	Σ1183	2004	328°	30.6″	6.2	7.8	*B9.5IV*		60-mm, 25×: An attractive wide pair for low power. It's a bright lemon-white star with a tiny green-white companion; both are sharply seen in lovely contrast.
08ʰ08.6ᵐ	−02° 59′	ζ Mon								125-mm, 50×: Nice effect! This is a broad boomerang-shaped triple in three different shades — brilliant orangish banana yellow, sky blue, and nebulous gray. Smyth: "Light yellow; gray; pale blue."
		AB	1998	105°	33.0″	4.4	10.1	*G2I*		
		AC	2002	247°	64.7″	4.4	9.7			

Musca

RA	Dec.	Name	Year	P.A.	Sep.	m_1	m_2	Spec.	Status	Comments
11ʰ 23.4ᵐ	−64° 57′	h 4432	1991	308°	2.4″	5.4	6.6	*B5V*	binary	Showcase pair. Hartung: "This pale yellow pair makes a good object for small apertures." Gould, 175-mm, 100×: A "bright, fairly close white pair in a faint field."
11ʰ 51.9ᵐ	−65° 12′	CorO 130	1991	159°	1.6″	5.0	7.3	*B4V*	binary	Gould, 175-mm: A "bright white star in an attractive field of faint and medium stars; at 180× it is an unevenly bright, nicely separated pair."
12ʰ 06.4ᵐ	−65° 43′	h 4498	2000	60°	8.9″	6.1	7.7	*G8/K*OIII		Gould, 175-mm, 100×: A "good easy pair, pale yellow with cream, in a quite rich field. There are some faint pairs in the field, 20′ northwest."
12ʰ 25.5ᵐ	−69° 29′	h 4522	2000	67°	12.7″	7.9	8.7	*B8*	optical	This pair is about 1° west-southwest of Alpha (α) Muscae. Gould, 175-mm, 100×: An "easy and moderately bright white pair, in a fairly starry but not bright field."
12ʰ 46.3ᵐ	−68° 06′	β Mus	2000	40°	1.1″	3.5	4.0	*B2.5V*	p=383 yr	Gould, 175-mm: "This brilliant white star is in a fairly rich but faint field; at 180× it is a tight even pair, just separated."
12ʰ 48.3ᵐ	−67° 08′	h 4550	2000	98°	13.5″	7.6	8.7	*A3*III/IV	optical	The data and location suggest an easily noticed wide pair, about 1° north of Beta (β) Muscae (above). Gould, 175-mm, 100×: A "fairly wide, moderately bright white pair in a mostly faint field."
12ʰ 49.0ᵐ	−65° 36′	Gli 185	2000	8°	8.7″	7.3	9.7	*B6*IV	binary	Gould, 175-mm, 100×: An "easy, moderately bright uneven pair, in a field quite rich in patches and streams of dim stars interleaved with dark lanes. Both stars are white."
12ʰ 59.3ᵐ	−70° 38′	CorO 144	2000	275°	4.0″	8.3	9.9	*G1V*	binary	Gould, 175-mm, 100×: "This yellow star has a less bright companion close to it, and an 11th-magnitude companion wide to the west. Nearby to the south is globular cluster NGC 4833."
13ʰ 08.1ᵐ	−65° 18′	θ Mus	2000	187°	5.3″	5.7	7.6	*B0I*	binary	Showcase pair. Hartung: "This elegant yellow and white pair ornaments a field sown profusely with stars . . . an attractive pair for small instruments."
13ʰ 15.2ᵐ	−67° 54′	η Mus AC	2000	331°	58.3″	4.8	7.2	*B8V*		Gould, 175-mm: This "bright white star has a very wide bright companion, and two fainter companions were also seen." [These are not listed in the WDS.].
13ʰ 28.4ᵐ	−67° 52′	h 4586	1991	141°	2.9″	7.3	9.1	*A5*III/IV	binary	Gould, 175-mm, 100×: "Good pair. It is attractive, close and uneven, and lies in a not very starry field [1¼° east of Eta (η) Muscae, above]. White."

Norma

RA	Dec.	Name	Year	P.A.	Sep.	m_1	m_2	Spec.	Status	Comments
$15^{h}32.7^{m}$	−57° 24′	h 4777	2001	295°	5.5″	7.5	9.1	*F2IV*	binary	The data suggest a fairly close binary star. Gould, 175-mm, 100×: "An easy uneven pair with a cream-white primary." Barker, 100-mm: "A pretty pair . . . white and yellow-white."
$15^{h}45.0^{m}$	−50° 47′	Hld 124	1991	194°	2.2″	6.6	8.5	*A3V*	binary	Gould, 175-mm, 100×: A "bright but close and quite uneven pair." Hartung: "The stars are pale yellow in a well sown field . . . the companion is a small point, close."
$15^{h}45.3^{m}$	−58° 41′	Δ 191 AB-C	1999	296°	32.5″	7.8	8.1	*A3/5V*		The data and location suggest a dim but easily noticed pair in a field packed with faint stars. Barker, 100-mm: "Beautiful. A nice wide pair, yellow and blue."
$15^{h}51.1^{m}$	−55° 03′	Δ 193	1999	12°	16.5″	5.8	9.1	*B2II*		Gould, 175-mm, 100×: "Quite good effect. A bright white star with a fairly wide companion, and several wider companions." Barker, 100-mm: A "gorgeous wide pair, white and orange."
$15^{h}54.8^{m}$	−50° 20′	Δ 195	1999	10°	12.0″	6.8	7.5	*A2*		Gould, 175-mm, 100×: A "fairly wide easy pair, with a wide, faint third companion. Some 2″ N is a little triangle of 11th-magnitude stars." Barker, 100-mm: A "lovely pair. Yellow and white."
$15^{h}55.5^{m}$	−60° 11′	h 4813	2000	100°	4.4″	5.9	8.4	*G5II-III*	binary	Hartung: "This bright yellow star has a small white point well clear [of it] which 75-mm shows easily. It dominates a field sown with stars, many in linear sequences."
$16^{h}03.5^{m}$	−57° 47′	ι Nor AB	1994	333°	0.3″	5.2	5.8	*A7IV*	p=26.9 yr	Hartung calls this triple a "close yellow pair" and a wider star that "looks reddish in contrast." Gould, 175-mm, 100×: AB is "an easy uneven pair . . . pale yellow and dull orange."
		AB-C	2000	243°	10.9″	4.6	8.0			
$16^{h}27.2^{m}$	−47° 33′	ε Nor	1999	334°	22.8″	4.5	6.1	*B2V*		Showcase pair. Hartung: "In a field sprinkled with stars is [this] fine wide unequal pair, pale yellow and bluish." Barker, 100-mm: "Easy and bright. White and yellow green."
$16^{h}31.0^{m}$	−46° 29′	h 4857 AC	1999	62°	6.0″	7.8	9.4	*B2II*		Barker, 100-mm: This "pretty easy pair is yellow and white; at low power, a bright white star is in the field."

Octans

RA	Dec.	Name	Year	P.A.	Sep.	m_1	m_2	Spec.	Status	Comments
$01^{h}37.4^{m}$	−82° 17′	Gli 14	2000	54°	5.4″	7.6	8.4		binary	Gould, 175-mm, 100×: A "neat little yellow pair; adjacent is a broad trapezium of 10th-magnitude stars. The field is middling starry."
$03^{h}42.5^{m}$	−85° 16′	R 38	1993	252°	1.9″	6.6	8.1	*B9.5IV*	binary	Gould, 175-mm: "A fine double. The field has a scatter of fairly bright stars, and this pair is part of a triangle of stars . . . it is a neat and close uneven pair at 180×. "
$09^{h}33.3^{m}$	−86° 01′	Δ 82	2000	275°	15.4″	7.1	7.6	*F3/5IV*	optical	This pair is 45′ east-southeast of Zeta (ζ) Octantis. Gould, 175-mm, 100×: A "wide and fairly bright easy pair, pale yellow in color. The field is faint."

Octans *(continued)*

RA	Dec.	Name	Year	P.A.	Sep.	m_1	m_2	Spec.	Status	Comments
10h 05.6m	−84° 05′	h 4310	2000	262°	3.9″	7.7	8.4	*F8V*	p. binary	Gould, 175-mm, 100×: A "neat, easy, and fairly close pair in a moderately starry but not bright field."
12h 02.3m	−85° 38′	h 4490	2000	146°	24.8″	6.2	9.0	*K3III*	optical	Gould, 175-mm, 100×: This "bright orange star has an easy wide companion; the field is good and starry."
12h 55.0m	−85° 07′	ι Oct	1991	239°	0.7″	5.9	6.9	*K0III*	p. binary	Gould, 175-mm: This "bright orange star . . . is elongated at 330×. Not a clean split, but definitely looks double. There is a 9th-magnitude field star nearby; the field thin."
20h 41.7m	−75° 21′	μ² Oct	2000	9°	16.7″	6.5	7.1	*G5III*		Gould, 175-mm, 100×: A "wide and easy yellow pair in a rather sparse field; the nearby stars are all faint."
21h 06.7m	−80° 42′	Gli 263	2000	245°	4.6″	7.3	9.6	*F0/2IV*	binary	Gould, 175-mm, 100×: This "fairly bright yellow star in a rather bare field has a faint companion quite close, but surprisingly easy."
21h 50.9m	−82° 43′	λ Oct	1999	62°	3.3″	5.6	7.3	*G8-K0III*	p. binary	Hartung: "This bright elegant close pair, deep yellow and white, is very easy with 75-mm." Gould, 175-mm, 100×: A "lovely and quite close uneven pair."
22h 03.1m	−76° 07′	h 5306	1999	71°	34.7″	6.0	10.6	*F3III*		Gould: 175-mm, 100×: This "bright pale yellow star has an easy wide companion. There is also a 9th-magnitude star about 90″ SW [which probably looks like a wide companion]."
22h 25.9m	−75° 01′	Δ 238	1999	80°	20.8″	6.2	8.9	*G3V*	optical	Gould, 175-mm, 100×: "Quite good. A wide uneven pair, deep yellow and dull yellow; the field is not very starry."

Ophiuchus

RA	Dec.	Name	Year	P.A.	Sep.	m_1	m_2	Spec.	Status	Comments
16h 25.6m	−23° 27′	ρ Oph AB	2001	340°	2.9″	5.1	5.7	*B2IV B2V*	p. binary	Showcase pair. 60-mm, 120×: A bright pair of touching stars, amber yellow in color. Smyth: "There are two other companions [besides B] . . . the whole forming a pretty group."
16h 28.8m	−08° 08′	Σ2048	1999	297°	5.6″	6.6	9.7	*F4V*		Smyth: "A very delicate double star . . . yellow; dusky," and there is a star "of a deep orange tinge" also in the field. Webb: "yellowish, no color given."
16h 30.9m	+01° 59′	λ Oph	2004	34°	1.6″	4.2	5.2	*A0V+A0V*	p=129 yr	125-mm, 200×: Stunning pair! A brilliant star touched by a bright star; both look lemon yellow.
16h 47.2m	+02° 04′	19 Oph	2003	88°	23.6″	6.1	9.7	*A3V*		325-mm, 54×: Dramatic contrast! A brilliant yellow-white star with a little smoke puff beside it, quite wide apart.
16h 51.1m	+09° 24′	Σ2106	2003	174°	0.7″	7.1	8.2	*F7IV*	p=1,270 yr	This binary star is probably resolvable with 250-mm until at least 2025.
16h 56.8m	−23° 09′	24 Oph	1994	301°	1.0″	6.3	6.3	*A0V*	p. binary	Hartung: A "bright yellow pair . . . in 1960 [when the separation was 0.9″] 150-mm clearly resolved the stars."

Ophiuchus *(continued)*

RA	Dec.	Name	Year	P.A.	Sep.	m_1	m_2	Spec.	Status	Comments
$16^h57.1^m$	−19° 32′	SHJ 240	2003	232°	4.6″	6.6	7.6	*B*6V+*B*7V	binary	125-mm, 83×: Nice effect — a close bright pair that forms a T-shaped asterism with four field stars. It's a straw-yellow star with a smaller arctic blue split by a tiny gap.
$17^h02.0^m$	+08° 27′	Σ2114	2003	192°	1.3″	6.7	7.6	*A*4V	p. binary	125-mm, 200×: A beautifully tight pair. 200× shows a lop-sided figure-8, whitish yellow and yellowish ash in color. Smyth: This "pair greatly resembles Sigma [σ] Coronae Borealis."
$17^h06.9^m$	−01° 39′	Σ2122	2003	279°	19.9″	6.4	9.7	*A*9V		125-mm, 50×: A wide pair with beautiful colors, which forms the corner of star triangle. It's a bright whitish lemon star with a mint-green speck beside it.
$17^h10.4^m$	−15° 44′	η Oph	2004	239°	0.6″	3.1	3.3	*A*1IV+*A*1IV	p=87.6 yr	Hartung: In 1961, "this brilliant white star . . . [showed] an elliptical image with 300-mm." [The separation has been widening since then.]
$17^h15.3^m$	−26° 36′	36 Oph	2002	146°	4.7″	5.1	5.1	*K*0V *K*1V		Showcase pair. 60-mm: Very pretty! A pair of bright citrus-orange twins that are split by a hair at 77×. Smyth: "Ruddy; pale yellow." Webb: [both stars] "Golden yellow."
$17^h16.6^m$	−00° 27′	41 Oph	2004	8°	1.0″	4.9	7.5	*K*1IV	binary	Hartung: A "fine orange pair . . . the stars could be plainly seen with 200-mm."
$17^h17.7^m$	−26° 38′	H I 35	1993	340°	5.3″	6.9	9.1	*B*9.5V	binary	325-mm, 100×: Striking contrast for the separation! A bright amber-yellow star almost touched by a little ghostly speck.
$17^h18.0^m$	−24° 17′	39 Oph	2002	355°	10.0″	5.2	6.6	*G*8III	optical	Showcase pair. 60-mm, 25×: Beautiful combination — the stars of this pair look tangerine orange and silvery white. They are attractively close while wide enough to be easy.
$17^h19.9^m$	−17° 45′	β 126								Hartung: "This dainty white triplet needs 150-mm to show
		AB	2003	263°	2.3″	6.4	7.6	*A*2V	binary	the third star [C], but the close pair [AB] is clear with
		AC	1998	138°	12.0″	6.4	11.3			75-mm." Webb: Dembowski calls AB "white, azure."
$17^h21.0^m$	−21° 07′	ξ Oph	1989	40°	4.4″	4.4	8.9	*F*1III-IV		Gould, 175-mm, 180×: This "bright pale yellow star in a moderately starry field has a tiny speck of light quite close to it."
$17^h27.9^m$	+11° 23′	Σ2166	2002	282°	27.7″	7.2	8.6	*A*5V	optical	125-mm, 50×: A wide, easy pair with a fairly bright primary. It's a white or peach-white star and a smaller bluish silver.
$17^h30.4^m$	−01° 04′	Σ2173	2001	305°	0.3″	6.1	6.2	*G*5V	p=46.4 yr	This very fast-moving binary is predicted to be closer than 0.5″ until 2007 and wider than 0.5″ from 2008 to 2017. Webb: "Very yellow, gold."
$17^h34.4^m$	+13° 10′	54 Oph	2002	66°	23.0″	6.7	11.6	*G*8III		Smyth: "A most delicate double star . . . pale straw color; blue; several other stars in the field." Webb: "Yellow, no color given."

Ophiuchus *(continued)*

RA	Dec.	Name	Year	P.A.	Sep.	m_1	m_2	Spec.	Status	Comments
$17^h 34.6^m$	+09° 35′	53 Oph	2004	190°	41.5″	5.8	7.5	*A2V*		60-mm, 25×: Nice combination of colors. An easy wide pair, bright white and fainter silvery blue, beside an orange field star.
$17^h 39.8^m$	−04° 58′	Σ2191	2002	267°	26.2″	7.8	8.5	*F2V*		150-mm: An easy wide pair, comfortably bright at 36×. It's a white or peach-white star and a smaller silvery blue.
$17^h 44.6^m$	+02° 35′	61 Oph	2004	93°	20.6″	6.1	6.5	*A1IV-V*	optical	60-mm, 25×: Beautiful combination. A wide, bright and easy pair of straw-yellow stars next to bright lemon yellow Gamma (γ) Ophiuchi. It's a nearly equal pair. Smyth: "Both [stars] silvery white."
$17^h 52.1^m$	+01° 07′	S 694	2003	237°	79.0″	6.7	7.3	*K0 A0*		125-mm, 50×: Interesting pair — the two stars are similar in brightness, but one is pumpkin orange and the other greenish white. Rachal, 70-mm: "Yellowish orange, bluish white."
$17^h 53.0^m$	−07° 55′	Σ3128	1997	57°	1.1″	7.8	10.0	*G0IV-V*	p=259 yr	This relatively fast-moving binary is now probably resolvable with 300-mm; it's separation is slowly widening.
$17^h 57.1^m$	+00° 04′	Σ2244	2003	98°	0.6″	6.9	6.6	*A3V*	p=368 yr	250-mm should split this binary until about 2021.
$18^h 00.6^m$	+02° 56′	67 Oph AC	2003	142°	54.3″	4.0	8.1	*B5I*		60-mm, 25×: Grand sight! A bright lemon-yellow star with a little silvery dot beside it. The pair looks about as wide as Albireo. Webb: "Yellowish, blue." Smyth: "Straw color; purple."
$18^h 03.1^m$	−08° 11′	69 Oph	2003	283°	1.5″	5.3	5.9	*F4IV F5V*	p=257 yr	80-mm, 150×: Grand sight! A bright but barely split pair that forms a triangle with two field stars. It's a pure lemon star with a definite bump on its edge. Smyth: "Both [stars] pale white."
$18^h 05.5^m$	+02° 30′	70 Oph	2004	139°	4.7″	4.2	6.2	*K0V K4V*	p=88.4 yr	125-mm, 200×: A grand sight with a good sky — a bright pair of tangerine-orange stars encased in beautiful diffraction rings. But it looks single in an average sky. Smyth: "Pale topaz; violet."
$18^h 05.7^m$	+12° 00′	Σ2276	2002	257°	6.9″	7.1	7.4	*A7*	binary	60-mm: An attractively close little pair. It's a coppery white star and a smaller ash white split by a hair at 25×. Easily found from 71 and 72 Ophiuchi, which point northward to it. Webb: "White, ruddy."
$18^h 09.6^m$	+04° 00′	73 Oph	2004	293°	0.6″	6.0	7.5	*F2V*	p=294 yr	Smyth: "A close double star . . . silvery white; pale white; and the two point nearly upon a dusky telescopic star."
$18^h 20.1^m$	−07° 59′	Σ2303	1991	240°	1.6″	6.6	9.3	*F2V*	binary	Webb, listed under Scutum Sobieski: "Yellowish, no color given."
$18^h 33.4^m$	+08° 16′	OΣ 355	2003	248°	37.5″	6.4	10.3	*B8III*		The data suggest a bright star with a small companion. Harshaw, 200-mm: "White and orange."
$18^h 37.3^m$	+07° 32′	Σ2346	2003	298°	29.1″	7.9	10.0	*G0*	optical	125-mm, 50×: An ordinary but easily noticed wide pair about 2° north of the giant cluster IC 4756. The stars are whitish grapefruit orange and dim bluish silver.

Orion

RA	Dec.	Name	Year	P.A.	Sep.	m_1	m_2	Spec.	Status	Comments
$04^h 49.8^m$	+06° 58′	π^3 Ori	1999	165°	74.9″	3.2	11.3	*F6V*		125-mm, 50×: Profound contrast! A bright lemon-yellow star with a tiny speck beside it; the pair looks no wider than Albireo. Castle, 150-mm, 171×: "There are several close field stars."
$04^h 59.0^m$	+14° 33′	SHJ 49								125-mm, 50×: Interesting sight — a bright white star between two dim companions. The trio forms a neat boomerang shape. Castle, 150-mm, 171×: The A and B members are "yellowish and bluish."
		AB	2002	305°	39.5″	6.1	7.4	*B7V*		
		AC	2002	89°	53.3″	6.1	9.6			
$05^h 00.6^m$	+03° 37′	Σ627	2004	260°	21.0″	6.6	7.0	*B9V B9V*		125-mm, 62×: Very nice! A wide and obvious pair of bright white twins about 2° northwest of Pi[5] (π^5) Orionis.
$05^h 01.8^m$	+11° 23′	S 463	2002	29°	32.6″	7.2	10.1	*B8*		Castle, 150-mm, 171×: An "extremely wide pair with a blue-white primary, in a field without stars." [This makes it sound like the pair jumps out of its field!]
$05^h 02.0^m$	+01° 37′	Σ630								This pair is inside a neat little trapezium of field stars. Castle, 150-mm, 171×: A "bright white star with a bluish companion; very wide." Webb saw the same colors.
		A-BC	2004	50°	14.4″	6.5	7.7	*B8V*	optical	
$05^h 07.9^m$	+08° 30′	14 Ori	2004	312°	0.7″	5.8	6.7	*A*	p=199 yr	Hartung calls it "deep and pale yellow" and says that "200-mm separates the stars cleanly."
$05^h 11.8^m$	+01° 02′	Σ652	2003	180°	1.6″	6.3	7.4	*A+G2III*	binary	The data suggest a pair of fairly bright stars. Hartung: "This deep yellow star has a close companion, which 75-mm will show in good conditions."
$05^h 13.3^m$	+02° 52′	ρ Ori	1998	64°	7.1″	4.6	8.5	*K2II*	binary	125-mm, 62×: Striking contrast for the separation. A brilliant tangerine star with a little speck of azure blue close beside it; these colors are vividly seen. Webb: "Very yellow, blue."
$05^h 13.5^m$	+01° 58′	OΣ 517	2004	241°	0.7″	6.8	7.0	*A5V*	p=530 yr	A 200-mm should be able to resolve this binary; the separation is slowly widening.
$05^h 14.5^m$	−08° 12′	β Ori								*(Rigel)* 125-mm, 62×: Grand sight! This brilliant star has a tiny speck within its glow, in stunning and beautiful contrast. The primary is also lovely — it's brilliant white with a hint of violet.
		A-BC	2004	204°	9.4″	0.3	6.8	*B8I*		
$05^h 14.7^m$	−07° 04′	Σ667	2004	316°	4.2″	7.2	8.8	*K2*	binary	This pair is about 1° north of Beta (β) Orionis. Castle, 150-mm: A "close, very colorful, and pretty pair. The stars are orange and blue." Webb: "Very yellow, ash."
$05^h 15.2^m$	+08° 26′	Σ664	2003	176°	4.9″	7.8	8.4	*A9IV*	binary	This binary star is about 2° east of 14 Orionis. Castle, 150-mm, 171×: "A close, very cute, nearly matched blue-white pair."
$05^h 17.6^m$	−06° 51′	τ Ori								Smyth: "An elegant and extremely delicate triple star . . . pale orange; blue; lilac."
		AB	1998	251°	33.7″	3.6	11.0	*B5III*		
		AD	1998	60°	35.9″	3.6	10.9			

Orion *(continued)*

RA	Dec.	Name	Year	P.A.	Sep.	m_1	m_2	Spec.	Status	Comments
$05^h20.4^m$	−08° 02′	Σ692 AB-C	2002	5°	34.6″	7.6	8.6	*F*2		This pair should be easy to spot in the low-power field with 29 Orionis about 15′ to its south. Webb: "Yellowish, white. Fine wide object."
$05^h22.8^m$	+03° 33′	23 Ori	2002	29°	31.6″	5.0	6.8	*B*1V		Showcase pair. 60-mm, 25×: A wide pair of bright stars, azure white and ashy blue in color. The stars are very unequal but both shine vividly. Webb: "Pale yellow, fine blue."
$05^h23.1^m$	+01° 03′	Σ700	2002	5°	4.9″	7.7	7.9	*B*9V *B*9.5V	binary	Castle, 150-mm, 171×: "A close and pretty pair of bluish stars, well matched in brightness and color, in a field with many stars." Webb: "Yellowish, bluish."
$05^h23.3^m$	−08° 25′	Σ701	2003	139°	6.0″	6.1	8.1	*B*8III	binary	This bright star is about 40′ south of 29 Orionis. Castle, 150-mm: "A wide pair with 171×, blue white and blue."
$05^h23.5^m$	+16° 02′	Σ697	2002	286°	26.2″	7.3	8.1	*B*7V		The data and location suggest a wide pair in a large asterism of seven bright stars — an 8 × 50 finderscope should show the whole view. Castle, 150-mm: "Blue white and blue."
$05^h23.9^m$	−00° 52′	Wnc 2 A-BC	2003	159°	3.0″	6.9	7.0	*F*6V		Castle, 150-mm: A "beautiful very close pair, cleanly split at 171× . . . with 27 Ori at the field's edge. Both stars look 'warmish,' without an obvious color."
$05^h24.5^m$	−02° 24′	η Ori	2003	78°	1.7″	3.6	4.9	*B*1V+*B*2	p. binary	125-mm, 200×: Grand sight! A kissing pair of bright stars, slightly unequal, that are straw yellow and silvery yellow in color. Webb: "White, purplish."
$05^h26.5^m$	+02° 56′	Σ712	2002	65°	3.3″	6.7	8.6	*B*9.5V	binary	The data suggest a fairly bright star with a close companion about 20′ southwest of 30 Orionis. Ferguson, 150-mm: "Bluish white and blue."
$05^h26.8^m$	+03° 06′	ψ² Ori	1991	327°	2.9″	4.6	8.6	*B*2IV		Webb: "Yellow, fine blue . . . well seen by Burnham with 3.5 inches [90 mm] of 6-inch achromat. In a grand region."
$05^h29.6^m$	+03° 09′	Σ721 A-BC	2002	149°	25.2″	7.1	9.1	*B*5V		The data suggest a wide easy pair about 30′ southwest of 33 Orionis. Webb: "White, no color."
$05^h29.7^m$	−01° 06′	31 Ori	2001	87°	12.8″	4.7	9.7	*K*5III		The data suggest a bright star with a dim companion. Webb: "Very gold, no color given," [but] Franks and Sadler call the companion "blue."
$05^h30.8^m$	+05° 57′	32 Ori	2004	46°	1.2″	4.4	5.8	*B*5V	p=614 yr	This binary should be resolvable by 125-mm; its separation is slowly widening. Webb calls both stars "yellowish."
$05^h31.2^m$	+03° 18′	33 Ori	2003	26°	1.9″	5.7	6.7	*B*1.5V	binary	125-mm: Grand! A bright pair of deep kissing stars, mildly uneven, that are almost apart at 200×. Both stars are yellow-white.
$05^h32.0^m$	−00° 18′	δ Ori Aa-C	2003	0°	52.8″	2.4	6.8	*B*4		Showcase pair. 60-mm, 25×: A brilliant star with a vivid little companion, yelllow-white and very bluish white in color. Smyth and Webb call the companion "violet."

Orion *(continued)*

RA	Dec.	Name	Year	P.A.	Sep.	m1	m2	Spec.	Status	Comments
05h 33.1m	−01° 43′	Σ734								125-mm, 50×: AC is a wide easy pair, in a splendid group of bright stars, just outside the field of brilliant white Epsilon (ε) Orionis. It's pale lemon and khaki white in color.
		AB	2003	356°	1.5″	6.7	8.2	B4V		
		AC	2004	244°	29.6″	6.7	8.4			
05h 35.1m	+09° 56′	λ Ori								Showcase pair. 60-mm, 120×: A bright star almost touching a much smaller one, in the lovely combination of lemon white and ashy blue-violet. Ferguson: A quadruple; C and D are "very faint blue."
		AB	2003	44°	4.3″	3.5	5.5	O8 B0.5V	binary	
		AC	1957	182°	28.0″	3.5	10.7			
		AD	1991	271°	78.0″	3.5	9.6			
05h 35.3m	−05° 23′	θ1 Ori								Showcase system — the Trapezium. 60-mm at 25×: A tight quadruple of bright stars at the heart of the Great Orion Nebula, a beautiful and striking effect. To the author, all the stars look white, but Smyth calls them "pale white [A]; faint lilac [B]; garnet [C]; reddish [D]."
		AB	2004	31°	8.8″	6.6	7.5	O7 B1V		
		AC	2002	132°	12.7″	6.6	5.1			
		CD	2002	61°	13.3″	5.1	6.4			
05h 35.4m	−05° 25′	θ2 Ori	2002	93°	52.2″	5.0	6.2	O9.5V		This is a wide pair of bright white stars beside Theta1 (θ1) Orionis. They're a fine sight for binoculars or a finderscope.
05h 35.4m	−05° 25′	θ1 - θ2 Ori	2000	315°	134.9″	5.0	5.1			8 × 50 finderscope: This super-wide pair consists of the unresolved Theta1 (θ1) Orionis and the brightest member of Theta2 (θ2) Orionis. It's another nice object for very low power.
05h 35.4m	−04° 50′	42 Ori	1995	205°	1.1″	4.6	7.5	B1V		Hartung: "The field shows a large straggling star group involved in luminous clouds. This very bright star needs good definition to see the close companion with 105-mm."
05h 35.4m	−05° 55′	ι Ori	2002	141°	11.3″	2.9	7.0	O9III		Showcase pair. 60-mm, 42×: Iota (ι) Orionis is a pair with fantastic contrast. It's a brilliant yellow-white star with a tiny ghostly speck inside its glow. The wide white pair Σ747 is in the same view.
05h 35.0m	−06° 00′	Σ747	2003	224°	36.0″	4.7	5.5	B0.5V B1V		
05h 35.5m	−04° 22′	Σ750	2003	61°	4.2″	6.4	8.4	B2.5IV	binary	125-mm: An attractively close pair in the large bright cluster NGC 1981. It's a bright white star and a silvery opal split by hairs at 83×. Webb: "White, ash."
05h 38.7m	−02° 36′	σ Ori								Showcase system. 60-mm, 25×: A bright straw-yellow star with two dim companions (D and E). The trio is in the shape of a tight fishhook. 125-mm shows a third companion.
		AB-C	2002	238°	11.5″	3.7	8.8	O9.5V		
		AB-D	2002	84°	12.7″	3.8	6.6			
		AB-E	2003	62°	41.5″	3.8	6.3	B2V		
05h 40.3m	+15° 21′	Σ766	2004	275°	10.1″	7.0	8.4	F0	optical	125- mm, 50×: Very nice; a miniature Beta (β) Cephei. A bright white star and a dim blue-green that are close with great contrast. Both shine vividly.
05h 40.7m	−01° 57′	ζ Ori								125-mm, 200×: Grand! A brilliant star touching a fainter star. Both stars look sharp and vivid, but the companion looks small by contrast. Yellow, silvery yellow.
		AB	2002	166°	2.6″	1.9	3.7	O9.5I	p=1,509 yr	
		BC	2002	8°	59.8″	4.2	8.5	B0		
05h 41.7m	−02° 54′	β 1052	2000	190°	0.6″	6.7	8.2	A9IV-V	p=109 yr	A 300-mm should now be able to just separate this pair.

RA	Dec.	Name	Year	P.A.	Sep.	m_1	m_2	Spec.	Status	Comments
05ʰ44.7ᵐ	+03° 50′	Σ788								125-mm, 50×: Nice "double-double." There are two pairs, about 15′ apart, dominated by a bright straw-yellow star (the primary of Σ789). Each couple is a wide pair of unequal stars.
		AB	1999	89°	7.5″	7.6	10.1	*B9*	binary	
		AC	2003	148°	35.6″	7.6	10.4			
05ʰ45.0ᵐ	+04° 00′	Σ789	1999	149°	14.3″	6.1	10.2	*F0*		
05ʰ46.0ᵐ	−04° 16′	Σ790	2003	89°	7.1″	6.4	9.0		binary	The data suggest a bright star with a close dim companion. Webb: "Reddish yellow, blue."
05ʰ48.0ᵐ	+06° 27′	52 Ori	2003	216°	1.1″	6.0	6.0	*A5V*	p. binary	125-mm, 200×: A beautifully tight pair! It's a perfect figure-8 made by identical yellow-white stars. Webb: "Yellowish, pale yellowish. Excellent test [of sky and instrument]."
05ʰ48.2ᵐ	−08° 23′	Σ798	2002	182°	20.7″	7.3	9.5	*B9*	optical	125-mm, 50×: An ordinary but conspicuous wide pair about 1¼° north of Kappa (κ) Orionis. It's an ash-white star and a little silvery nebulous star.
05ʰ54.2ᵐ	+10° 15′	OΣ 123	1995	186°	2.1″	7.3	9.1	*G5*	binary	The data suggest a close but not extremely difficult binary star. Webb: "Yellowish, ash."
05ʰ54.7ᵐ	+13° 51′	S 502	2001	131°	46.1″	7.9	8.3	*O6*		Brydges, 60-mm, 35×: "These two pairs are in the same view, in a rich starry field. S 502 is white and fairly equal, and S 503 is white and blue with a nice magnitude spread. Each is wide."
05ʰ56.1ᵐ	+13° 56′	S 503	2002	322°	76.0″	6.7	8.4	*G5IV*		
05ʰ54.9ᵐ	+05° 52′	Σ816	2002	289°	4.3″	6.9	9.3	*B9*	binary	Brydges, 60-mm: "A nice [small scope] challenge, with a delicate companion, nicely split with 90×. White, gray."
05ʰ55.2ᵐ	+07° 24′	α Ori								*(Betelgeuse)* Smyth, referring to AE: This bright naked eye star has a "distant companion." It is "orange tinge; bluish, and the two point nearly upon a pale small star."
		AE	2000	154°	176.4″	0.9	11.0			
05ʰ58.4ᵐ	+01° 50′	59 Ori	1999	206°	36.5″	5.9	10.4	*A5*		Smyth: A star "with a very minute companion . . . white; blue . . . delicate though wide. William Herschel remarked that the small individual 'is a point requiring some attention to be seen.'"
05ʰ58.9ᵐ	+12° 48′	OΣ 124	2004	298°	0.5″	6.1	7.4	*K2III+A5V*	p=140 yr	200-mm should be able to split this fast-moving binary, but its separation is slowly closing.
05ʰ59.7ᵐ	+22° 28′	OΣ 125	2000	0°	1.4″	7.9	8.9	*A0*		Webb included this pair in his catalog and gave the same separation as listed. He calls both members "red." The data suggest a close but not too difficult pair.
06ʰ05.1ᵐ	+00° 53′	Σ838	2002	328°	40.5″	7.1	10.4	*G3II*		Brydges, 60-mm, 90×: This is a "difficult [for 60-mm] but very rewarding pair that needs averted vision. The colors are amber and blue."
06ʰ06.5ᵐ	+10° 45′	Σ840								This pair is about 1° east-northeast of the faint cluster NGC 2141. Brydges, 60-mm, 90×: "A nice pair, white and grey in color, with a good medium separation."
		A-BC	2000	248°	21.5″	7.2	9.5	*A0V F0*		

Orion *(continued)*

RA	Dec.	Name	Year	P.A.	Sep.	m_1	m_2	Spec.	Status	Comments
$06^h 08.5^m$	+13° 58′	Σ848	2001	108°	2.3″	7.3	8.2	*B*1V *B*2V	binary	125-mm, 200×: Nicely placed! An attractively close pair on the edge of bright cluster NGC 2169. It's a straw-yellow star and a smaller pearly white just a small gap apart.
$06^h 09.0^m$	+02° 30′	Σ855								Brydges, 60-mm, 35×: A "bright, easy and pretty pair, white and greenish, that makes a nice triple with a wider third star." Webb: The "third star makes a beautiful group."
		AB	2003	114°	29.4″	5.7	6.7	*A*3V *A*0V		
		AC	2002	107°	118.5″	5.7	9.7			
$06^h 10.3^m$	+15° 54′	H VI 114	2002	107°	54.3″	7.2	10.4	*G*5		This pair is about 30′ west-southwest of 69 Orionis. Brydges, 60-mm, 100×: "A difficult unequal pair, with the pretty colors of yellow and blue.
$06^h 14.3^m$	+14° 30′	S 509	2002	199°	170.8″	7.3	8.0	*A*1V		Brydges, 60-mm: "A 'double-double.' S 509 looks very ordinary, but Σ877 was the finest sight of the night; it is a close pair of white stars, very closely alike in brightness."
$06^h 14.7^m$	+14° 35′	Σ877	2003	263°	5.7″	7.6	8.0	*B*9.5V	binary	
$06^h 14.5^m$	+11° 48′	ΟΣΣ 71	1999	312°	90.6″	7.2	7.6	*K*0		60-mm, 25×: Interesting and lovely sight. A pair of equal stars with different colors — one is citrus orange, the other blue-white — in the field with two bright white stars (73 and 74 Orionis).
$06^h 19.4^m$	+13° 26′	ΟΣΣ 73	1997	44°	72.5″	6.9	7.7	*F*6V *K*1V		The data suggest a very wide and conspicuous pair whose stars should look similiar in brightness but different in color.
$06^h 19.7^m$	+12° 18′	Σ891	1998	294°	21.9″	7.7	11.3	*B*8V		This pair is 48′ east of 74 Orionis. Smyth, under Monoceros: "A most delicate double star . . . both [stars] dull yellow."
$06^h 25.0^m$	+10° 31′	Σ901	2002	248°	19.8″	7.8	10.4	*B*9IV		Brydges, 60-mm, 90×: "A nice couple, fairly close, whose companion needs averted vision. White and blue."

Pavo

RA	Dec.	Name	Year	P.A.	Sep.	m_1	m_2	Spec.	Status	Comments
$18^h 23.2^m$	−61° 30′	ξ Pav	1988	156°	3.4″	4.4	8.1	*K*4III		Hartung: "This very bright orange star pair shines like a jewel in a field of scattered stars; it has a white companion which 75-mm will show." Gould: "Probably a test object for 100-mm."
$18^h 34.2^m$	−66° 17′	MlbO 5	1991	292°	4.7″	7.0	9.1	*G*5III *A*5	binary	Gould, 175-mm, 100×: "An easy uneven pair in a sparse field, the brighter star yellow. It has a nice contrast in brightness."
$18^h 49.7^m$	−73° 00′	R 314	1991	271°	1.9″	6.2	8.1	*B*9.5IV-V	binary	Gould, 175-mm: An "uneven pair, close but clear: a little point nestling beside the brighter one . . . easy at 180×. It lies in a scatttered field."
$19^h 04.1^m$	−63° 47′	h 5075	1991	113°	1.7″	7.7	7.7	*A*0	binary	Gould, 175-mm: This is "one of two fairly bright stars seen in the field [that becomes] a close, even pair at 100×; white."
$19^h 10.6^m$	−60° 03′	h 5085	1991	239°	2.7″	7.6	9.1	*B*9II/III	binary	Gould, 175-mm: This pair is "among the outliers of globular NGC 6752 . . . a beautiful combination. A fine uneven pair, just separated with 100×."

Pavo *(continued)*

RA	Dec.	Name	Year	P.A.	Sep.	m_1	m_2	Spec.	Status	Comments
$19^h 17.2^m$	−66° 40′	Gale 3	1999	329°	0.5″	6.1	6.4	*A5V+A8V*	p=157 yr	A 250-mm should resolve this fast-moving binary until at least 2025.
$19^h 29.8^m$	−67° 18′	h 5109								Gould, 175-mm, 100×: This triple is "a flattened triangle of stars in a sparse field. The main star is orange."
		AB	2000	142°	27.8″	7.8	9.4	*G8/K0III*		
		AC	2000	13°	36.8″	7.8	10.3			
$19^h 42.6^m$	−59° 01′	I 119	2001	151°	2.2″	7.9	9.0	*G5V*	p. binary	Gould, 175-mm: "An uneven, moderately bright pair in a starry field; close at 100×. An 11th-magnitude star is 90″ SE, and a coarse wide pair of the 9th magnitude is wide to the E."
$19^h 49.1^m$	−61° 49′	h 5141								Gould, 175-mm, 100×: This "modestly bright yellow star has a fainter wide companion. The field is thin."
		AB-C	2000	341°	13.9″	7.4	10.5			
$19^h 50.7^m$	−59° 12′	I 121	1999	151°	0.8″	5.6	7.2	*A0IV*	p. binary	Gould, 175-mm: This "bright white star has a wide, faint companion . . . I found difficult to see. At 180× the bright star is not quite single; at 330× there is still no clean split."
$20^h 05.1^m$	−63° 04′	h 5163	1991	249°	1.3″	7.7	8.4	*A3V*	binary	Gould, 175-mm: "A test [object] for 100-mm. A nearly even, very close, pale yellow pair . . . cleanly split with 180×. The field is thin."
$20^h 14.6^m$	−64° 26′	h 5171								Gould, 175-mm, 100×: "Quite a nice triple in an arc, though the companions are not bright. There is a scatter of stars to the north, in a thin field."
		AB	2000	306°	17.5″	7.0	9.8	*A2/3IV*		
		AC	2000	336°	33.8″	7.0	10.0			
$20^h 14.9^m$	−56° 59′	Rmk 25	1999	28°	7.2″	8.0	8.0	*F6/8+F*	binary	Gould, 175-mm, 100×: "Quite good. An easy even pair, pale yellow in color. There are some stars in the field to the north, in a sparse area."
$20^h 25.6^m$	−56° 44′	α Pav								Gould: 175-mm, 90×: "This very bright white star dominates a faint field and the easy though not bright pair BC is clear nearby."
		AB	1991	81°	249.0″	1.9	9.1	*B2IV*		
		BC	1999	332°	16.7″	9.2	10.5			
$20^h 51.6^m$	−62° 26′	Rmk 26	2002	82°	2.4″	6.2	6.6	*A2-3IV-V*	p. binary	Showcase pair. Gould, 175-mm, 100×: A "neat and close, pale yellow pair . . . gemlike at 180×. The field is sparse and faint."
$21^h 09.4^m$	−73° 10′	HdO 305								Hartung: "The companion of the bright star may be glimpsed with 250-mm on a clear dark night of good seeing, a real test of instruments and conditions."
		AB-C	1982	123°	7.2″	5.8	13.5	*F8-G0V*		

Pegasus

RA	Dec.	Name	Year	P.A.	Sep.	m_1	m_2	Spec.	Status	Comments
$21^h 22.1^m$	+19° 48′	1 Peg	2003	312°	35.9″	4.2	7.6	*K0.5III*		60-mm, 25×: The best pair in Pegasus for a small aperture. It's a bright yellow star with a tiny glimpse star wide beside it. Smyth: "Pale orange; purplish." Webb: "Orange, blue."
$21^h 26.7^m$	+13° 41′	Σ2797	1998	217°	3.5″	7.4	8.8	*A2V*	binary	125-mm: An easy and attractively close binary star. It's a bright yellow-white star and a little silver star split by a small sliver of space.

Pegasus *(continued)*

RA	Dec.	Name	Year	P.A.	Sep.	m_1	m_2	Spec.	Status	Comments
$21^h 28.9^m$	+11° 05′	Σ2799	2004	264°	1.9″	7.4	7.4	F4V	p=619 yr	125-mm: A modestly dim but attractively close binary star. It's a pair of amber-yellow twins split by a hair at 200×.
$21^h 29.0^m$	+22° 11′	2 Peg	1996	177°	41.1″	6.1	10.2	M4-5III		125-mm, 50×: A very wide pair that shows beautiful colors. It's a bright orange star, exactly the color of tangerine, and a small sapphire blue.
$21^h 33.0^m$	+20° 43′	Σ2804	2004	357°	3.6″	7.7	8.0	F5V	p. binary	125-mm: A pretty pair of amber-yellow stars, nearly equal in brightness, that are bright, sharp, and attractively close at 83×.
$21^h 37.7^m$	+06° 37′	3 Peg	2003	349°	38.6″	6.2	7.5	A2V		125-mm, 50×: Nice effect — a wide pair of bright stars that dominate their field. It's a yellow-white star and a smaller ash white. The pair OΣ 443 is in the field. Smyth: "White; pale blue."
$21^h 44.2^m$	+09° 53′	ε Peg AC	2000	318°	144.0″	2.5	8.7			The data suggest a very bright star with a small companion. Webb: "Pale yellow, blue."
$21^h 44.6^m$	+25° 39′	κ Peg AB-C	2004	288°	14.5″	4.1	10.8	F5IV		325-mm, 98×: Shocking contrast, glorious. A brilliant lemon-yellow star with a tiny point of light next to it; this companion is the faintest speck still possible to see.
$21^h 46.5^m$	+22° 10′	Ho 465	2001	246°	44.6″	7.1	9.1	A7IV		125-mm, 83×: Very nice contrast. This is a bright white star with a misty speck beside it. These stars are wide apart but look much closer than their listed separation.
$21^h 54.3^m$	+19° 43′	Σ2841 A-BC	2004	111°	22.6″	6.5	8.0	K0III+F7V	optical	125-mm, 50×: A pretty pair. It's a bright star with a wide little companion in the lovely combination of yellowish peach and Atlantic blue.
$21^h 58.0^m$	+05° 56′	Σ2848	2002	56°	10.9″	7.2	7.7	A F2V	optical	60-mm, 25×: An easy, sharp, and attractively close pair at 25× — a fine pair for a small aperture. The stars are grapefruit orange in color and nearly alike.
$22^h 01.1^m$	+13° 07′	20 Peg	1999	323°	59.1″	5.6	11.1	F4IIIV		Smyth: "A most delicate double star . . . lucid white; blue; several glimpse stars in field."
$22^h 04.4^m$	+13° 39′	Σ2854	2004	84°	1.7″	7.8	7.9	F6V	binary	125-mm, 200×: A dim but sharply defined and attractively close pair. It's a pair of straw-yellow stars, nearly alike, that are split by a hair.
$22^h 06.2^m$	+10° 06′	Σ2857	2003	112°	19.8″	7.1	9.8	A2		325-mm, 54×: A wide but pretty pair. It's an amber-yellow star with a little dot of silver beside it, and it forms a large right triangle with two field stars.
$22^h 10.4^m$	+14° 38′	Σ2869	2001	252°	21.3″	6.3	12.4	K0III		The data suggest a bright star with a very faint companion. Smyth: "A most delicate double star in a barren field . . . white; blue." Webb: "Very yellow, no color given."
$22^h 14.3^m$	+17° 11′	Σ2877	2003	310°	22.2″	6.7	9.2	K4IV		125-mm, 50×: A modestly wide pair with beautiful colors. It's a vivid grapefruit-orange star with a little dot of powder blue beside it.

Pegasus *(continued)*

RA	Dec.	Name	Year	P.A.	Sep.	m_1	m_2	Spec.	Status	Comments
22h 14.5m	+07° 59′	Σ2878	2003	117°	1.5″	6.9	8.1	B9IV	binary	The data suggest a fairly bright star with a close companion. Hartung: "The companion may be seen surely with 105-mm, [and it is] whiter [than the primary]."
22h 14.6m	+29° 34′	Σ2881	2004	79°	1.3″	7.7	8.2	F6III	binary	125-mm, 200×: An attractively bright and close but difficult pair. It's a yellow-white star that becomes a pair of touching stars for scattered moments. Webb: "Yellowish, bluish white."
22h 20.5m	+35° 07′	OΣ 469	2002	292°	27.7″	7.5	9.7	A9III		125-mm, 50×: A conspicuous wide pair in a field of dim stars. The colors are peach white and dim ivory, and the companion is easily seen despite its listed magnitude.
22h 23.7m	+20° 51′	33 Peg AC	2003	308°	89.9″	6.3	8.5	F7IV		125-mm, 50×: This super-wide pair is included because of its pretty primary star. It's a bright Sun-yellow star with a beige-white companion, and it seems to jump out of its thin field.
22h 26.6m	+04° 24′	34 Peg	1990	224°	3.6″	5.8	12.5	F8IV-V	p=420 yr	Hartung: "The faint companion of this bright yellow star is close . . . and not easy, but 250-mm will show it in good conditions." [This was said when the separation was 3.2″.]
22h 28.2m	+17° 16′	Σ2908	2001	114°	9.1″	7.7	9.7	G9III	binary	325-mm: Splendid contrast for the separation! A bright citrus-orange star with a pale dot beside it, split by hairs at 98×.
22h 34.5m	+04° 13′	Σ2920	2004	143°	13.5″	7.6	8.9	B9.5V	optical	125-mm, 50×: A pretty pair that's easily spotted about 1¼° east of 37 Pegasi. It's a gloss-white star and a smaller powder blue fairly wide apart.
22h 43.0m	+30° 13′	η Peg Aa-BC	2001	339°	93.8″	3.0	9.9	G8III+F0V		The data suggest a very bright star with a wide companion. Webb: "Like [Epsilon] ε Peg, [it] has a bluish 10th-magnitude companion, but the large star is pale yellow."
22h 46.7m	+12° 10′	ξ Peg	1998	97°	11.2″	4.2	12.3	F6V		Smyth: "A most delicate double star . . . pale yellow, blue . . . the companion is only to be caught under intense gazing, with the most favorable circumstances of atmosphere."
22h 54.2m	+28° 01′	Σ2952	1997	138°	17.4″	7.7	10.5	F8V	binary	Webb: "White, blue."
22h 59.2m	+11° 44′	52 Peg	2004	349°	0.5″	6.1	7.3	A8III	p=216 yr	Hartung: "A yellow pair . . . 200-mm showed [the stars] just in contact in 1961." [The separation then was approximately 0.9″.]
23h 00.7m	+31° 05′	Σ2968	2003	93°	3.3″	6.7	9.5	B9	binary	125-mm, 200×: Striking contrast for the separation! A bright straw-yellow star with a vaporish shadow beside it, just a small space apart.
23h 07.5m	+32° 50′	Σ2978	2003	145°	8.3″	6.4	7.5	A3V	binary	125-mm, 50×: A bright pair with vivid colors — the stars look pure yellow and arctic blue. Webb: "Several little pairs similiar to each other lie dispersed in this region."

Pegasus *(continued)*

RA	Dec.	Name	Year	P.A.	Sep.	m_1	m_2	Spec.	Status	Comments
23h09.5m	+08° 41′	57 Peg	2002	198°	32.6″	5.1	9.7			325-mm, 98×: The wide little companion looks almost like a field star, but the pair shows the lovely combination of brilliant citrus orange and powder blue. Smyth: "Orange; greenish."
23h10.0m	+14° 26′	Σ2986	2004	270°	31.6″	6.6	8.9	*G0V*		325-mm, 98×: A wide and easy pair with pretty colors, easily spotted about 1½° east-southeast of Alpha (α) Pegasi. The stars are bright lemon yellow and powder blue in lovely contrast.
23h13.4m	+11° 04′	Σ2991	2001	358°	33.1″	6.0	10.2	*G8III*		125-mm, 50×: This wide pair is pretty for its profound contrast. It's a brilliant lemon-yellow star with a little ghostly spot next to it.
23h22.8m	+20° 34′	Σ3007	2002	91°	5.8″	6.7	9.8	*G2V K6*	binary	Hartung: "This dainty unequal pair, deep yellow and ashy, is clearly shown by 75-mm in a black field."
23h30.4m	+30° 50′	Σ3018								125-mm, 50×: A conspicuous wide pair with a pretty primary easily spotted about 1° southwest of 72 Pegasi. It's a yellowish peach star with dim silver companion.
		AB-C	2002	203°	18.8″	7.4	9.8	*F7V*		
23h34.0m	+31° 20′	72 Peg	2003	95°	0.5″	5.7	6.1	*K4III*	p=246 yr	Gould, 175-mm, 90×: "A bright orange star in a thin field, with the hint of an elongated image at 83×."
23h44.0m	+29° 22′	78 Peg	1999	268°	0.8″	5.1	8.1	*G8III*	p=630 yr	Hartung: "This bright deep yellow star forms a large triangle with two stars less bright . . . 200-mm will show the [close] companion in good conditions."
23h53.0m	+11° 55′	Σ3044	2001	282°	19.8″	7.3	7.9	*F0*	optical	125-mm, 50×: An easy wide pair conspicuous in the field 1° north of 82 Pegasi. The colors are gloss white and ash white. Webb: Franks calls it "yellowish, bluish."
23h54.9m	+29° 29′	OΣ 252	1999	145°	111.1″	6.8	8.4	*K2 K0*		125-mm: A surprisingly lovely pair, given its wide separation. The colors are bright whitish lemon and brownish cherry, and this unusual combination was well seen.
23h58.1m	+24° 20′	Σ3048	2003	312°	8.9″	7.9	9.5	*G5*		Webb, listed under Andromeda: "Yellow white, no color," and years later the companion was "decidedly blue."

Perseus

RA	Dec.	Name	Year	P.A.	Sep.	m_1	m_2	Spec.	Status	Comments
01h49.3m	+47° 54′	Σ162								125-mm, 200×: An attractive and interesing triple. It's three stars in a cone shape — a touching pair of straw-yellow twins and a wide little bluish star.
		AB	2004	200°	1.9″	6.5	7.2	*A3V*	p. binary	
		AB-C	2003	178°	20.1″	6.5	8.4			
02h14.9m	+58° 29′	Σ230	2002	257°	25.0″	7.9	9.4	*B8III*		125-mm, 50×: A wide pair with pretty colors, with part of a giant star cluster (Stock 2) in the field. It's a very yellowish white star with a small green companion.
02h16.9m	+57° 03′	OΣΣ 25	1999	205°	102.9″	6.5	7.4	*B1I*		60-mm, 25×: Nicely placed! This is an easily noticed super-wide pair at the edge of the Double Cluster (NGC 869 and 884). The colors are gloss white and ash white.

Perseus *(continued)*

RA	Dec.	Name	Year	P.A.	Sep.	m_1	m_2	Spec.	Status	Comments
02ʰ 29.4ᵐ 02ʰ 30.8ᵐ	+55° 32′ +55° 33′	Σ268 Σ270	2002 2003	131° 305°	3.0″ 20.9″	6.7 7.0	8.5 9.7	*A*2ll *F*4V	binary	125-mm: These two pairs are just minutes apart. Σ270 is a wide pair with a tiny companion, and Σ268 is a white star with a little gray dot on its edge. It looks like a miniature Delta (δ) Cygni!
02ʰ 44.2ᵐ	+49° 14′	θ Per AB AC	 2004 1924	 305° 229°	 20.5″ 77.2″	 4.1 4.1	 10.0 9.5	 *F*8V	 p=2,720 yr	125-mm, 50×: Grand sight! A brilliant amber-yellow star with a tiny ember of light just beyond its glow. (Only one companion was seen.) Smyth: "Yellow; violet; gray."
02ʰ 47.6ᵐ	+53° 57′	Σ301	2002	17°	8.0″	7.9	8.7	*A*0	binary	125-mm: An easy and attractively close binary star. It's a Sun-yellow star and a small bluish gray just a small space apart.
02ʰ 50.6ᵐ	+38° 19′	16 Per	2003	148°	234.0″	4.3	10.4	*F*2III		60-mm, 25×: This pair is included because of its pretty primary star. It's a bright yellow star with a tiny glimpse star wide beside it. The little star looks more like a companion than a field star.
02ʰ 50.7ᵐ	+55° 54′	η Per	2002	301°	28.5″	3.8	8.5	*M*3I-II		Showcase pair. 60-mm, 25×: A brilliant star with a wide little companion in the lovely combination of apricot orange and cobalt blue. Webb: "Very yellow, very blue."
02ʰ 52.9ᵐ	+53° 00′	Σ314 AB-C	 2002	 314°	 1.5″	 7.0	 7.3	 *B*8III		125-mm: Nice! This is a bright yellowish white star 19′ northwest of Tau (τ) Persei. It becomes a pair of kissing twins at 200×.
02ʰ 53.7ᵐ	+38° 20′	20 Per AB-C	 1998	 236°	 13.9″	 5.0	 9.7	 *F*4IV		125-mm, 50×: Fantasic contrast for the separation! A very bright grapefruit-orange star with a ghostly presence close beside it, which is the faintest speck of light still possible to see.
03ʰ 00.9ᵐ	+52° 21′	Σ331	2002	85°	11.9″	5.2	6.2	*B*7V *B*9V	optical	Showcase pair. 60-mm: An easy pair of bright stars, lemon white and dusty blue-green, that are just a tiny gap apart at 25×. Webb: "White, bluish . . . form a triangle with γ and τ [Gamma and Tau Persei]."
03ʰ 01.5ᵐ	+32° 25′	Σ336	2002	8°	8.5″	7.0	8.3	*G*5IV	binary	60-mm, 40×: A dim but pretty pair. It's a whitish grapefruit-orange star with a little gray companion a small gap apart. A bright white star (21 Persei) is just outside the field.
03ʰ 04.8ᵐ	+53° 30′	γ Per	1998	329°	55.3″	3.1	10.8	*G*8III+*A*2V		The data suggest a very bright star with a very dim companion. Smyth: "A wide and unequal double star . . . flushed white, clear blue."
03ʰ 12.2ᵐ	+37° 13′	Σ360	2004	126°	2.8″	8.0	8.3	*G*0	p=871 yr	125-mm, 200×: A pretty but difficult binary star. It's a lucid peach-orange star that becomes a hair-split pair of stars for scattered moments.
03ʰ 17.1ᵐ	+40° 29′	Σ369	2002	29°	3.6″	6.8	7.7	*B*9V	binary	125-mm, 200×: Striking contrast for the separation! A bright white star with a shadowy greenish spot almost touching it.

RA	Dec.	Name	Year	P.A.	Sep.	m_1	m_2	Spec.	Status	Comments
03ʰ 17.7ᵐ	+38° 38′	Σ53	2004	248°	0.7″	7.7	8.5	G0	p=113 yr	A 300-mm should still be able to split this fast-moving binary, but its separation is rapidly closing.
03ʰ 24.5ᵐ	+33° 32′	Σ382	2002	153°	4.8″	5.8	9.3	A0V	binary	125-mm, 200×: Striking contrast for the separation! A bright yellowish white star with a small bump of powder blue on its edge and many stars in the field.
03ʰ 28.7ᵐ	+50° 26′	Σ388	2000	213°	2.8″	8.0	9.0	F0	binary	125-mm, 200×: A fine pair that's easily spotted at the north point of a perfect triangle formed with Alpha (α) and 34 Persei. It's a yellowish star with a little misty globe beside it, just a small gap apart.
03ʰ 29.3ᵐ	+45° 03′	Σ391	1999	93°	3.4″	7.6	8.3	B2IV	binary	125-mm, 83×: Most exotic! A white star hair-split from a red star, and they look similar in brightness.
03ʰ 34.2ᵐ	+48° 37′	β 787								This binary is probably resolvable with 200-mm until
		AB	1999	288°	4.5″	7.4	11.9	B9.5V	p=400 yr	at least 2025. It's 34′ northwest of Psi (ψ) Persei.
		AD	1999	180°	34.8″	7.4	10.3			
03ʰ 35.6ᵐ	+31° 41′	β 533	1998	42°	1.1″	7.7	7.6	F4V	binary	This tight binary is inside the large dark nebula Barnard 2. A 125-mm should be able to split it, but the author failed to resolve it with this aperture.
03ʰ 38.3ᵐ	+44° 48′	S 430	2003	96°	40.5″	7.2	7.5	A0		60-mm, 25×: A fine sight for low power. It's a wide pair of white stars, nearly alike, in a field filled with stars.
03ʰ 40.1ᵐ	+34° 07′	Σ425	2001	65°	2.0″	7.5	7.6	F9V	p. binary	125-mm, 200×: A splendid sight easily spotted about 30′ west of brilliant yellow 40 Persei. It's a kissing pair of bright twins, bluish white in color.
03ʰ 40.7ᵐ	+46° 01′	OΣ 59	1998	355°	2.7″	7.9	8.9	G5	binary	125-mm, 83×: Splendid view. A hair-split pair of stars, yellow-white in color, in a little trapezium of field stars. They're modestly unequal.
03ʰ 40.8ᵐ	+39° 07′	Σ426	1998	343°	19.7″	7.9	9.3	A7IV	optical	125-mm, 50×: An easy wide pair with pretty colors in a rich field of stars. It's a Sun-yellow star with a little greenish companion.
03ʰ 42.4ᵐ	+33° 58′	40 Per	2001	249°	25.6″	5.0	10.4	B1.5IV		125-mm, 50×: Striking contrast! A bright banana-yellow star with a shadowy blue dot beside it, just a modest gap apart.
03ʰ 44.0ᵐ	+38° 22′	Σ434	1998	83°	33.2″	7.8	8.3	K5III		60-mm, 25×: A wide easy pair in the field with a scattered cloud of dim stars. It's a pair of very orangy white stars nearly alike in brightness.
03ʰ49.5ᵐ	+52° 39′	Σ446	2002	254°	8.7″	6.9	9.9	B0.5III	binary	This pair is inside the cluster NGC 1444. 125-mm, 50×: Very fine view — a bright amber-yellow star with a blue-green speck beside it, within a cloud of dim pinpricks.
03ʰ 50.2ᵐ	+34° 49′	Es 277	2004	141`°	21.7″	6.8	9.8	F0		125-mm, 50×: An easy wide pair in a pretty concentration of bright and dim stars. It's a fairly bright white star with a little silvery companion.

Perseus *(continued)*

RA	Dec.	Name	Year	P.A.	Sep.	m_1	m_2	Spec.	Status	Comments
03h 54.1m	+31° 53′	ζ Per	2003	208°	12.7″	2.9	9.2	*B1I*		125-mm, 200×: A miniature Rigel, grand! It's a brilliant banana-yellow star with a speck of blue light at the edge of its glow. Smyth: "Flushed white, smalt blue." Webb: "Green white, ash."
03h 57.9m	+40° 01′	ε Per	1998	9°	9.0″	2.9	8.9	*B0.5V+A2V*		125-mm, 50×: A gorgeous contrasting pair. It's a brilliant whitish lemon star almost touching a little dot of rich cobalt blue; these colors are vivid and striking.
04h 01.6m	+38° 40′	Σ476	2002	290°	24.7″	8.0	9.2	*K2*		60-mm, 25×: A wide but pretty pair, grapefruit orange and dim arctic blue, in a broad asterism of modestly bright stars. Webb: "Yellow, blue."
04h 04.1m	+39° 31′	Σ483	2003	61°	1.3″	7.4	9.4	*G5*	p=440 yr	This binary is now probably resolvable with a 200-mm; its separation is increasing.
04h 07.6m	+38° 04′	OΣ 531	2004	358°	2.3″	7.3	9.7	*K1V*	p=590 yr	125-mm, 200×: A difficult but gratifying binary with a very pretty primary. It's a grapefruit-orange star with a faint, vaporish blur at its edge. Webb: "Yellow, red."
04h 08.0m	+43° 11′	AG 308	2002	71°	130.4″	7.2	6.6	*K0*		60-mm, 25×: A fine sight for low power. It's a very wide pair of stars that are similar in brightness with contrasting colors — one looks pure white, the other looks pale peach.
04h 20.9m 04h 21.0m	+50° 23′ +50° 15′	Σ519 S 445	1999 2002	348° 329°	17.9″ 71.1″	7.9 7.3	9.4 8.2	*K2* *K2*		125-mm, 50×: These pairs are part of the cluster NGC 1545, and they look like pairs within it. S 445 is the very wide pair that dominates the cluster.
04h 24.4m	+34° 19′	Σ533	2002	61°	19.4″	7.3	8.5	*B8V*	optical	60-mm, 25×: Splendid view. This wide pair lies in a straight line with two bright white stars (55 and 56 Persei). It's a fainter white star with a little gray companion.
04h 24.6m	+33° 58′	56 Per	2002	17°	4.4″	5.8	9.3	*F4IV-V*	binary	Hartung: "This is a bright golden star with a yellow companion . . . in a field well sprinkled with stars."
04h 31.4m	+40° 01′	Σ552	2003	116°	8.9″	6.8	7.2	*B8V*	binary	60-mm: Grand sight! A bright pair of identical stars, very deep white in color, that are split by a hair at 25×. It's about 1½° to the southwest of 58 Persei, which is just beyond the field.
04h 33.4m	+43° 04′	57 Per	2003	198°	120.5″	6.1	6.8	*F0V*		60-mm, 25×: Grand! A super-wide pair of bright stars, alike in brightness but contrasting in color — one is bluish white and the other pale peach orange. Webb: "Yellow, pale lilac."
04h 36.3m	+47° 22′	S 451	1998	202°	55.8″	7.6	7.9	*F5*		60-mm, 25×: Nice effect. A pair of white twins, very wide apart, that pop out of a desolate field — they look like a pair of eyes glowing in the dark.

Perseus *(continued)*

RA	Dec.	Name	Year	P.A.	Sep.	m_1	m_2	Spec.	Status	Comments
04ʰ36.7ᵐ	+41° 05′	Σ563	1998	32°	11.6″	8.0	10.5	*B*9I	optical	125-mm, 200×: A dim but pretty pair, in the field with brilliant orange 58 Persei. The pair looks much closer than its listed separation.
04ʰ38.1ᵐ	+42° 07′	Σ565	1998	169°	1.3″	7.7	9.1	*K*0		A 170-mm should split this star. It's about 1° north-northeast of 58 Persei. Webb: "Yellowish, bluish."

Phoenix

RA	Dec.	Name	Year	P.A.	Sep.	m_1	m_2	Spec.	Status	Comments
23ʰ27.2ᵐ	−50° 17′	Δ 250	2000	83°	28.4″	7.6	8.5	*K*2III		Gould, 175-mm, 100×: "The pattern is quite good. This is an easy wide pair at the apex of a flat triangle of three fairly bright stars."
23ʰ39.5ᵐ	−46° 38′	θ Phe	2002	276°	3.9″	6.5	7.3	*A*8V+*F*0V	p. binary	Hartung: "Small apertures find an attractive object in this elegant unequal white pair."
00ʰ15.3ᵐ	−40° 06′	Wg 2	1999	2°	13.9″	7.6	9.7	*G*8/*K*1III	optical	This pair is about 1° south of galaxy NGC 55 in Sculptor. Gould, 175-mm, 100×: A "fairly bright, easy, and well separated pair with a pale yellow primary. A fainter star is 2′ west."
00ʰ42.0ᵐ	−55° 47′	MlbO 1	1999	161°	6.3″	8.0	9.0	*F*7III/IV	binary	The data suggest a conspicuous pair 43′ north of Xi (ξ) Phoenicis. Gould, 175-mm, 100×: "Quite good. An easy, moderately bright and nicely separated pair; a few bright stars are about."
00ʰ43.3ᵐ	−45° 11′	h 3390	2000	313°	14.3″	7.1	10.3	*K*0III		Gould, 175-mm, 100×: An "easy wide pair. A bright deep yellow star with a rather faint companion that shows no particular color effect. Some fairly bright stars are wide in the field."
01ʰ06.1ᵐ	−46° 43′	β Phe	1999	258°	0.3″	4.1	4.2	*G*8III		Hartung: "This very bright yellow pair is clearly divided with 150-mm." [But the data suggest that it should require 400-mm.]
01ʰ08.4ᵐ	−55° 15′	ζ Phe AB-C	2000	242°	6.8″	4.0	8.2	*B*6V+*B*9V		Showcase pair. Gould, 175-mm, 100×: A "lovely white pair, fairly close and uneven, with some moderately bright stars in the field." Hartung: A "beautiful white pair."
01ʰ20.3ᵐ	−48° 41′	h 3428	1999	157°	20.5″	7.7	9.7	*F*6V		The data and location suggest a conspicuous star about 2° west of Delta (δ) Phoenicis. Gould, 175-mm, 50×: An "easy, uneven, and moderately wide pair with a pale yellow primary."
01ʰ20.5ᵐ	−57° 20′	h 3430	1991	224°	3.2″	7.2	9.5	*F*7V	binary	Gould, 175-mm, 100×: "Good effect. This close and uneven, pale yellow pair is in a rather nice grouping of six stars — a four-star trapezium plus two extra stars."

Pictor

RA	Dec.	Name	Year	P.A.	Sep.	m_1	m_2	Spec.	Status	Comments
$04^h 41.9^m$	−47° 50′	CorO 23	1993	229°	3.7″	7.5	9.6	*A8IV*	binary	Gould, 175-mm: "A moderately bright yellow star with a tiny close companion, easy and obvious at 180×. There are various fairly bright stars [surrounding it] over a 1° field."
$04^h 50.9^m$	−53° 28′	ι Pic	2002	59°	12.6″	5.6	6.2	*F0IV F4V*		Showcase pair. Hartung: "This beautiful bright yellow pair dominates a field sprinkled with scattered stars, an excellent object for small apertures."
$04^h 59.5^m$	−49° 27′	h 3715	1999	112°	9.9″	7.2	9.1	*F3V*	binary	This pair forms a wide triangle with Eta[1] (η^1) and Eta[2] (η^2) Pictoris, which lie to its east. Gould, 175-mm, 100×: "Quite good. An easy and moderately bright uneven pair, with some moderate stars wide in the field. Yellowish."
$05^h 24.8^m$	−52° 19′	θ Pic								Gould, 175-mm: A "bright, wide and obvious pair, white; a few bright stars are wide in the field."
		AB-C	2002	289°	38.1″	6.2	6.7	*A0V A2V*		
$05^h 30.2^m$	−47° 05′	Δ 21								Gould: An "extremely wide pair of bright yellow stars . . . a reasonable double for the 8 × 50 finder."
		AD	2000	271°	197.8″	5.5	6.7	*G3IV*		
$05^h 38.2^m$	−46° 06′	h 3784	1999	76°	4.7″	7.6	9.4	*G6V*	p. binary	Gould, 175-mm, 100×: A "rather nice little yellow pair, uneven but easy . . . which lies at the end of a bent line of three stars."
$05^h 57.2^m$	−53° 26′	h 3822								Gould, 350-mm, 110×: "A neat trapezium, with two bright and two faintish stars. The bright stars [A and B] are orange and yellow . . . and 250-mm should show all four stars."
		AB	2000	303°	55.6″	6.6	7.6	*K5-M0III*		
		BC	2000	117°	19.4″	7.6	13.0	*K1III*		
$06^h 16.3^m$	−59° 13′	Δ 27	1999	233°	34.4″	6.5	7.6	*G0V F8*		Gould, 175-mm, 100×: "Good pair. A bright, wide, and easy yellow pair, in a fairly starry field."
$06^h 32.0^m$	−58° 45′	μ Pic	1991	230°	2.5″	5.6	9.3	*B9V*	binary	Hartung: "A bright white star with a companion close . . . which 75-mm shows with close attention."

Pisces

RA	Dec.	Name	Year	P.A.	Sep.	m_1	m_2	Spec.	Status	Comments
$23^h 24.3^m$	+03° 43′	Σ3009	2004	229°	6.6″	6.9	8.8	*K2III*	binary	125-mm, 83×: Nice combination. A bright yellowish peach star with a little silvery lime companion attractively close but wide enough to be easy. It's about 2° south-southeast of 7 Piscium.
$23^h 30.7^m$	+05° 15′	Σ3019	2003	184°	10.9″	7.8	8.4	*A8III*	optical	125-mm: A dim but conspicous pair easily noticed 1¼° south-southeast of Theta (θ) Piscium. Both stars look white and nearly alike, and they're attractively close at 50×.
$00^h 10.0^m$	+11° 09′	34 Psc	2000	159°	7.6″	5.5	9.4	*B9V*		Smyth: A "neat double star . . . silvery white, pale blue; and they point to some small stars . . . from the delicacy of the companion it is excessively difficult to measure."
$00^h 15.0^m$	+08° 49′	35 Psc	2003	148°	11.4″	6.1	7.5	*A9V F3V*	optical	Showcase pair. 60-mm, 25×: A bright Sun-yellow star and a little sapphire-white star attractively close together. Their colors are well defined. Smyth: "Pale white, violet tinted."

RA	Dec.	Name	Year	P.A.	Sep.	m_1	m_2	Spec.	Status	Comments
00ʰ 17.4ᵐ	+08° 53′	38 Psc AB-C	2003	234°	4.0″	7.1	7.7	*F*7V		125-mm: A close bright pair with a pretty color. It's a pair of grapefruit-orange stars, nearly equal, that are split by a hair at 83×. Smyth: "Light yellow; flushed white."
00ʰ 22.4ᵐ	+13° 29′	42 Psc	2003	317°	28.7″	6.4	10.3	*K*3III *G*5		Smyth: A "delicate double star . . . topaz yellow; emerald green . . . [which has] an elegant aspect from the strong contrast of its colours in so barren a field of view."
00ʰ 32.4ᵐ	+06° 57′	51 Psc	2003	82°	27.1″	5.7	9.5	*B*9.5V		125-mm, 50×: Nice effect. This is a bright white star with a dot of arctic blue beside it. The companion is very tiny but easily seen. Webb: "White, ash." Smyth: "Pearl white; lilac tinted."
00ʰ 32.6ᵐ	+20° 18′	52 Psc	1997	299°	54.4″	5.5	11.7	*G*9III		Smyth: "Fine yellow; deep blue. This is a neat and most delicate object."
00ʰ 39.9ᵐ	+21° 26′	55 Psc	2003	195°	6.6″	5.6	8.5	*K*0III+*F*3V	binary	125-mm, 83×: Striking contrast! A beautifully bright citrus-orange star with a tiny blue dot just beyond its glow. Webb: "Very yellow, very blue." Hartung: "Orange yellow and ashy."
00ʰ 42.4ᵐ	+04° 10′	OΣ 18	2004	206°	1.9″	7.9	9.7	*F*8V	p=386 yr	This moderately fast-moving binary star is now probably resolvable with 150-mm; its separation is increasing.
00ʰ 49.9ᵐ	+27° 43′	65 Psc	2004	296°	4.3″	6.3	6.3	*F*5III		Showcase pair. 60-mm: A bright, vivid pair of identical stars, citrus orange in color, that are split by a hair at 45×. Webb: Both stars are "yellowish." Smyth: "Both [stars] pale yellow."
00ʰ 54.6ᵐ	+19° 11′	66 Psc	2000	190°	0.5″	6.1	7.2	*B*9V	p=275 yr	This bright binary should be splittable with 250-mm; the gap is very slowly widening. Webb: "Yellowish, bluish."
01ʰ 05.7ᵐ	+21° 28′	ψ^1 Psc	2004	161°	30.0″	5.3	5.5	*B*9.5V *A*0V	optical	Showcase pair. 60-mm, 25×: A bright, wide, and easy pair of identical stars, straw yellow in color. This star, along with Psi² (ψ^2), Psi³ (ψ^3), and Chi (χ) Piscium form a fine asterism for binoculars.
01ʰ 05.8ᵐ	+04° 55′	77 Psc	2003	83°	32.8″	6.4	7.3	*F*3V *F*5V		60-mm, 25×: Nice view! Three bright stars just barely in one field — a pumpkin-orange star (73 Piscium), a pearly white star (80 Piscium), and a straw-yellow star with a wide companion (77 Piscium).
01ʰ 09.7ᵐ	+23° 48′	β 303	2001	292°	0.7″	7.3	7.6	*F*0IV	binary	Hartung: A "close yellow pair in a thin field . . . 200-mm resolved them but 150-mm was not successful."
01ʰ 12.9ᵐ	+32° 05′	Σ98	2001	249°	19.8″	7.0	8.1	*A*0V *A*3IV	binary	125-mm, 50×: A wide pair that is easily noticed in the field with 82 Piscium, which is about 45′ to the south-southwest. It's a pure white star and a smaller bluish white.
01ʰ 13.0ᵐ	+30° 04′	OΣ 26	1999	258°	10.6″	6.3	10.5	*G*9III+*G*1V		The data suggest a bright star with a tiny close companion. Webb: "Yellow, no color given."
01ʰ 13.7ᵐ	+24° 35′	ϕ Psc	1980	226°	7.8″	4.7	9.1	*G*8III		Hartung: "The colours are orange yellow and white, and 75-mm will show the companion."

Pisces *(continued)*

RA	Dec.	Name	Year	P.A.	Sep.	m_1	m_2	Spec.	Status	Comments
$01^h 13.7^m$	+07° 35′	ζ Psc	2004	62°	22.8″	5.2	6.2	*A*7IV *F*7V		Showcase pair. 60-mm, 25×: A wide, bright, and easy pair of goldish white stars, mildly unequal in brightness. Smyth: "Silver white; pale grey." Webb: "Yellowish, pale lilac or rose."
$01^h 26.9^m$	+03° 32′	Σ122	2003	328°	5.9″	6.7	9.5	*B*9V+*A*8V	binary	The data and location suggest a fairly conspicuous pair in a thin field. Webb: "Very white, blue."
$01^h 28.4^m$	+07° 58′	S 398	2001	100°	69.1″	6.3	8.0	*K*1III		The spectral type suggests pretty colors. Webb, referring to it as ΟΣΣ 19: "Rosey and bluish."
$01^h 34.9^m$	+12° 34′	100 Psc	1998	77°	15.6″	7.3	8.3	*A*6V	optical	110-mm, 100×: Nice combination. A pure white star with an ashy nebulous companion attractively close but wide enough to be easy.
$01^h 36.0^m$	+07° 39′	Σ138	2002	56°	1.7″	7.5	7.6	*F*6V	binary	125-mm, 200×: A dim but very pretty binary star. It's a touching pair of whitish orange twins that are strikingly identical.
$01^h 41.3^m$	+25° 45′	Σ145	1997	31°	10.5″	6.3	10.9	*F*6V		Smyth: "A neat double star in a barren field . . . cream yellow; blue, and there is a small star near [in the field]."
$01^h 42.5^m$	+20° 16′	107 Psc	1910	248°	19.0″	5.3	11.7	*K*1V		Smyth: "Pale yellow; dusky. This is a very delicate but wide object in a barren field."
$01^h 44.3^m$	+09° 29′	Σ155	2002	325°	4.9″	7.9	8.0	*F*2	binary	80-mm, 60×: Splendid view! A hair-split couple in a little star group with the bright yellow star Omicron (ο) Piscium also in the field. The stars are pearly white in color and very nearly equal.
$02^h 02.0^m$	+02° 46′	α Psc	2004	270°	1.9″	4.1	5.2	*A*0 *A*3	p=933 yr	80-mm, 150×: An ideal pair for 80-mm — it shows a bright pair of stars that are just short of being fully apart. One is slightly fainter than the other, but both look lucid white.

Piscis Austrinus

RA	Dec.	Name	Year	P.A.	Sep.	m_1	m_2	Spec.	Status	Comments
$22^h 00.8^m$	−28° 27′	η PsA	1999	113°	1.8″	5.7	6.8	*B*8V	binary	Gould, 175-mm, 100×: A "bright white pair, moderately uneven and barely split, that lies in a thin field. 180× makes it an easy, attractive object." Hartung: Five "small wide pairs" are in the field.
$22^h 31.5^m$	−32° 21′	β PsA	1999	173°	30.4″	4.3	7.1	*A*0V		Showcase pair. Gould, 175-mm, 100×: A "wide, easy, and attractive bright white pair in a thin field." Hartung: "Pale yellow and white, the [companion] appearing sometimes reddish."
$22^h 36.6^m$	−31° 40′	Δ 241	1999	31°	93.3″	5.9	7.6	*K*1III		Gould, 175-mm, 100×: A "bright and extremely wide pair, orange and yellow . . . in a moderately starry field; a better double in the 8 × 50 finder."

Pisces Austrinus *(continued)*

RA	Dec.	Name	Year	P.A.	Sep.	m_1	m_2	Spec.	Status	Comments
22ʰ39.7ᵐ	−28° 20′	H VI 119								Gould, 175-mm: "The main star is deep yellow, and has a very wide companion that in turn is a double . . . this closer pair is a fairly bright and attractive, pale yellow pair."
		AB	1991	159°	86.5″	6.4	7.5	*K*0III		
		H N 117								
		BC	2000	69°	3.1″	7.5	8.6	*F*5IV	optical	
22ʰ52.5ᵐ	−32° 53′	γ PsA	1992	258°	4.0″	4.5	8.2	*A*0III	binary	Gould, 175-mm, 100×: This "bright, off-white star has a much less bright companion close, an effective contrast; not difficult." Hartung: "Beautiful pair . . . excellent object for small apertures."
22ʰ55.9ᵐ	−32° 32′	δ PsA	1966	243°	5.2″	4.2	9.2	*G*8III		Hartung: "In a sparingly sprinkled field this bright deep yellow star has a faint companion close . . . which 75-mm will show with care."
22ʰ57.8ᵐ	−26° 06′	h 5371	1998	344°	9.1″	7.7	9.2	*G*3V	binary	Gould, 175-mm, 100×: "Quite good. This fairly bright yellow star has an easy less bright companion, and lies at the east end of a group of fairly bright stars."

Puppis

RA	Dec.	Name	Year	P.A.	Sep.	m_1	m_2	Spec.	Status	Comments
06ʰ04.7ᵐ	−45° 05′	h 3834								Gould, 175-mm, 100×: This triple is "three bright stars dominating a thin field — a very wide yellow pair (AC), whose primary . . . has a neat little point near it."
		AB	1999	216°	5.7″	6.0	9.0	*F*5V	binary	
		AC	1999	321°	196.2″	6.0	6.4	*G*01V-V		
06ʰ04.8ᵐ	−48° 28′	Δ 23	1999	120°	2.6″	7.3	7.7	*G*6V	p=304 yr	Gould, 175-mm, 100×: "A nice object. A fine close and nearly even pair, in a not very starry field. Both stars are yellow."
06ʰ22.9ᵐ	−45° 38′	h 3856	1999	4°	34.0″	6.7	9.7	*K*1III		Gould, 175-mm, 100×: "This bright orange star has a faintish wide companion; a few 10th-magnitude stars are nearby, and some brighter ones are wide in the field to the south."
06ʰ29.8ᵐ	−50° 14′	R 65								The AB pair is a very fast-moving binary. Gould, 175-mm: An "easy uneven pair [referring to AB-CD], bright pale yellow and dull darker yellow; at 240× the primary star was split."
		AB	1999	87°	0.8″	6.0	6.1	*F*2V	p=53 yr	
		AB-CD	1999	312°	11.9″	6.0	8.0			
06ʰ35.9ᵐ	−48° 17′	CorO 39	1999	207°	9.5″	7.5	9.9	*K*0	optical	Gould, 350-mm, 120×: "An easy pair, orange and dull brown. Δ 31 [below] is close nearby [about 1/2° to the east]."
06ʰ38.6ᵐ	−48° 13′	Δ 31	1999	320°	12.9″	5.1	7.4	*G*8III	binary	Showcase pair. Gould, 350-mm, 120×: "An easy good pair, in a rambling line of fainter stars that ends in a trapezium. It's a bright yellow star with a less bright, bluish white companion."
06ʰ38.7ᵐ	−45° 04′	h 3882								Gould, 175-mm: "This moderately bright white star has a fainter, fairly wide companion; the field has a few moderate stars."
		AC	1999	331°	18.1″	7.7	9.2	*B*5V		
06ʰ41.2ᵐ	−40° 21′	h 5443	1999	108°	15.5″	6.1	9.4	*B*4V	binary	Gould, 175-mm, 100×: "Quite good. This bright white star has an easy, well separated companion, and there is a fainter star to the NW." [The trio forms a neat L shape.]

Puppis *(continued)*

RA	Dec.	Name	Year	P.A.	Sep.	m_1	m_2	Spec.	Status	Comments
$06^h 42.3^m$	−38° 24′	Δ 32	1999	277°	7.9″	6.6	7.7	*A*5	binary	Gould, 175-mm, 65×: A "bright and easy uneven double, cream white and dull off-white in color; the field has a few fairly bright stars."
$06^h 42.8^m$	−50° 27′	h 3889	1999	267°	42.1″	7.0	8.8	*B*5V		This pair is about 1° west of Tau (τ) Puppis. Gould, 350-mm, 120×: "A very wide and easy white pair, uneven in brightness, that lies in a thin field."
$06^h 46.7^m$	−47° 48′	h 3895	1999	66°	25.7″	7.1	10.9	*K*3III		The spectral type suggests a pretty primary star. J. Herschel: "Orange, blue."
$06^h 50.0^m$	−45° 27′	I 159	1966	323°	6.6″	6.5	11.0	*K*5-*M*0III		Gould: 175-mm, 100×: This "bright, deep-yellow star has a minute companion fairly close; perhaps more obvious at 180×. It lies in a plain field."
$06^h 58.3^m$	−35° 25′	I 66								Gould, 350-mm, 70×: "An easy and obvious white pair, in a nice field with four bright stars and scattered fainter ones in patterns. At 120× the main star is a close, uneven double star."
		AB	1991	253°	1.9″	7.8	9.3	*A*	binary	
		h 3905								
		AC	1999	270°	14.3″	7.8	9.7		binary	
$07^h 04.0^m$	−43° 37′	Δ 38	1999	126°	20.9″	5.6	6.7	*G*3V *K*0V	optical	Showcase pair. Gould, 175-mm, 100×: This "bright, deep golden yellow star has a well separated orange companion, and there is another middling bright orange star 3′ to the NW."
$07^h 05.5^m$	−34° 47′	h 3928								Gould, 175-mm,100×: "Good effect. In a plain field is this bright little golden pair, flanked on each side by a wide fainter companion."
		AB	1994	148°	2.7″	6.5	7.8	*F*2V	p. binary	
		AC	1999	291°	37.8″	6.5	9.7			
		AD	1999	126°	42.7″	6.5	10.8			
$07^h 06.0^m$	−42° 20′	h 3931								Gould, 175-mm: This broad triple is "three stars nearly in a line, with the brightest one in the middle. A bright star [C Puppis] is west."
		AB	1999	212°	54.7″	7.2	11.1	*A*0V		
		AC	1999	41°	72.2″	7.2	8.8	*F*5V		
$07^h 12.4^m$	−36° 33′	β 757	1993	69°	2.5″	6.0	8.4	*B*3V	binary	Hartung: "75-mm shows this bright pale yellow pair very well in the black field dotted with faint stars."
$07^h 17.1^m$	−37° 06′	π Pup	1999	213°	68.9″	2.9	7.9	*K*3Ib		Gould, 350-mm, 70×: "This very bright orange star has a wide white companion." [It is the greatest of four bright stars, including Jc 10 (below), inside the large dim star cluster Cr 135.]
$07^h 17.5^m$	−46° 59′	I 7	2001	208°	0.8″	7.1	8.4	*K*3V	p=94 yr	200-mm should be able to split this very fast-moving binary, but the separation is slowly closing.
$07^h 18.3^m$	−36° 44′	Jc 10								Gould, 350-mm, 120×: This triple is "two bright white stars and a less bright star making a triangle with them. It is part of the Cr 135 cluster, as is π Pup."
		AB	1991	98°	240.0″	4.7	5.1	*B*2V+*B*3IV		
		CD	1999	206°	2.3″	8.7	9.2	*A*0/1V		
$07^h 20.9^m$	−48° 31′	h 3956	2000	166°	7.1″	8.4	9.5	*B*9.5V	binary	These two pairs are just arcminutes apart. Gould, 175-mm, 100×: "Δ 45 is a bright and easy white pair, fairly wide, with the delicate pair h 3956 about 7′ to the W — a good contrast."
$07^h 21.4^m$	−48° 32′	Δ 45	2000	157°	22.6″	6.8	7.7	*B*9V	optical	

Puppis *(continued)*

RA	Dec.	Name	Year	P.A.	Sep.	m_1	m_2	Spec.	Status	Comments
$07^h 22.3^m$	−35° 55′	h 3957	1999	191°	7.4″	7.1	7.9	*F*8	p. binary	This pair is in a conspicuous triangle of three stars located 1.6° northeast of Pi (π) Puppis and 1.1° northeast of NV Puppis. Gould, 175-mm, 70×: This "bright yellow star has an easy less bright companion."
$07^h 24.8^m$	−37° 17′	h 3966	1999	323°	6.8″	6.9	6.9	*F*0V *F*0V		Gould, 175-mm, 100×: This "small even pair is quite bright, and both stars are cream white; the field [about 1½° east of Pi (π) Puppis] is nondescript."
$07^h 27.0^m$	−34° 19′	h 3969	1999	227°	17.7″	7.1	8.1	*F*6/7V *F*8	optical	Gould, 175-mm, 100×: A "wide easy pair, bright yellow and duller yellow, in a starry field."
$07^h 27.6^m$	−18° 30′	S 550	2003	116°	39.0″	6.9	7.6	*A*0 *B*8I		Gould, 350-mm, 120×: A "bright and very wide white pair, in a little gathering of fainter stars."
$07^h 27.9^m$	−11° 33′	β 332								Smyth, listed under Monoceros, referring to AC: "A delicate
		AB	2003	171°	0.7″	6.2	7.4	*G*8I-II	binary	double . . . yellow, violet." Webb, also under Monoceros
		Σ1097								and referring to AC: "Pale red, deep blue."
		AC	2003	313°	19.8″	6.0	8.5			
$07^h 28.9^m$	−31° 51′	Δ 49	1999	53°	9.0″	6.3	7.0	*B*3V *B*4V	binary	This pair is in an arc of three bright stars about 1° in size. Gould, 175-mm, 100×: "A fine white pair, easy and effective; there are some faint field stars close, and brighter stars well wide."
$07^h 29.2^m$	−43° 18′	σ Pup	1981	74°	22.2″	3.3	8.8	*K*5III		Showcase pair. Hartung: This "brilliant orange star with white (by contrast) companion makes a fine sight in the star sprinkled field, and 75-mm shows it well."
$07^h 29.4^m$	−15° 00′	Σ1104	2002	27°	1.9″	6.4	7.6	*F*7V	p=686 yr	Hartung: "Clearly divided by 75-mm, this beautiful golden yellow pair lies in a profuse starry field."
$07^h 31.9^m$	−20° 56′	h 3973	2000	38°	9.1″	7.9	9.9	*B*6III/IV		Gould, 175-mm, 100×: "A fine little pair, easy but not bright; the colors are white and red, and the field is fairly good."
$07^h 34.0^m$	−23° 42′	Howe 18	1991	204°	1.9″	8.1	8.9	*B*6/8III		Showcase pair. Gould, 175-mm: "H 19 is an even, bright
$07^h 34.3^m$	−23° 28′	H 19	1998	117°	9.8″	5.8	5.9	*F*5/7V		and easy pair of yellow stars; 15′ to the south is Howe 18: a quite close pair of white stars, just double at 100×."
$07^h 34.1^m$	−50° 50′	h 3986	1999	220°	48.5″	7.8	9.3	*M*3/4III		Gould, 175-mm: "This moderately bright, deep red star has a wide yellow companion, and a wider field star, all in a straight line."
$07^h 35.4^m$	−28° 22′	p Pup								Gould, 175-mm: "This bright white star has two faint
		AB	1999	150°	38.3″	4.6	9.1	*B*8V		companions, well separated southeast; the field is
		BC	1999	130°	43.2″	9.1	10.4			moderately starry."
$07^h 36.6^m$	−14° 29′	Σ1121	2004	305°	7.4″	7.0	7.3	*B*6V *B*6V		125-mm, 50×: Splendid view! This is the bright, wide pair of stars that dominates the cluster M47. They are straw-yellow twins that look like a pair within the cluster.

RA	Dec.	Name	Year	P.A.	Sep.	m_1	m_2	Spec.	Status	Comments
07ʰ38.8ᵐ	−26° 48′	κ Pup	2002	318°	9.8″	4.4	4.6	*B6V B6V*	p. binary	Showcase pair. 60-mm, 25×: Beautiful combination — a vivid set of bright white stars, exactly alike, that are attractively close while wide enough to be easy.
07ʰ38.9ᵐ	−20° 16′	Arg 47	2000	334°	2.9″	7.8	8.4	*A7III/IV*	binary	Gould, 175-mm: "Quite good. A neat and nearly even little white pair, fairly close but easy. It is in a faint cloud along with some 8th- and 9th-magnitude stars."
07ʰ44.2ᵐ	−50° 27′	Δ 55	1999	133°	51.9″	6.7	7.6	*F8V G0*		Gould, 175-mm, 100×: "This pair is probably best for small scopes at low power. It's a very wide pair, both bright and yellow, in a field with a scattering of moderately bright stars."
07ʰ45.5ᵐ	−14° 41′	2 Pup	2003	340°	16.6″	6.0	6.7	*A2V A8V*	optical	60-mm, 25×: Splendid field! The view shows a wide, easy pair of yellowish peach stars next to a bright white star (4 Puppis), with cluster M46 at the field's edge.
07ʰ47.8ᵐ	−16° 01′	Knott 4	2002	132°	129.4″	6.5	6.6	*M2II-III*		125-mm, 50×: Beautiful couple! It's a pair of stars that are super-wide apart but alike in brightness, and both have the same unusual color — a yellowish rose red.
07ʰ47.9ᵐ	−12° 12′	5 Pup	2003	349°	1.3″	5.7	7.3	*F5*	p. binary	Hartung: "A fine object in a starry field, and 75-mm shows it well." Smyth: "A close double star . . . pale yellow; light blue." Webb: "Pale yellow, ruddy."
07ʰ51.8ᵐ 07ʰ52.5ᵐ	−13° 54′ −13° 52′	9 Pup H 28	2001 2003	345° 220°	0.2″ 18.8″	5.6 6.6	6.5 10.5	*G1V* *K2III*	p=22.7 yr	Gould: 175-mm, 100×: "These two pairs are just 10′ apart, and both have bright deep yellow primaries. 9 Puppis was seen as a possibly elongated star in 1997."
07ʰ52.3ᵐ	−34° 42′	Howe 65	1994	270°	3.6″	5.1	8.6	*F5V*	binary	Gould, 175-mm, 100×: "A beautiful delicate pair — a bright pale yellow star with a tiny point beside it. A scattered field surrounds it."
07ʰ55.3ᵐ	−41° 50′	h 4019	1999	157°	5.4″	7.7	9.9	*G8/K0III*		Gould, 175-mm: "Not bright, but rather nice. This deep yellow star has a fairly close faint companion, and a wide 11th-magnitude star; the field is starry and effective."
07ʰ55.8ᵐ 07ʰ55.6ᵐ	−43° 51′ −43° 47′	See 91 CorO 64	1991 1999	340° 152°	0.7″ 5.1″	6.6 7.5	6.9 11.0	*B6V* *K3III*	p. binary	Gould, 150-mm: "The field shows two pairs in a fairly rich field — an elongated star (See 91) and an easy, uneven pair with a deep yellow primary (CorO 64)."
07ʰ58.8ᵐ	−47° 41′	I 30	1991	352°	1.7″	7.6	8.7	*B7V*	binary	Gould, 175-mm: "A close uneven pair, pale yellow in color (despite the spectral type); a neat pair at 180×. The field is faint with a few stars wide."
07ʰ59.2ᵐ	−49° 59′	Δ 59	1999	48°	16.5″	6.2	6.2	*B2IV-V*	optical	Gould, 175-mm, 100×: A "beautiful and evenly bright easy pair, with a little cloud of faint stars in the field. Despite the spectral typle, the stars look yellow."
07ʰ59.8ᵐ	−47° 18′	h 4032 AB-C	 1999	 351°	 28.8″	 6.7	 9.4	 *B8V*		Gould, 175-mm, 100×: "An easy wide and uneven pair of white stars, with some fairly bright stars in the field."

Puppis *(continued)*

RA	Dec.	Name	Year	P.A.	Sep.	m_1	m_2	Spec.	Status	Comments
$08^h01.3^m$	−22° 20′	β 333								60-mm, 25×: This wide pair shows unusual color contrast. The stars look alike in brightness, but one is pale yellow, the other ruddy. It's between two parallel rows of dim stars.
		AB	1991	45°	1.6″	7.2	9.2	A1V		
		AC	1998	71°	42.6″	7.2	7.6	G0		
$08^h02.0^m$	−27° 13′	β 202	1999	161°	7.7″	6.9	9.9	B7II	binary	This pair sits on the edge of a faint ring-shaped asterism. Gould, 175-mm, 100×: "This bright white star has an easy fainter companion near, and a wide faint one to the south-west."
$08^h02.6^m$	−27° 33′	h 4037								Gould, 175-mm: "Good object. A triple made of three stars in a straight line — a bright yellow star between two less bright companions. β 202 [above] is about 1/2° to the north-northwest."
		AB	1999	243°	7.2″	7.2	8.4	G2IV	binary	
		AC	1999	74°	63.8″	7.2	9.2			
$08^h02.7^m$	−41° 19′	h 4038	1993	347°	25.2″	5.5	9.0	B9		Hartung: "J. Herschel . . . noted as very remarkable the unusual colour combination of very pale yellow and red; this is visible with 75-mm, and a fine field makes this an attractive object."
$08^h03.1^m$	−32° 28′	h 4035	2000	135°	35.7″	6.0	9.0	K2.5III		Gould, 175-mm: "This bright orange star has a wide and easy fainter companion; the field is quite good, and includes lines and patterns of stars."
$08^h05.7^m$	−33° 34′	h 4046	1999	88°	21.9″	6.3	8.4	G1I		Hartung: "In this beautiful and varied field is a bright orange star with a white companion wide." The pair lies between a deep red field star and close dim pair.
$08^h06.9^m$	−27° 07′	Δ 61	1999	35°	69.5″	7.0	9.0	B7III K0		A tiny asterism is at this location — this wide pair and two close field stars. Gould, 175-mm: "White and orange."
$08^h08.2^m$	−22° 08′	β 334	2000	346°	3.1″	7.9	8.5	G0V		Gould, 175-mm: An "easy little yellow pair in a fairly starry but not bright field."
$08^h08.5^m$	−19° 52′	S 563	2002	57°	134.3″	7.0	7.6	B8V K0III		125-mm, 50×: Interesting! A pair of stars super-wide apart but alike in brightness, with contrasting colors. They look pure white and reddish copper and are in the field with 16 Puppis.
$08^h09.1^m$	−39° 31′	Gli 80	1999	249°	40.1″	7.7	8.7	K1III		Gould, 175-mm: "Nice effect. This is less a pair than an arc of three bright stars — two orange, the other ashy; another bright star [NS Puppis] is nearby, and a fainter triangle of stars is adjacent."
$08^h09.8^m$	−42° 38′	Δ 63	1999	80°	5.5″	6.6	7.5	B7V	binary	Hartung: "This elegant white unequal pair is in a very fine starry field with a pronounced curved line of stars coming in S and ending in a small wide pair."
$08^h11.0^m$	−37° 18′	DAW 43								Hartung: "This yellow star is the brightest in a beautiful field [that includes cluster NGC 2546] . . . the two faint companions are probably field stars . . . 200-mm shows them plainly."
		AB	1999	206°	16.8″	6.4	13.5	O9.5II		
		h 4051								
		AC	1999	263°	17.6″	6.4	13.3			

Puppis *(continued)*

RA	Dec.	Name	Year	P.A.	Sep.	m_1	m_2	Spec.	Status	Comments
08ʰ 11.3ᵐ	−12° 56′	19 Pup								125-mm, 50×: Splendid field! The view shows a wide, cone-shaped triple on the edge of the cluster NGC 2539. It's a bright orange star with two dim companions.
		AD	1899	276°	60.6″	4.7	8.9	*G*8III		
		AE	2002	256°	69.1″	4.7	9.4	*A*7V		
08ʰ 11.4ᵐ	−42° 59′	h 4057	1999	299°	26.3″	4.8	9.8	*A*5I		Gould: 150-mm, 75×: "Good effect. This bright yellowish-white star has a small, well separated companion, and a little pattern of faint stars is in the field to the south."
08ʰ 14.0ᵐ	−36° 19′	Δ 67	1999	176°	66.9″	5.0	6.0	*B*1.5IV		Gould, 175-mm, 100×: A "bright and wide uneven pair, both stars pale yellow, in a field with various fainter pairs about."
08ʰ 15.5ᵐ	−37° 22′	h 4063	1999	352°	16.6″	7.4	9.3	*B*4V	optical	Gould, 350-mm, 120×: "A fine pair, cream and dark red; cluster NGC 2546 is nearby [about ½° to the southwest]."
08ʰ 15.9ᵐ	−30° 56′	β 454	1997	2°	1.9″	6.5	8.2			Gould, 350-mm: "This deep yellow star dominates a fine field, with stars in wide pairs and patterns. At 240×, it is a fine and fairly close uneven pair. Just north is an arc of dim stars."
08ʰ 18.2ᵐ	−37° 22′	h 4073	1994	175°	1.9″	7.2	7.8	*A*0IV	binary	This binary is about 1° south of q Puppis. Gould, 350-mm: "A slightly uneven and fairly bright pale yellow pair, just split with 120×; more power makes it very easy."
08ʰ 25.0ᵐ	−42° 46′	Rst 4888	1991	109°	0.4″	6.6	6.8		p. binary	Gould, 175-mm: "A bright white star . . . with a suggestion of elongated image at 330×." [This much resolution can be gratifying for a binary as close as this one.]
08ʰ 25.2ᵐ	−24° 03′	S 568	2001	90°	42.2″	5.5	8.4	*K*4/5III		125-mm, 50×: A very wide but lovely pair. It's a bright apricot-orange star with a small ruddy star beside it, which so dominate their field that they make an obvious couple.
08ʰ 26.3ᵐ	−39° 04′	h 4093								Gould, 175-mm, 100×: "Good, easy pair. A fine, bright, and somewhat uneven pair, in a nice star field. Both stars are white."
		A-BC	1999	124°	8.2″	6.5	7.1	*B*9V *A*2V		
08ʰ 33.1ᵐ	−24° 36′	β 205	1997	318°	0.6″	7.1	6.8	*A*8IV	p=136 yr	This fast-moving binary is probably resolvable with 250-mm until at least 2025.

Pyxis

RA	Dec.	Name	Year	P.A.	Sep.	m_1	m_2	Spec.	Status	Comments
08ʰ 31.4ᵐ	−36° 42′	h 4106	1999	305°	4.4″	7.7	9.7	*K*1IV		Gould, 175-mm, 100×: A "pleasing, fairly delicate pair with an orange yellow primary. It lies in a rambling line of stars, with an asterism to the west."
08ʰ 33.1ᵐ	−24° 36′	β 205	1997	318°	0.6″	7.1	6.8	*A*8IV	p=136 yr	Hartung: A "difficult close pair . . . in 1962 [when the separation was 0.5″], I saw the star elongated and almost divided . . . 250-mm showed the elongation clearly."
08ʰ 38.7ᵐ	−19° 44′	β 207	2003	102°	4.4″	6.6	9.2	*K*5III	binary	Gould, 175-mm, 100×: "Nice pair, with good contrast. This fairly bright orange star has a faint companion close, which is easily seen. The field has some patches of stars."

RA	Dec.	Name	Year	P.A.	Sep.	m_1	m_2	Spec.	Status	Comments
08ʰ 39.1ᵐ	−22° 40′	β 208	2001	33°	1.3″	5.4	6.8	*G6IV*	p=123 yr	Gould, 175-mm: A "bright pale yellow star that becomes a very close pair at 180×. Probably a test object for 100-125-mm." Hartung: "Orange yellow and whitish."
08ʰ 39.7ᵐ	−29° 34′	ζ Pyx	1999	61°	52.4″	5.0	9.6	*G4III*		Gould, 175-mm, 100×: This "bright yellow star in a pleasantly starry field has an easy wide companion." Smyth, listed under Argo Navis: "A coarse double star . . . red; green."
08ʰ 50.4ᵐ	−35° 56′	h 4144	1991	315°	2.4″	7.0	9.0	*B5IV*	binary	Gould, 175-mm: "Quite good. An uneven close pair with a white primary that is just double at 100×. It lies in a moderately starry but not bright field."
09ʰ 03.3ᵐ	−33° 36′	h 4166 A-BC	1999	153°	13.7″	7.1	7.9	*A0V*	optical	Hartung: "Both stars are pale yellow and ornament a field sown profusely with stars. It is a fine object for small telescopes." Gould: "10′ SE is a bright star with a wide companion."
09ʰ 09.8ᵐ	−25° 48′	β 410	1991	157°	1.8″	7.3	8.8	*A0V*	binary	Gould, 175-mm: "A nice pair in a field partly bare, partly starry [about 1/2° east of Kappa (κ) Pyxidis]. A pale yellow star with a very close, less bright companion; easy at 180×."
09ʰ 09.9ᵐ	−30° 22′	ε Pyx A-BC	1999	147°	17.7″	5.6	9.9	*A4IV-V*		Gould, 175-mm: This "bright pale yellow star has a wide and easy fainter companion, and the field is moderately starry."
09ʰ 20.7ᵐ	−31° 46′	h 4200	1998	75°	3.0″	7.4	7.9	*B9.5V*	binary	Gould, 175-mm, 100×: "Fairly good pair. A neat and nearly even little white pair in an ordinary field."

Reticulum

RA	Dec.	Name	Year	P.A.	Sep.	m_1	m_2	Spec.	Status	Comments
03ʰ 15.2ᵐ	−64° 27′	Δ 12 A-BC	1999	104°	19.1″	6.7	9.0	*F5*		This star should jump out of the black field surrounding it. Gould, 350-mm: "This bright yellow star has an easy wide companion."
03ʰ 18.2ᵐ	−62° 30′	ζ Ret	1991	217°	309.3″	5.3	5.6	*G2V*		Gould: These "two bright yellow stars make an obvious pair in an 8 × 50 finderscope. It is a pair for very low power only."
03ʰ 38.2ᵐ	−59° 47′	Δ 14	1999	272°	57.3″	7.0	8.3	*F3V F5*		Gould, 175-mm,100×: A "very wide pair . . . both stars yellow. Probably best at very low power." The companion star has a minute companion (magnitude 12.8) of its own.
03ʰ 44.6ᵐ	−54° 16′	h 3592	1993	17°	5.0″	6.5	9.3	*K1III*	binary	Gould, 400-mm, 64×: "Nice object. This bright deep yellow star has a much less bright companion close; the field has a good scatter of fairly bright stars."
04ʰ 01.0ᵐ	−54° 24′	Δ 17	1999	142°	64.6″	7.7	8.2	*A3V*		Gould, 400-mm, 64×: "A fine field. Δ 17 is a very wide double that makes a triangle with a third star."
04ʰ 17.7ᵐ	−63° 15′	θ Ret	2000	2°	4.3″	6.0	7.7	*B9III-IV*	p. binary	Showcase pair. Gould, 350-mm, 110×: "A fine pair that is bright, easy and fairly close. The stars are yellow and deeper yellow, and uneven in brightness."

Reticulum *(continued)*

RA	Dec.	Name	Year	P.A.	Sep.	m_1	m_2	Spec.	Status	Comments
04ʰ33.6ᵐ	−62° 49′	h 3670	1998	100°	31.8″	5.9	9.3	*K*1III		Gould, 350-mm, 110×: "Good pair. It is a bright deep yellow star with a wide less bright companion of contrasting bluish color."

Sagitta

RA	Dec.	Name	Year	P.A.	Sep.	m_1	m_2	Spec.	Status	Comments
19ʰ02.0ᵐ	+19° 07′	h 2851								125-mm, 50×: A lovely citrus-orange star with a companion much closer and dimmer than the data suggest. It's a ghostly presence seen only with averted vision.
		AC	2003	292°	47.8″	6.9	9.2	*G*9III		
19ʰ12.6ᵐ	+16° 51′	β 139								125-mm, 50×: The AC pair is bright and pretty. It's a yellow-white and a smaller orange-white star super-wide apart. Hartung says 200-mm splits the AB pair.
		AB	1999	136°	0.6″	7.1	8.0	*B*9IV	binary	
		ΟΣΣ 177								
		AC	2002	278°	99.7″	7.1	8.0			
19ʰ14.3ᵐ	+19° 04′	Σ2484	2004	242°	2.1″	7.9	9.5	*F*8	p=2,316 yr	125-mm, 200×: A dim but attractively close binary star. It's a lemony white star with a dim green companion split by a hair at 200×.
19ʰ21.0ᵐ	+19° 09′	Σ2504	2003	282°	8.7″	7.0	9.0	*F*5V	binary	125-mm: An easy pair that's easily noticed about 1½° southwest of the Coathanger asterism. It's a lemon-white star and a small blue-green modestly wide apart.
19ʰ24.4ᵐ	+16° 56′	2-3 Sge	2002	79°	341.7″	6.3	6.9	*A*2III-IV		Webb included this very wide pair in his catalog. He says both stars are "very white" and "form a wide pair."
19ʰ37.3ᵐ	+16° 28′	ε Sge	2002	82°	87.4″	5.8	8.4	*G*9III		60-mm, 25×: This wide pair is pretty. It's a bright star with a conspicuous small companion, whitish Sun yellow and azure white in color. They seem closer than the data suggest.
19ʰ39.4ᵐ	+16° 34′	H N 84	2003	302°	28.2″	6.4	9.5	*K*4I		Webb notes that these two pairs are 15′ apart. 125-mm, 50× H N 84 is a bright orange star with a little nebulous companion very wide apart. W. Herschel: H N 84 is "red, blue."
19ʰ44.8ᵐ	+16° 49′	Σ2569	2000	357°	2.1″	8.4	9.1	*A*0	binary	
19ʰ49.0ᵐ	+19° 09′	ζ Sge								125-mm, 83×: Beautiful combination! A brilliant white star with a little dot of ocean blue beside it. These stars are attractively close while wide enough to be easy.
		AB-C	2002	310°	8.3″	5.0	9.0	*A*1V		
20ʰ09.6ᵐ	+16° 48′	Σ2634	2003	14°	4.0″	7.8	9.9	*K*0	binary	125-mm, 83×: Striking contrast for the separation. A very yellowish white star almost touched by a tiny cloud of vapor. W. Herschel: "Red, deeper red."
20ʰ09.9ᵐ	+20° 55′	θ Sge								This is the best sight in Sagitta for a small aperture. 60-mm, 25×: A bright, wide, and easy cone-shaped triple — a pair of white stars that are nearly alike and a little gray dot.
		AB	2004	330°	11.9″	6.6	8.9	*F*3V	optical	
		AC	2004	222°	89.0″	6.6	7.5	*K*2III		

Sagittarius

RA	Dec.	Name	Year	P.A.	Sep.	m_1	m_2	Spec.	Status	Comments
$17^h58.9^m$	−36° 52′	Δ 219	1999	254°	53.2″	5.8	7.7	*G8III*		These pairs are just arcminutes apart. Gould, 175-mm, 100×: "Δ 219 is very easy . . . and visible in an 8 × 50 finder . . . its main star is yellow. h 5000 is an easy, uneven, attractive white pair."
$17^h59.2^m$	−36° 56′	h 5000	2000	103°	7.5″	7.1	9.0	*A3III*	binary	
$17^h59.1^m$	−30° 15′	Pz 6	1994	107°	5.8″	5.4	7.0	*M1I G8II*	binary	Showcase pair. Gould, 175-mm, 100×: "An attractively bright pair, mustard yellow and deep yellow in color; a nice effect from uneven brightness and modest separation."
$18^h02.4^m$	−23° 02′	H N 40								Gould, 175-mm, 100×: "This multiple is in the Trifid Nebula. [This sounds like a nice view!] The AC pair is easy, the B member is faintish, and the CD pair is just visible."
		AB	1998	20°	6.1″	7.6	10.4	*O8V*		
		H N 6								
		AC	1998	214°	11.0″	7.6	8.7			
		CD	1934	280°	2.3″	8.5	10.5			
$18^h02.6^m$	−24° 15′	Arg 31								This pair and 7 Sagittarii are only about 60″ apart; at low power, does it look like a wide double companion of this bright star? Sawyer, 80-mm, 60×: Both stars are "bluish white."
		AC	1998	26°	35.6″	6.9	8.6	*B3*		
$18^h04.2^m$	−22° 30′	S 698	1999	316°	29.8″	7.2	8.5	*O*		The location suggests a very nice view — this pair is inside the cluster M21. Sawyer, 80-mm, 60×: "A bright star with a faint companion, and good separation."
$18^h05.7^m$	−36° 35′	Howe 88								Gould, 175-mm, 100×: A "neat little pair with no colour, in a trapezium of unevenly bright stars . . . in a good starry field."
		A-BC	1991	4°	3.2″	7.9	8.9	*B8*		
$18^h08.9^m$	−25° 28′	WNO 21	1999	65°	13.5″	6.8	8.8	*B3III*	optical	Gould, 175-mm, 100×: An "easy, moderately bright pair in a starry field. Despite the spectral type, the primary looks yellow. Two globular clusters (NGC 6553 and NGC 6544) are nearby."
$18^h10.1^m$	−30° 44′	β 245	1994	353°	3.9″	5.8	8.0	*K1II*	binary	Showcase pair. Gould, 175-mm, 100×: "A bright orange star with a close, less bright companion; a good contrast; the field is fairly rich . . . and there are some minor pairs nearby."
$18^h11.2^m$	−19° 51′	β 132	2001	190°	1.3″	7.0	7.1	*A4V*	binary	Gould, 175-mm, 100×: "A very close and fairly bright white pair, in a rich but not bright field. 180× makes it easy. A test for 100-mm."
$18^h13.8^m$	−21° 04′	μ Sgr								Smyth: "A multiple star on the north end of the Archer's bow . . . [A] pale yellow; [B] blue; [C and D] both reddish [and] point to a coarse double star."
		AB	1998	258°	16.8″	3.9	10.5	*B8I*		
		AC	1901	118°	25.6″	3.9	13.5			
		AD	1998	312°	48.7″	3.9	10.0	*B3*		
$18^h17.6^m$	−36° 46′	η Sgr	1988	107°	3.6″	3.1	7.8	*M3.5III*		Hartung: "This brilliant orange star dominates a field sown with scattered stars on a profuse faint ground, and the white companion close . . . is steadily visible with 75-mm."

RA	Dec.	Name	Year	P.A.	Sep.	m_1	m_2	Spec.	Status	Comments
$18^h 17.9^m$	−18° 48′	Sh 263	2002	12°	53.3″	6.8	9.3	*B8IV B9V*		Gould, 175-mm, 50×: "These three pairs are in a beautiful field that includes cluster NGC 6603 [a cluster within the giant star cloud M24] . . . a wonderful effect."
$18^h 18.0^m$	−18° 42′	Howe 42	1879	195°	20.2″	8.5	10.0	*F6I*		
$18^h 18.7^m$	−18° 37′	Sh 264								
		AB-C	2003	51°	17.1″	6.9	7.6	*B0I*		
$18^h 25.4^m$	−20° 33′	21 Sgr	1997	275°	1.5″	5.0	7.4	*A+K2III*	binary	Hartung: "This pair, orange and greenish (by contrast), lies in a field well sprinkled with stars . . . it may be resolved surely with 75-mm." [Is this really possible? 120-mm is more likely.]
$18^h 27.7^m$	−26° 38′	β 133	1996	243°	0.9″	6.6	8.5	*A8V+F2V*	p. binary	Hartung: A "pale yellow star . . . divided with 105-mm."
$18^h 29.0^m$	−26° 35′	WNO 6	1999	182°	41.8″	6.7	8.0	*A3III*		Gould, 175-mm, 100×: A "very wide, easy and bright pair that is just visible as a pair in an 8 × 50 finder. There is another very wide and bright pair 15′ SW of it."
$18^h 34.5^m$	−34° 49′	Stone 62	1996	118°	2.1″	7.6	7.8	*F0V*	binary	Gould, 175-mm: A "very neat and nearly even close pair, just split at 100×. The field is scattered with mostly faint stars, with a triangle of 8th- and 9th-magnitude stars just 12′ SE."
$19^h 02.6^m$	−29° 53′	ζ Sgr	2001	172°	0.2″	3.3	3.5	*A2III*	p=21.1 yr	At the listed separation, 250-mm may show an elongated star. It will be widest (0.6″) from 2014 to 2017.
$19^h 02.7^m$	−36° 06′	h 5080	1999	247°	5.5″	7.9	9.0	*A0IV/V*	binary	This binary is about 1° northeast of globular cluster NGC 6723. Gould, 175-mm, 100×: A "not bright, but fairly close, uneven pair; both stars are white. The field is moderately starry."
$19^h 03.1^m$	−19° 15′	h 5082								Gould, 175-mm, 100×: This "bright, deep yellow star has an obvious companion . . . and a fainter, wider one. South from it is a gathering of dim stars, including some pairs or triplets."
		AB	2002	88°	7.5″	6.2	9.0	*G5II-III*	binary	
		AC	1998	113°	19.9″	6.0	10.7			
$19^h 04.2^m$	−22° 54′	H N 129	1999	309°	8.0″	6.9	9.2	*A0V*	binary	This binary is about 1° south of Omicron (ο) Sagittarii. Gould, 175-mm, 100×: "Fairly good pair. It is quite bright and easy and lies in a moderately starry field. Uneven, but both stars are white."
$19^h 04.3^m$	−21° 32′	H N 126	2002	195°	1.2″	7.9	8.1	*F8V*	p=500 yr	A 180-mm should be able to split this star until at least 2025. Hartung: "This small yellow binary is the leader of a field sown profusely with stars."
$19^h 06.9^m$	−16° 14′	S 710	2003	0°	6.4″	6.1	8.4	*B8IV*	binary	Gould: 175-mm, 100×: "Good object. An easy uneven pair in a fairly starry field." Ferguson, 150-mm: "Bluish white and blue."
$19^h 08.5^m$	−30° 58′	h 5091	2000	213°	8.7″	8.0	10.2	*K0V*	binary	This binary heads a little triangle of three stars — probably a pretty asterism. Gould, 175-mm, 100×: An "easy uneven pair, with a yellow primary."

Sagittarius *(continued)*

RA	Dec.	Name	Year	P.A.	Sep.	m_1	m_2	Spec.	Status	Comments
$19^h 12.7^m$	−33° 51′	h 5094	1998	184°	28.4″	7.3	7.8	*F0V A0*		Gould, 175-mm, 100×: A "wide and easy pair of white stars, in a field where a circlet of stars loops around it to the north and west."
$19^h 17.7^m$	−15° 58′	S 715	2004	16°	8.4″	7.1	7.9	*A3V*	binary	These two pairs are just minutes apart, and each has the same position angle! Gould, 175-mm, 100×: "Good effect in a field quite starry."
$19^h 18.1^m$	−15° 57′	S 716	2004	195°	5.4″	8.4	8.6	*B9.5V*	binary	
$19^h 18.2^m$	−18° 52′	H V 77	2003	159°	36.9″	7.0	10.4	*B5V*		Gould, 175-mm, 50×: A "wide, uneven pair beside a bright white star (43 Sgr) . . . the primary is white, and the companion looks fainter than its listing."
$19^h 22.6^m$	−44° 28′	β Sgr	1999	76°	28.6″	4.0	7.2	*B8V F0V*		Showcase pair. Hartung: "Bright pale yellow and ashy white, this is a fine object for small telescopes." Gould: "White, yellowish."
$19^h 25.1^m$	−29° 19′	h 5113	1999	151°	10.4″	6.1	10.1	*K3III*		Gould, 175-mm, 100×: "This bright, deep dull orange star has a small, easy companion."
$19^h 28.4^m$	−43° 53′	h 5117	2000	259°	5.8″	7.9	9.4	*F8V*	binary	This binary should be easy to spot — it lies in an empty field $1\frac{1}{4}$° northeast of Beta^1 (β^1) and Beta^2 (β^2) Sagittarii (above). Gould, 175-mm, 100×: An "easy and moderately bright uneven pair, with a pale yellow primary."
$19^h 29.9^m$	−26° 59′	H N 119	1999	144°	7.2″	5.6	8.8	*K2III*	binary	Gould, 175-mm, 100×: "Very good pair with color contrast. A very bright orange star which has a small pale companion, fairly close." [About $1\frac{1}{2}$° north is a large cloud of scattered stars.]
$19^h 39.2^m$	−16° 54′	S 722	2002	236°	9.9″	7.2	7.5	*A8III*		Gould, 175-mm, 50×: An "elegant, nearly even, and moderately bright white pair, beside a fainter wide pair 15′ SE." 60-mm, 25×: The pair 54 Sagittarii (below) is wide in the field.
$19^h 40.7^m$	−16° 18′	54 Sgr AC	2003	42°	44.7″	5.4	7.7	*K2III+F8V*		60-mm, 25×: This is the easiest good pair in Sagittarius! It's a bright star with a very wide little companion — a bright citrus-orange star with a small dot of silver beside it.
$20^h 01.9^m$	−17° 33′	h 2918	1991	135°	15.7″	8.0	8.3	*A3V*		The data suggest a pair of equal stars split by a modest gap, like a pair of eyes. Gould, 175-mm, 50×: An "easy but not bright, fairly wide pair."
$20^h 13.7^m$	−34° 07′	h 5178	1991	9°	2.8″	7.1	8.2	*K1/2III*	binary	Gould, 175-mm, 135×: This "fairly bright yellow star has a less bright companion close . . . not many stars in the field."
$20^h 16.4^m$	−36° 27′	h 5183 AB	1999	229°	36.8″	6.4	11.2	*M4III*		The data suggest a fairly bright star with a dim companion. It's at the edge of a large starless patch.
$20^h 17.8^m$	−40° 11′	Δ 230	2000	117°	9.7″	7.4	7.7	*F8*	binary	Gould, 175-mm, 100×: A "fairly bright, easy double, with a yellowish primary star . . . a good pair for small telescopes. The field has a scatter of 9th- and 10th-magnitude stars."

Sagittarius *(continued)*

RA	Dec.	Name	Year	P.A.	Sep.	m_1	m_2	Spec.	Status	Comments
$20^h 20.5^m$	−29° 12′	h 5188								Gould, 175-mm: "Attractive multiple . . . all [members] visible at 100×. It's a bright white star which has a wide companion . . . a close companion . . . and one which is a neat little pair."
		AB	1998	42°	3.6″	6.7	10.1	*A2III*		
		AC	2002	321°	27.0″	6.7	7.6	*A2III*		
		DE	1998	187°	4.7″	10.1	10.0	*F0V*		
$20^h 26.9^m$	−37° 24′	R 321	1999	134°	1.5″	6.6	8.1	*K2IV K1V*	p=177 yr	This binary is probably resolvable with 200-mm until at least 2025.

Scorpius

RA	Dec.	Name	Year	P.A.	Sep.	m_1	m_2	Spec.	Status	Comments
$15^h 53.6^m$	−25° 20′	2 Sco	1991	269°	2.1″	4.7	7.0	*B2.5V*	binary	Hartung: "This bright yellow star has a companion close . . . which looks whiter, an attractive object for small apertures." [How small? In theory this pair can be split by 125-mm.]"
$16^h 02.9^m$	−25° 01′	β 38	1998	345°	4.4″	7.2	9.5	*B9.5V*	binary	Gould, 175-mm: This "bright white star has an easy, fainter companion at 100×; an attractive object in a field of fairly bright stars."
$16^h 04.4^m$	−11° 22′	ξ Sco								Showcase pairs. 125-mm, 83×: Two attractive pairs just arcminutes apart — a view as beautiful as Epsilon (ε) Lyrae. Xi (ξ) Scorpii is bright amber yellow and royal blue; Σ1999 is a wide and pretty pair of tangerine-orange stars.
		AC	2004	48°	7.5″	4.9	7.3	*G1V*		
$16^h 04.4^m$	−11° 27′	Σ1999	2003	98°	11.8″	7.5	8.1	*K0*	optical	
$16^h 05.4^m$	−19° 48′	β Sco								Showcase pair. 60-mm: With 25×, this bright naked-eye star becomes a brilliant white star beside a little globe of cobalt blue. Smyth: "Pale white; lilac tinge."
		AC	2003	20°	13.6″	2.6	4.5	*B2V*	optical	
$16^h 08.6^m$	−39° 06′	Δ 199								The data suggest a fairly bright pair of stars. Gould, 175-mm, 100×: "A very wide, easy pair . . . white and orange. A suitable pair for binoculars."
		AC	2000	184°	44.1″	6.6	7.1	*A0-3III*		
$16^h 09.5^m$	−32° 39′	BrsO 11	1998	83°	7.8″	6.7	7.2	*K0-2III*	binary	Gould, 175-mm, 100×: This binary is "a moderately bright and easy pair of yellow stars, at the end of a line of stars curving at the end to the double."
$16^h 12.0^m$	−19° 28′	ν Sco								Gould, 175-mm: "A southerly version of a 'double-double.'" It is a "bright, wide pair of stars," each member becoming a pair with 180×. "The CD pair is easy; the AB pair is uneven and just split."
		Aa-B	2003	2°	1.3″	4.4	5.3	*B3V*	binary	
		Aa-C	2003	337°	40.8″	4.2	6.6			
		CD	2003	54°	2.4″	6.6	7.2	*B8/9V*		
$16^h 12.3^m$	−28° 25′	12 Sco	1999	71°	3.8″	5.8	8.1	*B9V*	binary	Gould, 175-mm, 100×: A "beautiful white pair, moderately close, with a fine contrast in brightness. There are two bright stars wide in the field, to the north and east."
$16^h 14.3^m$	−10° 25′	Σ2019								This pair is about 40′ east–southeast of Psi (ψ) Scorpii. Gould, 350-mm, 110×: This "bright yellow star has an easy, fairly wide companion; the field is thin."
		AB-C	1999	153°	22.3″	7.4	9.8	*F7V*		
$16^h 19.5^m$	−30° 54′	BrsO 12	1998	319°	23.5″	5.6	6.9	*F6III F9V*		Showcase pair. Gould, 175-mm, 100×: A "bright yellow pair, wide and easy; the field has a broad scatter of 8th- to 12th-magnitude stars."

RA	Dec.	Name	Year	P.A.	Sep.	m_1	m_2	Spec.	Status	Comments
16h 20.1m 16h 20.5m	−20° 03′ −20° 07′	SHJ 225 SHJ 226 AB-C	2002 2002	333° 20°	46.9″ 12.8″	7.4 7.6	8.1 8.3	*B9*	 optical	Gould, 175-mm, 100×: "Two easy pairs at different separations and angles, both fairly bright, which compliment each other . . . both pairs are white." [An easy "double-double."]
16h 21.2m	−25° 36′	σ Sco Aa-B	 1999	 273°	 20.0″	 2.9	 8.4	 *BiV*		Showcase pair. 60-mm, 45×: A brilliant amber-yellow star with a tiny but conspicuous speck beside it — it looks like a star with a planet! Smyth: "Creamy white; lilac."
16h 23.9m	−33° 12′	h 4848	1998	153°	6.2″	6.9	7.3	A0V+A0V	binary	Gould, 175-mm, 100×: A "lovely white pair, evenly bright, with a wide 9th-magnitude star near it to the north, and a faint wide pair to the northeast."
16h 24.7m	−29° 42′	H N 39	1998	358°	4.4″	5.9	6.6	G0IV G0V	p. binary	Showcase pair. The data suggest a bright pair of stars that are close but easy to split. Gould, 175-mm, 100×: An "easy pair of pale yellow stars in a fairly starry field."
16h 29.4m	−26° 26′	α Sco	1997	274°	2.5″	1.0	5.4	*M1*	p=1,218 yr	*(Antares)* 275-mm, 120×: Breathtaking sight in this large aperture! It's a stunningly brilliant star, mulberry red in color, with a little green cloud sitting on its edge. Smyth: "Fiery red; pale."
16h 38.4m	−43° 24′	h 4867	1999	294°	16.3″	5.8	9.4	*B2V*		Gould, 175-mm, 100×: "An ordinary pair nicely placed. It's an easy uneven pair at the western end of a rambling line of stars, beside a faintish cluster (NGC 6192) 15′ to the E."
16h 42.5m	−37° 05′	R 283	1991	261°	0.5″	7.0	7.8	*G3III*	p=201 yr	This binary is probably now resolvable with 300-mm, but the gap is slowly closing.
16h 43.9m	−41° 07′	CapO 70	2000	258°	97.2″	6.1	6.2	*B6-7V*		Gould: "A finderscope or binocular double. With a 10 × 50, it is a pair of white stars, evenly bright. A star of similar brightness is nearby, and the brilliant cluster NGC 6231 is 2° SE."
16h 44.3m	−27° 27′	β 1116	1991	13°	2.0″	6.6	10.2	*A2V*	binary	Hartung: "105-mm will just show close north the companion of this star . . . the field stars are in linear order, especially following [meaning east]. "
16h 48.2m	−36° 53′	Δ 209	1999	139°	23.8″	7.5	8.4	*A5IV/V*		The data and location suggest an easily noticed pair, about 1.4° northwest of Mu (μ) Scorpii. Gould, 175-mm, 100×: A "wide and easy pair, both stars white . . . probably more attractive at very low power."
16h 51.0m	−37° 31′	h 4889	1999	4°	6.8″	6.2	7.8	*B9*	binary	The data and location suggest a good view — a bright star with a companion, about 1/2° north-northwest of Mu (μ) Scorpii. Gould, 175-mm, 100×: "An easy, nicely uneven pair of white stars."
16h 54.2m 16h 54.2m	−41° 50′ −41° 51′	See 297 WFC 183	1999 1999	132° 36°	13.5″ 30.5″	6.1 6.4	9.9 7.7	O7		These are pairs inside the bright cluster NGC 6231. The data suggest that they should look like double stars within it.

RA	Dec.	Name	Year	P.A.	Sep.	m_1	m_2	Spec.	Status	Comments
$17^h 07.4^m$	−44° 27′	CorO 208	1999	138°	4.8″	7.2	9.3	*A0V*		Gould, 175-mm, 100×: "A fairly good though not special pair. The primary is white, and the companion is a pinpoint quite close [to it]; it stands out in a bare section of a starry field."
$17^h 13.9^m$	−38° 18′	Howe 86	1991	146°	2.7″	6.9	9.0	*F5V*	binary	Gould, 175-mm, 100×: An "interesting setting. In a little asterism — shaped of stars in lines and rectangles — is a brighter pale yellow star with a little companion close southeast."
$17^h 14.5^m$	−39° 46′	h 4926								Hartung: "This orange red star shines like a gem in a field profusely scattered with stars; the two whitish companions . . . may be seen with 105-mm."
		AB	1999	334°	14.1″	6.7	10.1	*G5I*		
		AC	1999	209°	16.9″	6.6	11.3			
$17^h 18.3^m$	−32° 33′	Hdo 268	1998	173°	19.0″	6.4	11.9	*B5V*		Smyth, referring to it as 31 Scorpii: "A delicate double star, between the left foot of Ophiuchus and Scorpio's back . . . pale white; ash colored."
$17^h 19.0^m$	−34° 59′	MlbO 4								This is a very fast-moving binary for amateur telescopes, now slowly closing. Gould, 175-mm, 100×: "Deep yellow orange and lemon yellow . . . and a 3rd prominent star is wide in the field."
		AB	2002	234°	1.7″	6.4	7.4	*K3V+K5V*	p=42.1 yr	
		h 4935								
		AC	2002	139°	31.2″	6.1	9.9			
$17^h 29.0^m$	−43° 58′	Δ 217	1999	168°	13.4″	6.3	8.5	*B5III*	binary	The data suggest a bright star with a companion. Gould, 175-mm, 100×: "A white pair, rather wide . . . with a little group of stars some 20′ to the south."
$17^h 31.3^m$	−39° 01′	Howe 87	1991	232°	3.2″	7.4	9.0	*F6V*	binary	Gould, 175-mm, 100×: "Quite nice. A moderately bright and fairly close pair with a yellowish primary. Several small pairs are in the field, the most obvious [of these] is 10′ NE."
$17^h 34.7^m$	−32° 35′	h 4962								The data and location suggest a very nice view. This is the bright star that dominates the cluster NGC 6383 and should look like a triple within the cluster.
		Aa-B	1998	101°	5.1″	5.7	10.5			
		Ho 647								
		Aa-C	1998	83°	13.6″	5.7	10.9			
$17^h 37.5^m$	−37° 47′	CorO 216								Surely a pleasant view — this pair is at the edge of the large cluster Cr 338. Gould, 175-mm, 100×: "An easy yellowish pair . . . the bright yellow star 5′ south is I 247 — a difficult pair."
		AB-C	2000	194°	13.7″	7.7	9.0	*F2V*		
$17^h 51.2^m$	−30° 33′	Pz 5	1998	189°	10.1″	6.7	8.1	*A0V*	optical	Gould, 175-mm, 100×: "Quite attractive. An easy bright white pair in a middling starry field [with the cluster NGC 6451 just 22′ to the north-northeast]."
$17^h 57.9^m$	−36° 00′	R 306	1999	15°	3.4″	6.8	9.5	*A0III/IV*	binary	Gould, 175-mm, 100×: This "moderately bright white star has a small companion, fairly close . . . there is an asterism some 15′ E [and cluster M7 is about 1½° northwest]."

Sculptor

RA	Dec.	Name	Year	P.A.	Sep.	m_1	m_2	Spec.	Status	Comments
23h 37.1m	−31° 52′	Howe 93	1991	251°	5.5″	6.7	9.9	*K*1III	binary	The data suggest a bright star with a companion. Hartung: "This elegant unequal pair is not difficult for small apertures."
23h 44.5m	−26° 15′	h 5417	2002	319°	8.1″	6.3	9.4	*F*7V		Gould, 175-mm, 100×: "A good object. An easy, uneven pair with a yellow primary. The field has stars south and east but is sparse in the north."
23h 48.9m	−28° 08′	δ Scl AC	1998	296°	74.7″	4.6	9.4	*K*1V		The data suggest a very bright and deeply colored star, which has a wide little companion.
23h 54.4m	−27° 03′	Lal 192	2002	271°	6.4″	6.8	7.4	*A*2V *F*2V	binary	Gould, 175-mm, 100×: "Attractive pair. A fairly bright and nearly even pair in a thin field. Both stars are pale yellow (with a slight difference in shade)."
23h 59.5m	−26° 31′	Lal 193	1998	170°	10.5″	8.0	8.3	*F*0	optical	Gould, 175-mm, 100×: An "easy, well separated pair, but not bright. Both stars are yellow, and the field is ordinary."
00h 09.4m	−27° 59′	κ¹ Scl	2002	261°	1.4″	6.1	6.2	*F*4IV-V	binary	Hartung: "This bright yellow star in a rather thin field is cleanly resolved with 105-mm." Gould, 175-mm, 180×: "A neat and close even pair of stars."
00h 33.6m	−26° 06′	h 3377	1999	60°	21.1″	7.5	9.8	*K*3/4III	optical	Gould, 175-mm, 100×: An "easy, fairly wide uneven pair, orange and dull orange in color; the field is sparse."
00h 33.7m	−35° 00′	h 3375	1998	169°	4.5″	6.6	8.5	*G*3IV	binary	This is the first bright star south-southeast of Eta (η) Sculptoris. Gould, 175-mm, 100×: "Easy, attractive double. It is uneven, fairly close, and yellow in color. South is a little patch of stars."
00h 38.8m 00h 38.9m	−25° 06′ −25° 36′	h 1991 h 1992	1998 1998	95° 248°	47.1″ 45.9″	6.6 7.8	9.7 8.9	*K*0 *A*7V		These two wide pairs are just 1/2° apart. Gould, 175-mm, 100×: h 1991 is a "bright, deep yellow star which has a wide and easy, fairly faint companion; the field has a modest scatter of stars."
00h 42.7m	−38° 28′	λ¹ Scl	1996	17°	0.7″	6.6	7.0	*A*0V	binary	Gould, 175-mm: "A bright white star, with another bright star [Lambda² (λ²) Sculptoris] nearby; at 330× it is elongated, not quite split." Ferguson, 150-mm: Both stars are "bluish white."
01h 27.1m	−30° 14′	h 3436	1998	128°	9.7″	6.9	9.6	*K*0/1III	binary	Gould, 175-mm, 100×: "Fairly good. This is one of two bright stars in a widely scattered field — it is deep orange-yellow, and has a small, easy companion near [to it]."
01h 32.3m	−26° 33′	Arg 4	1998	72°	18.0″	8.0	9.1	*A*/*G*2V	optical	Gould, 175-mm, 100×: "An easy uneven pair, pale yellow and deeper yellow; a 10th-magnitude star is close by [forming a broad arc shape]; the field is thin."
01h 36.1m	−29° 54′	τ Scl	1999	169°	0.8″	6.0	7.4	*F*2V	p. binary	Hartung: A "close yellow pair . . . in 1960 [when the separation was 1.2″], 150-mm was needed to resolve it."
01h 39.7m	−37° 28′	h 3452	1998	277°	19.8″	7.4	8.9	*K*0	optical	This pair forms a neat triangle, about 1° in size, with two bright field stars. Gould, 175-mm, 100×: "A wide, easy and fairly bright pair, deep yellow in color."

Sculptor *(continued)*

RA	Dec.	Name	Year	P.A.	Sep.	m_1	m_2	Spec.	Status	Comments
$01^h45.6^m$	−25°03′	ε Scl	2002	24°	4.9″	5.4	8.5	*F2V*	p=1,192 yr	Showcase pair. Gould, 175-mm: A "moderately close and attractive uneven pair, in a fainter starry field. The primary is pale yellow." Hartung: "Easy bright yellow pair."

Scutum

RA	Dec.	Name	Year	P.A.	Sep.	m_1	m_2	Spec.	Status	Comments
$18^h20.1^m$	−07° 59′	Σ2303	1991	240°	1.6″	6.6	9.3	*F2V*	binary	The data suggest a tight binary with a fairly bright primary. Webb, listed under Scutum Sobieski: "Yellowish, no color given."
$18^h24.7^m$	−06° 36′	Σ2313	2002	196°	5.9″	7.5	8.7	*G0IV*	binary	125-mm: Nice contrast with attractive closeness. A lovely lemon-yellow star and a small green-white just a tiny gap apart.
$18^h31.4^m$	−10° 48′	Σ2325	2003	256°	12.3″	5.8	9.3	*B2V*	binary	The data suggest a bright star with a companion. Webb, listed under Scutum Sobieski: Franks calls it "white, bluish."
$18^h45.9^m$	−10° 30′	Σ2373	1999	337°	4.1″	7.4	8.4	*F2*	binary	125-mm: A bright, easy pair with attractive closeness. A very lemony white pair of stars, just mildly unequal, that are split by a hair at 50×.
$18^h48.7^m$	−06° 00′	Σ2391	2003	332°	37.9″	6.5	9.6	*A2II*		125-mm, 50×: This ordinary wide pair is easily spotted in the field with M11, the Wild Duck Cluster. It's a very lemony white star and a tiny bluish turquoise very wide apart.
$18^h49.7^m$	−05° 55′	H VI 50 AC	2001	171°	111.4″	6.2	8.2	*K1II*		60-mm, 25×: A fine pair for low power with a pretty primary star. It's a bright grapefruit-orange star and a small pearly white, super-wide apart.

Serpens

RA	Dec.	Name	Year	P.A.	Sep.	m_1	m_2	Spec.	Status	Comments
$15^h12.7^m$	+19° 17′	Σ1919	2003	10°	23.1″	6.7	7.4	*G1V G5V*	optical	60-mm, 25×: Ideal couple for low power. It's a pair of peach-white stars in the field with a bright red star (FL Serpentis). They're wide enough to be easy but still attractively close.
$15^h18.7^m$	+10° 26′	Σ1931	2001	168°	13.2″	7.2	8.1	*F7V G3V*	optical	125-mm: An easy pair with a pretty primary. It's a bright yellow star and a dimmer yellow star fairly wide apart.
$15^h19.3^m$	+01° 46′	5 Ser	2000	35°	11.4″	5.1	10.1	*F8V*		325-mm, 200×: Fantastic view! A brilliant yellow star with a close little speck beside it, and globular M5 is at the edge of the field. Smyth: "Pale yellow; light gray."
$15^h21.0^m$	+00° 43′	6 Ser	2000	25°	3.0″	5.5	8.8	*K2III*	binary	125-mm, 200×: Difficult but striking binary. It's a bright citrus-orange star almost touched by a tiny, shadowy vapor that takes great effort to see. Webb: Sadler calls it "yellow, grey blue."

Serpens *(continued)*

RA	Dec.	Name	Year	P.A.	Sep.	m_1	m_2	Spec.	Status	Comments
$15^h 34.8^m$	+10° 32′	δ Ser	2004	174°	4.0″	4.2	5.2	F0IV	p=1,038 yr	Showcase pair. 60-mm, 45×: A brilliant pair of unequal stars, both yellow-white, that are just kissing at 45×. Smyth: "[Sometimes] both have a bluish tinge . . . against the theory of contrast."
$15^h 40.2^m$	+12° 03′	OΣ 300	2003	260°	15.0″	6.3	10.1	G7.5III		125-mm, 200×: Striking contrast for the separation! A bright whitish peach star with a little vaporish bump on its edge. At 50×, a bright white star [Chi (χ) Serpentis] is wide in the field.
$15^h 44.0^m$	+02° 31′	ψ Ser AC	2002	208°	195.5″	6.0	9.2	G3V		60-mm, 25×: This pair is included because of its pretty primary star. It's a bright vivid star, very yellowish white in color, with a close field star beside it that looks like a companion.
$15^h 46.2^m$	+15° 25′	β Ser	1999	264°	30.9″	3.7	10.0	A2IV		125-mm, 83×: Grand! A brilliant whitish yellow star with a distant pinpoint beside it, which looks more like a companion than a field star. Webb: Franks calls it "pale yellow, lilac."
$15^h 55.9^m$	−02° 10′	Σ1985	2004	354°	6.0″	7.0	8.7	F8V	p=2,331 yr	125-mm, 50×: An easy binary star with attractive closeness. It's a yellow-white star and a little nebulous globe only a small gap apart. Webb: "Yellow white, ash."
$15^h 56.8^m$	+12° 29′	Σ1988	2004	251°	1.9″	7.6	7.8	F1V	p. binary	125-mm, 200×: This double is fairly bright and attractively close. It's a pair of amber-yellow stars, nearly equal in brightness, just a tiny gap apart.
$15^h 57.2^m$	+03° 24′	Σ1987	2002	323°	10.9″	7.3	8.7	A0V	optical	125-mm, 50×: A pretty pair that seems to jump out of an empty field. It's an amber-yellow star with a tiny blue companion just a small gap apart. Webb: "White, lilac."
$16^h 00.9^m$	+13° 16′	OΣ 303	2003	171°	1.5″	7.7	8.1	F7V	binary	125-mm, 200×: A dim but attractively close binary star. It's a pair of amber-yellow stars, very slightly unequal, that are just short of touching.
$16^h 06.0^m$	+13° 19′	Σ2007	2003	322°	37.8″	6.9	8.0	G8III		60-mm: A wide pair with pretty colors — an ideal object for low power. It's a Sun-yellow star with a little red companion.
$16^h 16.3^m$	−01° 39′	Σ2031	2003	229°	19.8″	7.2	8.7	F7V		125-mm: This pair is attractive for its contrast. It's a fairly bright orange star with ghostly speck of light beside it, widely separated. The little star looks much fainter than its listed magnitude.
$16^h 21.8^m$	+01° 13′	Σ2041	1991	359°	2.6″	7.5	10.5	K0		This pair is just a few arcminutes north-northwest of Sigma (σ) Serpentis. Webb: "Yellow, no color given." Sadler calls the companion "blue."
$17^h 20.8^m$	−12° 51′	53 Ser	2002	26°	45.9″	4.3	9.4	A2V		125-mm, 50×: Nice color contrast. A brilliant white star with a little red companion. They're very wide apart but look like a couple. Smyth: "Pale sea green; lilac."
$17^h 34.8^m$	−11° 15′	h 4964	2002	225°	54.3″	5.5	9.9	B8V		The data suggest a bright star with a wide little companion. Harshaw 200-mm: "White and blue."

Serpens *(continued)*

RA	Dec.	Name	Year	P.A.	Sep.	m_1	m_2	Spec.	Status	Comments
$17^h 56.3^m$	−15° 49′	h 2814	2003	157°	20.5″	5.9	9.2	*A1V*		125-mm, 50×: Pretty view. It shows a bright white star with a small ruddy companion, very wide apart, at the end of a long stream of flickering pinpoints.
$18^h 13.6^m$	−15° 36′	β 131								Hartung: A "beautiful little triplet . . . 75-mm shows the pair [AB] only, and 150-mm is needed to see the third star [C]."
		AB	2002	276°	3.1″	7.3	9.3	*F6V*		
		BC	1933	296°	4.8″	9.3	11.5			
$18^h 20.1^m$	−07° 59′	Σ2303	1991	240°	1.6″	6.6	9.3	*F2V*	binary	Hartung: "This fairly bright star has a companion close . . . which 75-mm shows with care . . . the field is almost blank [because of a dark cloud within the Milky Way]."
$18^h 25.0^m$	−01° 35′	AC 11	2003	355°	0.9″	6.7	7.2	*F0*	p=248 yr	Hartung: A "close yellow pair . . . in 1961 [when the separation was 0.7″], the stars could just be divided with 200-mm."
$18^h 27.2^m$	+00° 12′	59 Ser	2003	320°	3.9″	5.4	7.6	G0III *A6V*	binary	Showcase pair. 125-mm, 200×: A dazzling star with a close little companion, in the splendid combination of lucid yellow and lovely royal blue. J. Herschel: "Orange, green."
$18^h 28.6^m$	+04° 51′	OΣΣ 168	2004	163°	47.4″	7.7	8.8	*A0*		8 × 50: Beautiful field at extremely low power — this pair and clusters NGC 6633 and IC 4756 are all in one view! It's a pair of touching stars, straw yellow and bold silver.
$18^h 35.6^m$	+04° 56′	Σ2342								125-mm, 50×: Very pleasant view. A fairly bright, Sun-yellow star with a wide little companion — just beyond a dense swarming cloud of stars (the edge of the giant cluster IC 4756).
		AC	2003	1°	33.3″	6.5	9.6	*A1V*		
$18^h 45.5^m$	+05° 30′	Σ2375								125-mm, 200×: Splendid combination! Attractively bright twins, deep pure white in color, which are just wide enough to be easy.
		Aa-Bb	2004	119°	2.5″	6.3	6.7	*A1V+A1V*		
$18^h 56.2^m$	+04° 12′	θ Ser	2004	104°	22.3″	4.6	4.9	*A5V A5V*	optical	Showcase pair. 60-mm, 25×: Brilliant white twins that are generously apart but still attractively close. They seem like a pair of eyes dominating a rich starry field.

Sextans

RA	Dec.	Name	Year	P.A.	Sep.	m_1	m_2	Spec.	Status	Comments
$09^h 43.5^m$	+02° 38′	Σ1377	2000	137°	4.2″	7.5	10.5	*F7V*		The data suggest a modestly bright primary and good contrast for the separation. Smyth: "A delicate double star . . . yellowish white, blue."
$09^h 52.5^m$	−08° 06′	γ Sex	2004	57°	0.6″	5.4	6.4	*A1V*	p=77.5 yr	A 250-mm should be able to split this fast-moving binary until about 2018. It's probably the best pair in Sextans, for those who can resolve it.
$09^h 54.1^m$	+04° 57′	9 Sex	2004	287°	52.5″	6.9	8.4	*M0 K0*		125-mm, 50×: This wide pair shows pretty colors. It's a bright tangerine star with a little blue companion super-wide apart. Smyth: "Both [stars] blue, and well defined." Webb: "Red, blue."

Sextans *(continued)*

RA	Dec.	Name	Year	P.A.	Sep.	m1	m2	Spec.	Status	Comments
10h 24.2m	+02° 22′	h 2530								60-mm, 25×: Splendid sight for low power. A bright pair of lemon-white twins, super-wide apart, that pop out of a nearly black field.
		AC	2002	64°	202.4″	6.4	6.7	G9V		
10h 26.2m	+03° 56′	β 1280	2002	192°	116.5″	6.7	9.4	A3V		125-mm, 50×: This pair is included because of its pretty primary star. It's a bright white star with a distant small companion in a hauntingly black background.
10h 31.0m	−07° 38′	Σ1441	2003	167°	2.8″	6.5	8.8	K5III+F6V	binary	The data suggest a bright star with a close companion. Webb: "Gold, no color given."
10h 38.7m	+05° 44′	Σ1457	1998	332°	2.1″	7.7	8.2	F5V	p. binary	125-mm, 200×: A modestly dim but attractively close pair. It's a hair-split pair of pearly white stars nearly equal in brightness.
10h 43.3m	+04° 45′	35 Sex	2004	241°	6.7″	6.2	7.1	K3III+K0III	binary	Showcase pair. 60-mm, 42×: A bright star with a close little companion in the splendid combination of citrus orange and blended blue-green. Smyth: "Topaz yellow; smalt blue."
10h 49.3m	−04° 01′	40 Sex	2004	16°	2.4″	7.1	7.8	A2IV	p. binary	125-mm, 200×: Grand sight — stars split by a hair with colors that are pure white and silver white.
10h 50.3m	−08° 54′	41 Sex	1999	307°	27.9″	5.8	11.7	A3		The magnitudes suggest a bright star with a very dim companion. Smyth: "White; dusky . . . it lies . . . nearly alone in its glory . . . like an oasis in the desert."

Taurus

RA	Dec.	Name	Year	P.A.	Sep.	m1	m2	Spec.	Status	Comments
03h 31.1m	+27° 44′	Σ7	2004	234°	43.3″	7.4	7.8	B9		60-mm, 25×: A lovely double double. It's two bright pairs are just a few arcminutes apart — a wide white pair (Σ7) and a close yellow pair (Σ401). Each is a pair of nearly equal stars.
03h 31.3m	+27° 34′	Σ401	2004	269°	11.2″	6.6	6.9	A2V		
03h 34.4m	+24° 28′	7 Tau								Smyth: "A triple star . . . white; pale yellow; bluish . . . this is a fine and very difficult object."
		AB	2004	356°	0.7″	6.6	6.9	A2V	p=522 yr	
		AB-C	2004	55°	22.0″	5.9	9.9			
03h 36.8m	+00° 35′	Σ422	2004	271°	6.7″	6.0	8.9	G8V	p=2,101 yr	125-mm, 83×: Splendid view — a pumpkin-orange star with a close little companion. A brilliant yellow star (10 Tauri) is in the field. Webb, listed under Eridanus: "Gold, blue."
03h 40.5m	+05° 08′	Σ430								The data suggest a bright and pretty star with two dim companions. Webb: "Very yellow, no color given [for either companion]."
		AB	2002	56°	25.8″	6.7	9.6	K2III		
		AC	2003	301°	34.6″	6.7	10.5			
03h 40.6m	+28° 46′	Σ427	2004	207°	7.0″	7.4	7.8	A1V A2V	binary	125-mm: Splendid combination — a close pair of bright stars just wide enough to be very easy. They're deep white in color, nearly identical, and split by a tiny gap at 50×.
03h 43.1m	+25° 41′	Σ435	1999	3°	13.2″	7.2	8.9	F3V		125-mm, 125×: This good pair is easily spotted about 2° northwest of the Pleiades. It's a star with a tiny wide companion in the pretty combination of deep lemon white and sapphire blue.

Taurus *(continued)*

RA	Dec.	Name	Year	P.A.	Sep.	m₁	m₂	Spec.	Status	Comments
03ʰ44.6ᵐ	+27° 54′	ΟΣΣ 38	2003	51°	133.8″	6.8	6.9	F4V		60-mm, 25×: Nice view at low power — it shows a super-wide pair of twins that seem to jump out of a rich field of dim stars. They're whitish gold in color.
03ʰ45.2ᵐ	+24° 28′	19 Tau	2003	329°	72.1″	4.3	8.1	B6V		*(Taggeta)* This is one of the six naked-eye stars that form the Pleiades. J. Herschel and Smyth saw it as a pair within the cluster. Smyth: "Lucid white; violet tint." Harshaw, 200-mm: "Blue white, blue."
03ʰ45.8ᵐ	+23° 09′	Σ444	1991	335°	3.6″	6.9	10.1	A0V	binary	The magnitudes and location suggest an easily spotted star just south of the Pleiades. Smyth: "Bright white; fine blue."
03ʰ47.4ᵐ	+23° 55′	Σ450	1997	264°	6.2″	7.1	9.1	A2V		125-mm, 50×: Nicely placed! An attractively close pair within the Pleiades, just south of Eta (η) Tauri. It's a pearly white star with a little nebulous companion split by a tiny gap.
03ʰ48.3ᵐ	+11° 09′	30 Tau	2000	59°	9.2″	5.1	9.8	B3V+B3V		Smyth: "A delicate double star . . . pale emerald; purple . . . elegant but difficult." Webb: "Bluish green; no color given."
03ʰ49.4ᵐ	+24° 23′	ΟΣΣ 40	2000	309°	86.5″	6.6	7.5	B9.5V		This is another pair within the Pleiades (just north of 28 Tauri). The data suggest a good sight for very low power. Harshaw, 200-mm: "White, white."
03ʰ50.0ᵐ	+23° 51′	ΟΣ 64								This triple is at the southeastern edge of the Pleiades; the position angles suggest three stars in a straight line. Harshaw: "White, orange."
		AB	2003	234°	3.2″	6.8	10.2	B9.5V		
		AC	2004	235°	10.0″	6.8	10.5			
03ʰ52.0ᵐ	+06° 32′	31 Tau	2002	208°	0.7″	6.3	6.6	B5V	p. binary	A 200-mm should be able to split this bright double.
04ʰ00.9ᵐ	+23° 12′	Σ479								125-mm, 50×: This triple forms an attractive little asterism. It's a close pair of white stars and a very wide brown-cherry star in a neat L shape.
		AB	2004	127°	7.4″	6.9	7.8	B9V	binary	
		AC	2004	242°	57.8″	6.9	9.5			
04ʰ02.2ᵐ	+28° 08′	Σ481								125-mm, 50×: This pair is included because of its pretty primary star. It's a deeply amber-yellow star with a ghostly dot beside it, split by just a small gap.
		AC	2002	325°	14.8″	7.5	10.1	G6V		
04ʰ07.7ᵐ	+15° 10′	Σ495	2002	221°	3.7″	6.1	8.8	F3V	binary	The data suggest a bright star with a close little companion. Hartung: An "easy yellow pair."
04ʰ08.0ᵐ	+17° 20′	ΟΣ 72	2002	328°	4.7″	6.1	9.7	K5III		The data suggest a bright star with a close companion. Ferguson, 150-mm: "Yellow and faint blue. Difficult."
04ʰ 08.9ᵐ	+23° 06′	Σ494	2000	188°	5.1″	7.5	7.7	A8IV A8IV	binary	110-mm: A bright binary star with attractive closeness. It's a yellow-white star and a smaller silvery white just a small gap apart at 32×.
04ʰ12.2ᵐ	+00° 44′	Σ510	2000	301°	10.9″	6.7	10.1	G8III	binary	The data suggest a bright star with a dim companion. Webb, listed under Eridanus: "Very yellow, no color given."
04ʰ13.9ᵐ	+09° 16′	47 Tau	2000	342°	1.3″	5.1	7.3	G5IV+A8V	binary	Hartung: A "deep yellow pair . . . 200-mm shows the small star lying close."

Taurus *(continued)*

RA	Dec.	Name	Year	P.A.	Sep.	m_1	m_2	Spec.	Status	Comments
$04^h 15.5^m$	+06° 11′	H IV 98	2002	316°	64.2″	6.4	7.0	G0IV G3V		125-mm, 50×: A fine sight for low power. It's a very wide pair of bright stars, vivid lemon yellow and fainter gloss white, that seem to jump out of their thin field.
$04^h 16.0^m$	+00° 27′	Σ517	1997	10°	3.3″	7.4	9.3	A1V	binary	125-mm, 83×: A binary with striking contrast for the separation. It's a fairly bright white star and a ghostly dot split by hairs at 83×. In an 8 × 50 finderscope, OΣΣ 49 (below) is in the same field.
$04^h 18.9^m$	+01° 46′	OΣΣ 49	2000	144°	102.8″	7.9	7.8	G5		125-mm, 50×: A bright pair of identical stars, super-wide apart, in a starless background. They're gloss white in color and make a fine sight at low power.
$04^h 19.8^m$	+23° 44′	Σ523	1997	163°	10.2″	7.6	9.9	B9.5V		125-mm, 83×: An attractive dim pair, easily spotted about 2° northwest of a bright asterism (65, 67, 69, and 72 Tauri). It's a yellow-white star with a little speck beside it, modestly wide apart.
$04^h 20.4^m$	+27° 21′	φ Tau	1999	256°	49.2″	5.1	7.5	K2III		125-mm, 50×: Grand! A brilliant star with a small companion, Sun yellow and bluish beige in color; they're very wide apart but look like a couple. Smyth: "Light red; cerulean blue."
$04^h 22.4^m$	+20° 49′	β 87	1991	169°	1.9″	6.2	8.6	B3V+K3II	binary	The data suggest a bright star with a close dim companion. Webb: "Red, blue." Sadler calls it "very lovely."
$04^h 22.6^m$	+25° 38′	59 Tau	2004	25°	19.1″	5.4	8.5	B9V F8V		Showcase pair. 110-mm, 100×: An easy pair with pretty colors. It's a bright yellow star with a little smoke puff beside it, fairly wide apart.
$04^h 22.7^m$	+15° 03′	OΣ 82	2004	340°	1.2″	7.3	8.6	F9V	p=241 yr	This binary is probably resolvable with 200-mm until at least 2025.
$04^h 23.3^m$	+11° 23′	Σ535	2004	276°	1.1″	7.0	8.3	A5III	p=1,128 yr	This binary is probably resolvable with 200-mm until at least 2025.
$04^h 24.0^m$	+24° 18′	62 Tau	2002	290°	28.7″	6.4	7.9	B3V A1		Showcase pair. 60-mm, 25×: A bright white star with a tiny white companion. Both stars are seen easily in dramatic contrast. Smyth: "Silvery white; purple." Webb: "Fine field."
$04^h 25.4^m$	+22° 18′	65-67 Tau	2002	174°	339.7″	4.2	5.3	A7IV/V		60-mm: A fine sight for low power. It's a brilliant pair of white stars super-wide apart. They're mildly unequal and look like a couple at 25×.
$04^h 27.1^m$	+18° 12′	Σ545	2004	58°	18.7″	6.9	8.8	A0V	optical	125-mm, 50×: Splendid view — this easy, wide pair is at the edge of the Hyades, about 1/2° northeast of Delta³ (δ³) Tauri. It's a gloss-white star and a dim silver wide apart.
$04^h 28.7^m$	+15° 52′	θ¹ - θ² Tau	2002	348°	336.7″	3.4	3.9	A7III		60-mm, 25×: Grand sight for low power! This is the bright, wide pair of twins that dominate the Hyades cluster; they look like a couple within it. Webb: "White, yellowish . . . pair to naked eye."

Taurus *(continued)*

RA	Dec.	Name	Year	P.A.	Sep.	m_1	m_2	Spec.	Status	Comments
$04^h 28.9^m$	+30° 22′	Σ548	2004	36°	14.6″	6.4	8.0	*F7V*	optical	125-mm, 50×: A bright, easy pair that seems to jump out of the field. It's a wide pair of white stars, modestly unequal, in a background without bright stars.
$04^h 30.1^m$	+15° 38′	80 Tau	2002	16°	1.6″	5.7	8.1	*F0V+G0V*	p=180 yr	Smyth: "A close double star . . . yellow; and dusky. This beautiful object . . . is of no easy measurement."
$04^h 31.1^m$	+06° 47′	OΣ 84	2002	253°	9.4″	7.2	8.1	*K0IV*	binary	125-mm, 50×: Nice combination. A fairly bright star and a fairly dim star attractively close while wide enough to be easy. Yellow-white, bluish silver.
$04^h 33.5^m$	+18° 01′	Σ559	2004	277°	3.0″	7.0	7.0	*B9IV*	binary	125-mm: Grand! A bright pair of bluish white stars, exactly alike, that are split by a hair at 83×. It's a conspicuous star about halfway between Alpha (α) and Epsilon (ε) Tauri.
$04^h 35.7^m$	+10° 10′	88 Tau	2003	300°	69.1″	4.3	7.8	*A5*		125-mm, 50×: Grand! A brilliant star with a wide little companion in the lovely combination of vivid lemon yellow and silvery cherry. Webb: "Yellow white, yellow red."
$04^h 35.9^m$	+16° 31′	α Tau AC	1997	31°	133.2″	0.8	11.3	*K5III*		*(Aldebaran)* Smyth: "Pale rose tint; sky blue . . . I was surprised on readily seeing it [the companion] with my 5-foot telescope of 3.75-inches [95-mm] aperture."
$04^h 37.4^m$	+00° 34′	OΣΣ 53	2002	173°	78.0″	7.6	7.6	*G5*		125-mm, 50×: Nice sight — a pair of equal stars that are different in color in the field with a bright white star. They're white and coppery red and are super-wide apart.
$04^h 38.5^m$	+26° 56′	Σ572	2003	192°	4.6″	7.4	7.2	*F2V F2V*	p. binary	125-mm: Beautiful couple. A bright pair of identical stars, gloss white in color, that are split by a hair at 62×. They completely dominate a nearly pitch black background.
$04^h 39.3^m$	+15° 55′	σ¹-σ² Tau	2002	195°	437.5″	4.7	5.1	*A5V A4*		125-mm, 50×: Dazzling! Two brilliant yellow stars, super-wide apart, appearing so alike that they make an obvious couple.
$04^h 42.2^m$	+22° 57′	τ Tau	1999	214°	63.0″	4.2	7.0	*B3V*		125-mm, 50×: This very wide pair shows beautiful colors. It's a brilliant lemon-yellow star and a pale mint green. These colors are seen vividly in striking contrast.
$04^h 59.9^m$	+27° 20′	Σ623	2004	206°	20.2″	7.0	8.7	*B7V+A0V*	p. binary	125-mm, 50×: A bright, wide, and easy pair that seems to jump out of a desolate field. Bright yellowish white and silvery blue-green.
$05^h 05.5^m$	+19° 48′	OΣ 95	2004	298°	0.9″	7.0	7.6	*A5*	p=760 yr	This binary is probably resolvable with 150-mm until at least 2025.
$05^h 08.1^m$	+24° 16′	103 Tau AC	1977	198°	36.0″	5.5	8.6	*B2V*		The data suggest a bright star with a companion. Smyth, referring to the pair as 295 P. IV. Tauri: "Pearly white; pale blue . . . and though a widish object in a bare field, it is fine and delicate."
$05^h 09.8^m$	+28° 02′	Σ645 A-BC	2002	29°	11.3″	6.0	9.1	*A5*		125-mm, 50×: A pretty pair whose primary jumps out of a thin field. It's a bright yellow and a pale mint green, modestly wide apart. Webb, listed under Auriga: "White, ash."

Taurus *(continued)*

RA	Dec.	Name	Year	P.A.	Sep.	m_1	m_2	Spec.	Status	Comments
05^{h}16.7^m	+18° 26′	Σ670	2002	166°	2.5″	7.7	8.3	*B*3V		125-mm: A modestly dim but attractively close pair in a rich field of dim stars. It's a green-white star and an ash white, modestly unequal, that are split by a hair at 83×.
05^{h}17.5^m	+20° 08′	Σ674	2002	149°	9.9″	6.8	9.7	*F*7V+*F*5IV	binary	The data suggest a pretty sight — two pairs, 1/2° apart, that sound alike in every way but color. Smyth: Σ674 is "bright white; bluish" and Σ680 is "deep yellow; bluish."
05^{h}19.2^m	+20° 08′	Σ680	2002	203°	9.1″	6.2	9.7	*K*0III	binary	
05^{h}29.3^m	+25° 09′	118 Tau	2002	207°	4.7″	5.8	6.7	*B*8.5V	p. binary	Showcase pair. 60-mm: A bright pair of peach-orange stars, mildly unequal, that are split by a hair at 45×. Smyth: "White; pale blue." Webb: "White, bluish."
05^{h}29.8^m	+18° 25′	h 3275								8 × 50: Beautiful sight for a finderscope! A close pair of white stars, nearly alike, in the field with a giant asterism of bright stars (110, 111, 113, 115, 116, and 117, 119, and 120 Tauri).
		AC	2002	21°	56.3″	7.7	8.2	*A*0		
05^{h}32.2^m	+17° 03′	Σ730	2002	141°	9.6″	6.1	6.4	*B*7III	binary	Showcase pair. 60-mm, 25×: A pair of bright stars, very nearly equal, that are attractively close while wide enough to be very easy. Blue-white and yellow-white.
05^{h}36.4^m	+22° 00′	Σ742	2004	274°	4.0″	7.1	7.5	*F*8	p=2,959 yr	125-mm, 50×: Lovely view. It shows two pairs in the field with the brilliant white star Zeta (ζ) Tauri. They are a close pair of bright stars (Σ742) and a wide pair of dim stars (Σ740).
05^{h}36.4^m	+21° 11′	Σ740	2004	121°	21.7″	9.0	9.9	*B*2III		
05^{h}37.1^m	+26° 55′	Σ749	2002	323°	1.1″	6.5	6.6	*B*9IV-V	p. binary	125-mm, 200×: Grand sight — a bright pair of beautifully close stars! They're a set of identical stars, gloss white in color, that form a perfect hourglass shape.
05^{h}47.9^m	+24° 41′	OΣΣ 66	2001	166°	94.3″	7.0	7.7	*K*2		125-mm, 50×: This super-wide pair is easily noticed in the field with 132 Tauri. It's a red-white star with an ash-white companion beside brilliant yellow 132 Tauri.
05^{h}49.0^m	+24° 34′	132 Tau								The data suggest a bright star with a close little companion. Although a 120-mm should split this pair, the author failed to resolve it. The pair OΣΣ 66 (above) is in the field.
		AC	1991	230°	3.8″	5.0	9.1	*G*8III		

Telescopium

RA	Dec.	Name	Year	P.A.	Sep.	m_1	m_2	Spec.	Status	Comments
18^{h}15.4^m	−48° 51′	h 5033								Gould, 175-mm, 100×: This "bright orange star has two fairly wide companions, and a third much fainter one. The field is quite attractive."
		AB	1999	114°	17.2″	6.8	9.8	*G*8III/IV		
		AD	1999	69°	25.5″	6.8	10.7			
18^{h}16.2^m	−46° 01′	h 5034	1994	98°	2.2″	7.5	8.6	*A*5IV/V		Gould, 175-mm: A "beautiful little close pair. It is moderately bright and uneven, and a fine sight against a mixed starry field. Easy at 180×."
18^{h}25.8^m	−53° 38′	h 5041	1991	260°	3.0″	7.3	9.2	*F*		Gould, 175-mm, 100×: This pair is "not unlike h 5034 [above] . . . a fine moderately bright pair, with a yellow primary. About 8′ SE is a little patch of stars, including an easy faint pair."

Telescopium *(continued)*

RA	Dec.	Name	Year	P.A.	Sep.	m_1	m_2	Spec.	Status	Comments
18ʰ30.9ᵐ	−48° 01′	h 5045	2000	21°	8.1″	6.7	9.7	*F6V*	binary	Gould, 175-mm, 100×: "A bright yellow star with a faintish companion, moderately separated."
18ʰ54.0ᵐ	−47° 16′	I 112								Gould, 175-mm: "Nice effect. A quite bright, very wide pair, yellow and white in color . . . the brighter star has a very close, fainter companion that is plainly seen at 180×."
		AB	1991	190°	1.8″	7.1	9.1	*F5V*	binary	
		Δ 224								
		AC	1998	62°	86.7″	7.1	7.3	*A0IV/V*		
18ʰ58.9ᵐ	−48° 30′	I 113	1991	227°	3.2″	6.8	10.7	*K3III BII*		Gould, 175-mm: "180× shows the small close companion of this bright orange star, and there are some fairly bright stars wide in the field." Hartung: "Fine sight in a field well sprinkled with stars."
19ʰ03.1ᵐ	−45° 43′	R 317								Gould, 175-mm: An "easy wide white pair at 100×; at 180×, the brighter star becomes a rather attractive close pair, somewhat uneven in brightness."
		AB	1991	284°	1.5″	8.0	8.8	*B9III*		
		h 5078								
		AC	1999	211°	18.5″	8.0	8.8			
19ʰ12.4ᵐ	−51° 48′	Δ 225	1999	250°	70.1″	7.2	8.4	*K5III*		Gould, 175-mm: A "very wide pair, orange and pale yellow, with a much fainter star between the two members. It lies in a moderately starry field."
19ʰ27.8ᵐ	−54° 20′	h 5114								The data suggest a bright star with two wide companions. Gould, 175-mm, 100×: This is a "very broad triple in the form of an arc: member A is orange and B is yellow."
		AB	1999	238°	75.1″	5.9	8.2	*K4III*		
		AD	1999	94°	72.9″	5.9	10.9			
19ʰ42.3ᵐ	−52° 57′	CorO 238	1992	47°	3.4″	7.7	9.3	*A6/7V*	binary	Gould, 175-mm, 100×: "Good effect. A delicate uneven pair in a modest gathering of faint stars. Both members are white."
19ʰ52.6ᵐ	−54° 58′	Δ 227	2000	148°	23.0″	5.8	6.4	*K0III A2V*	optical	Hartung: "This beautiful pair is an excellent object for small telescopes; both stars are bright and are in fine contrast, orange yellow and white." Gould: "Deep yellow and cream."

Triangulum

RA	Dec.	Name	Year	P.A.	Sep.	m_1	m_2	Spec.	Status	Comments
01ʰ55.1ᵐ	+28 47′	Σ183								125-mm: A fairly bright and obvious pair about 1° south-southeast of Alpha (α) Trianguli. It's a pearly white star and a smaller silvery blue star just a small gap apart at 50×.
		AB-C	2001	163°	5.2″	7.7	9.3	*F2*		
02ʰ03.0ᵐ	+33° 17′	ε Tri	1990	119°	4.2″	5.4	11.4	*A2V*		Smyth: "A most delicate double star . . . bright yellow; dusky . . . marked 'difficilis' in the Dorpat Catalog [of W. Struve]. It lies . . . [near] a deep orange coloured [star]."
02ʰ10.3ᵐ	+33° 22′	Σ219	2003	183°	11.4″	8.0	8.9	*A0*	optical	125-mm, 50×: This rather ordinary pair forms a nice asterism. It's a wide pair of yellow stars, slightly unequal, that create a neat butterfly shape with three field stars.
02ʰ12.4ᵐ	+30° 18′	ι Tri	2002	69°	3.9″	5.3	6.7	*G0III*	binary	Showcase pair. 60-mm, 80×: A bright grapefruit-orange star almost touched by a little gray globe; both stars are seen vividly in striking contrast. Smyth: "Topaz yellow; green."

Triangulum *(continued)*

RA	Dec.	Name	Year	P.A.	Sep.	m_1	m_2	Spec.	Status	Comments
02^h 14.7^m	+30° 24′	Σ232	2002	66°	6.6″	7.9	7.8	*A*0V	binary	60-mm, 35×: Splendid view — a touching pair of white twins at the corner of a star triangle, which is beside the bright orange star Iota (ι) Trianguli (previous page).
02^h 17.4^m	+28° 45′	Σ239	2001	212°	13.7″	7.1	7.8	*F*7V *F*9V	optical	60-mm, 17×: Splendid view at 17× — a pearly white star with a close companion. They are in the field with a bright white star (10 Trianguli) and a deep orange star.
02^h 18.7^m	+34° 29′	Σ246	2002	123°	9.9″	7.8	9.3	*G*0IV	binary	125-mm, 50×: Splendid view. A close and unequal pair — bright white and nebulous gray — in the field with the bright yellow star Delta (δ) Trianguli and the brilliant white star Gamma (γ) Trianguli.
02^h 28.2^m	+29° 52′	Σ269	2001	345°	1.6″	7.6	9.0	*G*0	binary	This binary star is very easy to locate — it's between the bright stars 12 and 13 Trianguli, which are separated by 18′. Smyth: "A close double star . . . yellow; gray . . . exquisite and difficult."
02^h 35.8^m	+34° 41′	15 Tri	2003	16°	142.2″	5.6	6.8	*M*3III		125-mm, 50×: A very wide pair with superb colors. It's a bright pumpkin-orange star and a bright pearly white. They look like a pair, though a very wide one.
02^h 38.8^m	+33° 25′	Σ285	1998	165°	1.8″	7.5	8.1	*K*2III	binary	125-mm: A difficult but lovely binary. 200× shows a figure-8 with color contrast — an amber-yellow star with an ash-white star coming out of it. Clearly resolved for only moments.

Triangulum Australe

RA	Dec.	Name	Year	P.A.	Sep.	m_1	m_2	Spec.	Status	Comments
15^h 20.7^m	−67° 29′	I 332	1991	107°	1.1″	6.4	8.2	*B*3V	binary	Hartung: "In a field sown profusely with stars is [this] bright pale yellow star with a companion very close . . . which is clearly separated with 200-mm."
15^h 47.9^m	−65° 27′	Rmk 20	1991	147°	1.8″	6.2	6.4	*A*5III-IV	p. binary	Barker, 100-mm: "Lovely. A tight but obvious blue white pair at 100×, which lies in a rich field with other bright stars."
15^h 54.9^m	−60° 45′	Slr 11								Jaworski, 100-mm: "Very attractive multiple. Three stars [A, C, D] in a boomerang shape, with the white primary at the apex." Hartung: "The B member is just visible with 105-mm."
		AB	1991	95°	1.1″	6.3	8.1	*B*9II	binary	
		Δ 194								
		AC	2000	48°	44.4″	6.4	10.0			
		AD	2000	256°	48.3″	6.4	9.0			
16^h 28.0^m	−64° 03′	ι TrA	2000	1°	16.7″	5.3	9.4	*F*4IV		Hartung: "The stars are deep yellow and white and lie in a field sown with stars. The small star is easy with 75-mm."
16^h 33.1^m	−60° 54′	Δ 203	2000	277°	22.2″	7.9	8.2	*A*3III	optical	A pair of equal stars with different colors — one is blue and the other yellow. Barker, 100-mm: "The field is full of fainter stars."

Tucana

RA	Dec.	Name	Year	P.A.	Sep.	m_1	m_2	Spec.	Status	Comments
$22^h 27.3^m$	−64° 58′	δ Tuc	2000	280°	6.9″	4.5	8.7	*B*9.5V		Showcase pair. Hartung: "A bright white star with a companion well clear . . . [that looks] definitely reddish." Gould, 175-mm, 100×: "Bright pale yellow and dull yellow."
$22^h 44.7^m$	−60° 07′	h 5358	2000	91°	31.1″	8.0	10.0	*K*0III		Gould, 175-mm, 100×: A "very wide and easy yellow pair in a scattered ordinary field."
$23^h 08.6^m$	−59° 44′	Δ 245	1999	290°	13.6″	7.5	9.4	*F*5V	optical	Gould, 175-mm, 100×: A "wide and easy uneven pair; with a few stars in the field."
$23^h 18.0^m$	−61° 00′	Δ 247	2002	293°	50.0″	6.9	8.2	*K*1III		Gould: A "very wide and bright pair, yellow and cream in color; in a field with fainter stars. It is just visible as a pair in an 8 × 50 finder."
$23^h 18.6^m$	−58° 18′	h 5392	1999	319°	45.1″	7.6	9.2	*G*2/3V		Gould, 175-mm, 100×: This pair is "in the field with [Gamma] γ Tuc. It is a very wide and easy uneven pair, with a deep yellow primary."
$23^h 31.0^m$	−69° 05′	h 5402	2000	198°	36.4″	7.2	9.1	*G*0V		Gould, 175-mm, 100×: A "wide, easy and fairly bright uneven pair in a scattered field, yellow and dark yellow orange in color."
$23^h 35.2^m$	−64° 41′	h 5403	1999	43°	35.4″	7.2	10.0	*F*2IV/V		Gould, 175-mm, 100×: "In a middling starry but faint field is this pale yellow star, which has a wide and easy, but fairly faint companion."
$00^h 00.6^m$	−66° 41′	Gli 289	1999	274°	3.3″	7.7	9.2	*G*2IV/V	binary	Gould, 175-mm, 100×: "Easy and good. A fairly bright and moderately close pair of pale yellow stars, uneven in brightness, in a field of faint stars."
$00^h 31.5^m$	−62° 57′	β Tuc AC	1999	168°	26.9″	4.3	4.5	*B*9V		Showcase pair. Gould, 175-mm, 100×: "A brilliant wide pair, pale yellow and cream yellow, with a third bright star some 12′ to the SE. A faint pair is also in the field, about 12′ to the S."
$00^h 44.5^m$	−62° 30′	CorO 3	1999	66°	2.3″	6.3	8.0	*F*5III-IV	binary	Gould, 175-mm, 100×: "Beautiful effect. An elegant close pair, yellow in color, with a bright star wide in the field to the SE, and a broad, faint trapezium [of stars] to the N."
$00^h 52.4^m$	−69° 30′	λ Tuc	1998	81°	20.4″	6.7	7.3	*F*7IV-V	optical	Gould, 175-mm: A "wide bright and easy yellow pair in a mostly faint field. Probably best with small telescopes at very low power."
$01^h 03.3^m$	−60° 06′	h 3416	1999	129°	5.1″	7.6	7.7	*F*5V	binary	Gould, 175-mm, 100×: "Fairly good. An easy and moderately close pair of yellowish stars, even in brightness. Some 20′ E is a gathering of moderately bright stars, including a wide pair."
$01^h 10.0^m$	−72° 57′	Gli 9	1998	357°	3.4″	7.2	10.2	*A*1/2V		Gould, 175-mm, 100×: This "fairly bright pale yellow star has a faint companion moderately close. The field [which is part of the Small Magellanic Cloud] includes globular NGC 419."

Tucana *(continued)*

RA	Dec.	Name	Year	P.A.	Sep.	m_1	m_2	Spec.	Status	Comments
01^h 15.8^m	−68° 53′	κ Tuc								The AB pair is a showcase binary. Hartung: "This pair [AB] is particularly beautiful in a moonlit field, the stars yellow and orange . . . 150-mm should split the CD pair [which is now widening]."
		AB	1999	322°	5.0″	5.0	7.0	*F6IV*	p=857 yr	
		A-CD	1991	310°	319.5″	5.0	7.9			
		I 27								
		CD	1991	229°	0.9″	7.8	8.4	*K2V*	p=85 yr	
01^h 17.1^m	−66° 24′	h 3426	1999	330°	2.4″	6.4	8.3	*A0V*	p. binary	The data suggest a bright star with a faint companion. Gould, 175-mm: "A fine object. An uneven and quite close pair, white in color."

Ursa Major

RA	Dec.	Name	Year	P.A.	Sep.	m_1	m_2	Spec.	Status	Comments
08^{h}20.7^m	+72° 24′	Σ1193	2001	90°	42.7″	6.2	9.7	*K4III F0*		125-mm, 50×: A fine pair for low power. It's a bright pumpkin-orange star with a little blue-green companion super-wide apart. Webb, listed under Camelopardalis: "Very yellow, no color given."
08^h 43.5^m	+48° 52′	Σ1258	2002	331°	9.9″	7.7	7.9	*F0*	p. binary	125-mm, 83×: Attractive double; a pair of identical stars, whitish peach in color, unusually wide for a probable binary.
08^h 59.2^m	+48° 03′	ι UMa								The data suggest a very bright star with a close dim companion. Ferguson, 150-mm: "White and very faint blue."
		A-BC	1983	24°	4.5″	3.1	9.2	*A7IV*		
09^h 10.4^m	+67° 08′	σ^2 UMa	2004	352°	4.0″	4.9	8.9	*F7V*	p=1,141yr	The data suggest a bright star with a close dim companion. Webb: "Greenish, no color given . . . companion very difficult with 3.7-inch [94-mm]."
09^h 12.8^m	+61° 41′	Σ1315	2000	27°	24.7″	7.3	7.7	*A3IV*	optical	125-mm, 50×: Splendid sight at low power. It's a wide and easy pair of nearly equal stars, deep white and green-white, in the field with a bright straw-yellow star (16 Ursae Majoris).
09^h 14.4^m	+52° 41′	Σ1321	2002	94°	17.7″	7.8	7.9	*M0V M0V*	p=975 yr	Interesting object – a pair of cool M stars that are wide apart but form a binary system. 60-mm, 25×: An easy wide pair of identical stars, peach white in color. Webb: "Yellow."
09^h 20.7^m	+51° 16′	37 Lyn								This pair was in Lynx but is now part of Ursa Major. Gould, 175-mm, 135×: A "bright yellow star which has a distant companion and a close companion making a triangle."
		AB	1988	138°	5.6″	6.2	10.0	*F5V*		
		AC	1998	7°	130.0″	6.2	10.7			
09^h 22.5^m	+49° 33′	Σ1340	1998	319°	6.2″	7.1	9.0	*B9.5V*	binary	125-mm, 50×: Lovely contrast for the separation. A bright white star with a little arctic-blue companion just a small gap apart. It seems to jump out of a desolate field.
09^h 24.9^m	+51° 34′	ΟΣ 200	2003	335°	1.3″	6.5	8.6	*G0IV*	binary	The data suggest a fairly bright star with a close dim companion. Webb: "Yellow, no color given."
09^h 25.6^m	+54° 01′	21 UMa	2002	314°	5.9″	7.7	8.6	*A2V*	binary	125-mm: An attractively close binary star in a pretty field of many scattered dim stars. It's a gloss-white star and a small silvery white star split by hairs at 83×.

Ursa Major *(continued)*

RA	Dec.	Name	Year	P.A.	Sep.	m_1	m_2	Spec.	Status	Comments
09ʰ 28.7ᵐ	+45° 36′	41 UMa	1999	161°	71.8″	5.5	7.8	*K*0III-IV		60-mm, 25×: A fine pair for low power. It's a bright star with a super-wide companion in the pretty combination of deep lemon yellow and bluish turquoise.
09ʰ 30.9ᵐ	+44° 41′	Σ1358	2002	174°	23.9″	7.9	9.3	*M*		The data suggest an easy wide pair with a deeply colored primary star.
09ʰ 31.2ᵐ	+67°32′	Σ1349	2002	168°	19.0″	7.5	9.0	*A*3V	optical	125-mm, 50×: This ordinary pair stands out conspicuously in a desolate region of the sky. It's a pair of unequal stars, gloss white and beige white, that are spread wide apart.
09ʰ 31.5ᵐ	+63° 04′	23 UMa	2003	269°	23.2″	3.7	9.2	*F*0IV		125-mm, 50×: Grand sight! It's a brilliant star with a very wide little companion in the beautiful combination of pure Sun yellow and greenish sapphire. Smyth: "Pale white, gray."
09ʰ 51.0ᵐ	+59° 02′	υ UMa	1999	296°	11.8″	3.8	11.5	*A*9IV		The data suggest a bright star with a very dim companion. Webb: "Pale yellow, no color given."
10ʰ 04.9ᵐ	+55° 29′	Σ1402	2003	105°	32.3″	7.9	8.9	*K*5		125-mm, 50×: A wide pair of stars with a pretty color contrast — terra cotta orange with a small blue companion. Webb: "Yellow, bluish."
10ʰ 17.8ᵐ	+71° 04′	Σ1415	2003	167°	16.5″	6.7	7.3	*A*7		125-mm, 50×: Grand sight! A bright star with a fairly wide companion in the pretty combination of Sun yellow and pearly white. It's a conspicuous pair in a barren field.
10ʰ 26.0ᵐ	+52° 37′	Σ1428	2004	88°	2.8″	8.0	8.4	*F*6V	p. binary	125-mm, 125×: A dim but attractively close pair. It's a yellow-white star and a silvery white, slightly uequal, just a tiny gap apart. Smyth: A "pretty but minute object."
10ʰ 48.0ᵐ	+41° 07′	ΟΣ 229	2004	266°	0.7″	7.6	7.9	*A*5IV	p=320 yr	This binary is probably resolvable with 300-mm until at least 2025.
10ʰ 59.8ᵐ	+58° 54′	Σ1495	2000	37°	34.1″	7.3	8.8	*K*2II-III		125-mm, 50×: An easy wide pair, effortlessly found; it's about halfway between the westernmost stars of the Big Dipper's bowl (Merak and Dubhe). The colors are pretty — it's an amber-yellow and a peach-white star.
11ʰ 03.7ᵐ	+61° 45′	α UMa	1991	204°	381″	2.0	7.0	G9III *F*8		Showcase pair *(Dubhe)*. 60-mm × 25: Striking contrast — a brilliant orange star with a tiny dot of almond brown beside it. They look like a pair, though an extremely wide one.
11ʰ 03.7ᵐ	+44° 20′	h 2554	2002	291°	40.0″	7.4	9.3	*G*5		125-mm, 50×: This wide pair is included because of its pretty primary star. It's a bright lemon-yellow star with a small dot of arctic blue beside it. Easily spotted about 1° west of Psi (ψ) Ursae Majoris.
11ʰ 08.0ᵐ	+52° 49′	Σ1510	1998	329°	5.3″	7.7	9.0	*F*8V	p. binary	125-mm, 83×: This pair shows nice contrast for the separation. It's a yellow star almost touching a little white star. Webb: "White, ashy."

Ursa Major *(continued)*

RA	Dec.	Name	Year	P.A.	Sep.	m_1	m_2	Spec.	Status	Comments
11ʰ16.1ᵐ	+52° 46′	Σ1520	2003	344°	12.3″	6.5	7.8	*F6V F9V*	optical	125-mm, 50×: Grand sight! A bright yellow star with a little white star beside it. The stars are attractively close while wide enough to be very easy. Webb: "White, bluish."
11ʰ18.2ᵐ	+31° 32′	ξ UMa	2004	245°	1.7″	4.3	4.8	*F9V G9V*	p=59.9 yr	Showcase pair — the fastest easy one. In 2001, 60-mm showed a yellow-white couple, about halfway apart — a star with half a star sticking out of it. They're now widening.
11ʰ18.5ᵐ	+33° 06′	ν UMa	1996	147°	7.1″	3.5	10.1	*K3III B0*		Smyth: "A delicate double star . . . immediately above [Zeta] ξ . . . orange tint; cerulean blue, preceded exactly on the equatorial line by a 7th-magnitude star." Webb: "Very yellow, no color given."
11ʰ29.1ᵐ	+39° 20′	57 UMa	2003	354°	5.5″	5.4	10.7	*A2V*	binary	125-mm, 83×: Striking contrast for the separation! A brilliant white star with a tiny blue dot beside it split by a tiny gap. Smyth: "Lucid white; violet." Webb: "White, ash."
11ʰ31.3ᵐ	+59° 42′	Σ1544	1999	91°	12.3″	7.3	8.0	*A3 A*		125-mm, 50×: Nice effect — a bright easy pair in a black background. A peach white and a blue-white star, mildly unequal, that are split by a modest gap.
11ʰ32.3ᵐ	+61°05′	ΟΣ 235	2004	2°	0.7″	5.7	7.6	*F8V*	p=72.7 yr	A 300-mm should be able to split this very fast-moving binary; the separation is increasing.
11ʰ36.6ᵐ	+56° 08′	Σ1553	2003	166°	6.1″	7.7	8.2	*G5*	p. binary	125-mm: A fairly bright and attractively close pair. A straw-yellow star and a smaller pearly white split by a tiny gap at 50×.
11ʰ38.7ᵐ	+45° 07′	Σ1561	2004	249°	9.0″	6.5	8.2	*G0V*	p=2,050 yr	125-mm: Grand sight! A bright gold star with a little peach-white companion; both stars are seen vividly in striking contrast. Webb: "Yellowish white, ash."
11ʰ38.8ᵐ	+64° 21′	Σ1559	1997	322°	1.8″	6.8	8.0	*A5IV*	p. binary	125-mm, 200×: Striking contrast for the separation! A deep white star with a shadowy presence beside it, just wider than touching. Webb: "White, ashy."
11ʰ55.1ᵐ	+46° 29′	65 UMa								125-mm, 83×: A colorful triple. It's a fairly bright white star (D), beside a bright yellow star (A), that's almost touching a little blue star (C). Webb, referring to AB-C: "Very white, blue."
		AB-C	2003	41°	3.9″	6.7	8.3	*A3V*	binary	
		AB-D	2002	114°	63.2″	6.7	7.0			
11ʰ58.1ᵐ	+32° 16′	β 918	1999	236°	6.9″	6.4	12.5	*A9V*		Webb incuded this extremely unequal pair in his catalog, so he must have seen the companion. Webb: "3 faint nebulae [galaxies] in low-power field [with this pair]."
12ʰ02.1ᵐ	+43° 03′	67 UMa	2002	62°	278.5″	5.2	6.7	*A7*		125-mm, 50×: A fine pair, despite the width. It's a beautifully bright couple of stars, Sun yellow and lemon yellow, that are just close enough together to look like a pair.
12ʰ05.6ᵐ	+51° 56′	Σ1600	2002	94°	7.6″	7.6	8.3	*G8III*	binary	125-mm: A bright, easy, and attractively close star. A straw-yellow star and a smaller whitish blue split by a tiny gap at 50×.

Ursa Major *(continued)*

RA	Dec.	Name	Year	P.A.	Sep.	m_1	m_2	Spec.	Status	Comments
$12^h 08.1^m$	+55° 28′	Σ1603	2002	83°	22.3″	7.8	8.3	*F8V F9V*	optical	60-mm: A haunting sight — a pair of white stars in a pitch-black field, like a pair of eyes in a dark forest. They look close at 25×.
$12^h 56.3^m$	+54° 06′	Σ1695	2003	281°	3.8″	6.0	7.8	*A5*	p. binary	125-mm, 50×: Grand sight! It's a bright star almost touching a modest star in the beautiful combination of Sun yellow and pure violet. Webb: "White, ash."
$13^h 00.7^m$	+56° 22′	78 UMa	2004	89°	1.2″	5.0	7.9	*F2V*	p=106 yr	275-mm: Striking contrast for the separation! A dazzling gold-white star with a little ember of bluish turquoise sitting on its edge, and both stars look bold and sharp.
$13^h 23.9^m$	+54° 56′	ζ UMa								Showcase system — the easiest to find. This is the middle star of the Big Dipper's handle. 60-mm, 25×: Two brilliant stars are in the field (Mizar and Alcor), and the brighter one (Mizar) is a vivid pair of green-white stars just a small gap apart. Smyth, referring to Mizar: "Brilliant white; pale emerald."
		Mizar & Alcor	1991	71°	708.5″	2.2	4.0	*A1V A5V*	optical	
		Mizar	2003	153°	14.3″	2.2	3.9	*A1V*	binary	
$13^h 37.7^m$	+50° 43′	Σ1770	2004	122°	1.7″	6.9	8.2	*M2III*	binary	125-mm, 200×: Difficult but striking. A bright yellowish peach star touched by a ghostly, shadowy presence that was very hard to see. The wide pair h 2676 (below) is nearby.
$13^h 43.0^m$	+50° 02′	h 2676	2002	126°	27.7″	7.8	10.2	*K0*		125-mm, 50×: This easy but ordinary pair is about 35′ from Σ1770 (above). It's a gloss-white star with a tiny nebulous companion, very wide apart.
$13^h 58.9^m$	+53° 06′	Σ1795	2003	3°	7.9″	6.9	9.8	*A3IV-V*		Webb: "Well seen [with] 3.7-inch [94-mm]. In a string of stars [81, 83, 84, and 86 Ursa Majoris] reaching from ζ [Zeta UMa, above] towards a coarse group [χ, ι, θ] in Boötes." Smyth: "Bright white; pale blue."
$14^h 16.1^m$	+56° 43′	Σ1831								125-mm: Beautiful view. A bright white star almost touched by a little smokeball.
		AB	2000	137°	5.8″	7.2	9.6	*A7IV*	binary	
		AC	2003	220°	109.7″	7.2	6.7	*F8V*		

Ursa Minor

RA	Dec.	Name	Year	P.A.	Sep.	m_1	m_2	Spec.	Status	Comments
$02^h 31.8^m$	+89° 16′	α UMi	2005	233°	18.6″	2.1	9.1	*F7I-II*		*(Polaris)* Showcase pair. 60-mm, 42×: Striking contrast! A brilliant amber-yellow star that has a ghostly, shadowy presence just beyond its glow. Smyth: "Topaz yellow; pale white."
$13^h 40.7^m$	+76° 51′	h 2682								80-mm, 22×: A fine pair for low power. It's a bright white star with a dot of silvery blue beside it, attractively close for its contrast. Only one companion was seen.
		AB	2003	282°	25.6″	6.7	10.3	*A7V*		
		AC	2003	318°	43.8″	6.7	9.0			
$13^h 55.0^m$	+78° 24′	Σ1798	1999	11°	7.5″	7.7	9.7	*F2*	binary	125-mm, 83×: Difficult but striking. An amber-yellow star with a ghostly presence beside it, which floated in and out of view.

Ursa Minor *(continued)*

RA	Dec.	Name	Year	P.A.	Sep.	m_1	m_2	Spec.	Status	Comments
$14^h 27.5^m$	+75° 42′	5 UMi								125-mm, 83×: This pair is included because of its pretty primary star. It's a bright grapefruit-orange star with a ghostly patch of light beside it, very wide apart. Smyth: "Fine yellow; plum."
		AC	1999	132°	58.6″	4.3	9.9	*K4III*		
$15^h 29.2^m$	+80° 27′	π^1 UMi	2000	78°	31.7″	6.6	7.3			60-mm, 25×: Splendid view — a very conspicuous pair in a bright asterism within a barren region of the sky. It's a wide pair of white stars, nearly equal, in the field with three bright stars.
$15^h 48.7^m$	+83° 37′	Σ2034	2001	111°	1.1″	7.7	8.0	*A3*	binary	125-mm, 200×: This binary is ideally bright for a close pair. It's a pair of gloss-white stars in a figure-8, easily seen but absent of glare, which show razor-sharp definition.

Vela

RA	Dec.	Name	Year	P.A.	Sep.	m_1	m_2	Spec.	Status	Comments
$08^h 09.5^m$	−47° 20′	γ Vel								Showcase system. Gould, 150-mm, 45×: "A brilliant and beautiful object. A very bright white pair [AB], with a much less bright wide pair [members C and D] offset at right angles."
		AB	2002	219°	41.0″	1.8	4.1	*O7.5*		
		AC	2000	152°	2.6″	1.8	7.3			
		AD	2000	142°	93.9″	1.8	9.4			
$08^h 12.5^m$	−46° 16′	See 96	1997	276°	0.6″	6.2	7.7	*B5V*	binary	Gould: 175-mm: A "bright white star . . . in a field of mixed bright and faint stars. It is only elongated at 330× — egg shaped, without separation." Two faint pairs are in the field.
$08^h 13.6^m$	−47° 00′	Gli 87	1999	343°	35.0″	5.1	10.1	*B2.5V*		Gould, 175-mm, 100×: A "bright yellow star with an easy wide companion, in an attractive field of mixed bright and faint stars." [The pair is 46′ northeast of Gamma (γ) Velorum.]
$08^h 13.6^m$	−45° 58′	CorO 69	1999	46°	8.7″	8.1	8.6	*F2*		Gould, 175-mm, 100×: The view shows a "bright yellow star [h 4069] with an easy wide companion . . . and a neat little pale yellow pair [CorO 69] nearby."
$08^h 14.4^m$	−45° 50′	h 4069								
		AB-C	1999	249°	32.1″	5.8	8.7	*B2IV*		
$08^h 22.5^m$	−48° 29′	B Vel	1998	137°	0.7″	5.1	6.1	*B2III*	binary	Hartung: A "close pale yellow pair . . . in 1960 [when the separation was 0.8″] 200-mm showed the stars clearly apart."
$08^h 25.5^m$	−51° 44′	Δ 69	1998	219°	25.7″	5.1	9.6	*B2V*		This pair lies a few arcminutes southwest of a rich stream of mixed stars. Gould: 175-mm, 65×: A "bright white star with an easy, well separated companion."
$08^h 29.1^m$	−47° 56′	A Vel								Gould, 175-mm, 100×: "Very fine triple . . . inside a gathering of fainter stars. The main star is pale yellow, with a close companion; the third star is fainter and well separated, nearly in a line."
		AB	1999	250°	3.0″	5.5	7.2		binary	
		AC	1999	40°	18.9″	5.5	9.2	*B5*		
$08^h 29.5^m$	−44° 43′	Δ 70	2002	349°	4.3″	5.2	7.0	*B2IV*	binary	Showcase pair. Gould, 175-mm, 100×: "Good double. A bright and easy uneven pair, white in color . . . with an asterism [the cluster vdB-Ha 34] NE." Hartung: "Pale and deep yellow."

RA	Dec.	Name	Year	P.A.	Sep.	m_1	m_2	Spec.	Status	Comments
$08^{h}30.6^{m}$	−40° 31′	Δ 71	1999	51°	63.6″	7.0	7.7	B8V K4III		Gould, 175-mm, 65×: The view shows three stars nearly in a line — this "very wide pair and a third fainter star." About 12′ north is "a little triangle of stars, the SE one being a small pair."
$08^{h}31.4^{m}$	−39° 04′	h 4107								Gould, 175-mm, 100×: "A fine triple. Its stars are uneven in brightness and separation with contrasting colors — white, bronze and ashy brown. Four bright stars are in the field."
		AB	1999	330°	4.3″	6.5	8.2	B4V	binary	
		AC	1999	99°	30.1″	6.5	9.1			
$08^{h}32.1^{m}$	−53° 13′	Slr 8	1991	285°	0.8″	6.1	7.1	K0III+A3	binary	Hartung: A "fine pair, orange yellow and whitish, which 200-mm just separates."
$08^{h}36.7^{m}$	−47° 30′	h 4116	2000	3°	7.6″	7.8	8.8	A5/7V	binary	Gould, 175-mm, 100×: An "easy, fairly bright little pair, in a moderately starry field. The primary star is cream white, and there is a faint companion about 30″ away."
$08^{h}40.0^{m}$	−53° 03′	h 4126	1998	31°	16.8″	5.2	8.7	B5		Showcase pair. Gould: This "easy pair has a nice contrast in brightness, and lies in the middle of a broad loose cluster (IC 2391), with a brilliant white pair [BSO 18 AB, below] in the field."
$08^{h}40.3^{m}$	−40° 16′	CorO 74	1999	66°	4.1″	5.2	9.1	B9V	binary	Gould, 150-mm: "A delicate pair . . . in a quite starry field [at the edge of the giant star cluster Ru 64]. A bright pale yellow star with a much fainter companion, fairly close at 75×."
$08^{h}42.4^{m}$	−53° 07′	BSO 18								Gould, 175-mm, 100×: "A very wide triple, on the eastern side of the broad loose cluster IC 2391. The main stars are white, and the third star is relatively faint."
		AB	1998	311°	76.1″	4.8	5.5	B3IV		
		BD	1998	267°	60.4″	5.5	9.9	F5V		
$08^{h}44.7^{m}$	−54° 43′	δ Vel	1999	344°	1.1″	2.1	5.1	A1V	p=142 yr	Gould, 175-mm: This "very bright cream yellow star . . . becomes a very tight and very uneven double star with 180×. This one is a testing object — and the separation is decreasing."
$08^{h}52.7^{m}$	−52° 08′	CapO 9	1991	83°	2.9″	6.6	8.2	A0V	binary	This pair is about 1° northwest of H Velorum. Gould, 175-mm: A "rather nice little pair. It is quite bright but unequal, with a cream colored primary. Two 11th-magnitude stars form a triangle with the pair."
$08^{h}53.9^{m}$	−41° 50′	h 4150	1999	265°	17.5″	7.3	10.0	B5V	binary	Gould, 175-mm: This "moderately bright white star has an easy, well separated companion; the field [just beyond the edge of the nebula Gum 17] has some fairly bright stars in it."
$08^{h}56.3^{m}$	−52° 43′	H Vel	1995	335°	2.6″	4.7	7.7	B5V	binary	Hartung, referring to the pair as R 87: "In a field sown with stars this bright pair, pale and deep yellow, is an attractive object, easy for 75-mm."
$08^{h}57.1^{m}$	−43° 15′	See 108								This multiple star dominates the star cluster vdB-Ha 56. Gould, 150-mm, 75×: "This quite bright white star has a close companion, and two wide ones. The field . . . has a number of faint wide pairs."
		AB	1991	47°	3.1″	7.4	9.9	B2IV	binary	
		Gli 102								
		AC	1999	3°	42.9″	7.4	10.0	M		
		AD	1999	238°	46.3″	7.4	8.8	B8		

Vela *(continued)*

RA	Dec.	Name	Year	P.A.	Sep.	m_1	m_2	Spec.	Status	Comments
$09^h00.0^m$	−49° 33′	Gli 104	2000	307°	9.0″	7.1	9.7	*K*1/2II	binary	Gould, 175-mm, 100×: This "bright orange star has a fairly faint but not difficult companion. The field is quite starry, with mixed faint and middling bright stars."
$09^h01.7^m$	−52° 11′	h 4165	1995	136°	0.7″	5.6	6.6	*B*9	p. binary	Hartung: A "pale yellow pair . . . 150-mm resolved it plainly in 1961 [when the separation was 0.9″]."
$09^h04.5^m$	−56° 20′	h 4177								Gould, 175-mm, 100×: A "nice little mutiple star, whose main star is white. The field is moderately starry, with a loose astersim of 9th- to 11th-magnitude stars about 20′ to the north."
		AB-C	2000	225°	13.0″	7.2	8.8			
		AB-D	2000	294°	33.6″	7.5	9.7	*B*5V		
$09^h12.5^m$	−43° 37′	h 4188	1992	281°	2.8″	6.0	6.8	*B*8V	binary	These bright pairs are only 31′ apart. Each is a close pair. Gould, 150-mm, 75×: "h 4188 is off-white and nearly even" and z Velorum is a "bright creamy star with a tiny companion."
$09^h14.4^m$	−43° 14′	z Vel	1999	14°	5.7″	5.3	9.2	*B*4V		
$09^h15.2^m$	−45° 33′	I 11	1997	290°	0.8″	6.6	7.7	*B*8V	p. binary	Hartung: "In a fine starry field, this bright yellow star needs 150-mm to resolve it, the stars just in contact with no perceptible colour difference."
$09^h29.3^m$	−44° 32′	Δ 77	1999	77°	108.4″	7.1	7.0	*F*8V *F*8V		The data and location suggest a very wide pair in a lovely field full of stars — probaby a fine sight for binoculars or a finderscope.
$09^h30.7^m$	−40° 28′	ψ Vel	1997	226°	0.5″	3.9	5.1	*F*0IV+*F*3IV	p=34 yr	This very fast binary is sometimes resolvable by amateurs. Hartung: A "bright yellow binary . . . in 1961 [when the separation was 0.5″], 300-mm resolved the pair."
$09^h33.6^m$	−49° 45′	Δ 79	1999	33°	140.4″	7.5	7.6	*K*0III		This very wide pair is in a 1° field with two bright stars (M Velorum and a 5th-magnitude star). The spectral type suggests a deeply colored primary star. Probably a fine view with binoculars.
$09^h33.7^m$	−49° 00′	h 4220	1995	218°	2.0″	5.5	6.2	*B*5III	binary	Showcase pair. Hartung: "Dominating a field well sprinkled with stars, this fine pale yellow pair is clearly divided with 75-mm . . . [the companion] looks deeper yellow than the primary."
$09^h36.4^m$	−48° 45′	R 125	2000	188°	3.0″	6.3	10.3	*A*		This pair forms a triangle, about 1/2° in size, with M Velorum and a 5th-magnitude star. Gould: 175-mm, 100×: This "bright pale yellow star has a small companion, close."
$09^h37.2^m$	−53° 40′	See 115	1997	188°	0.7″	6.3	6.1	*A*2/3V	p. binary	Gould, 175-mm: A "good though close pair, bright pale yellow in color. At 330×, it is a figure-8 pair — two disks in contact. It is in a good field [about 1/2° east-southeast of cluster NGC 2925]."
$09^h42.7^m$	−55° 50′	R 129	1994	297°	3.1″	7.9	8.2	*G*2/5V+*F*I	p. binary	Gould, 175-mm, 100×: A "neat and fairly bright little pair, close and fine; both stars look yellow. The field is fairly rich in 9th-magnitude and fainter stars."

Vela *(continued)*

RA	Dec.	Name	Year	P.A.	Sep.	m_1	m_2	Spec.	Status	Comments
09ʰ 45.7ᵐ	−41° 40′	h 4242	1999	357°	7.9″	7.7	9.3	*B9IV*	binary	Gould, 175-mm, 100×: An "easy uneven pair . . . in a not very starry field. White."
09ʰ 46.1ᵐ	−45° 55′	h 4245	1999	217°	9.3″	6.8	9.6	*G8II/III*	binary	Hartung: An "elegant pair, orange yellow and bluish, which shines in a lovely field profusely sown with stars." Gould: This is the "most easterly star in a line of three [prominent] stars."
09ʰ 50.2ᵐ	−49° 37′	CorO 92	2000	24°	5.9″	8.0	9.2	*F5/6V*	binary	Gould, 175-mm, 100×: An "easy little pair in a field partly starry, partly vacant."
09ʰ 54.3ᵐ	−45° 17′	Δ 81	1999	240°	5.2″	5.8	8.2	*B5V*	binary	Showcase pair. Hartung: "This bright unequal pair, pale yellow and bluish, ornaments a lovely field." Gould, 175-mm, 100×: A "bright white star with an easy, fairly close companion."
09ʰ 57.7ᵐ	−48° 25′	h 4269	1999	321°	13.8″	6.1	10.1	*B3V*		Gould, 175-mm, 100×: This "bright white star has an easy small companion; the field has some fairly bright stars."
10ʰ 02.1ᵐ	−54° 59′	Δ 83	2000	227°	113.6″	7.9	7.9	*K2III*		The data and location suggest an easily noticed wide pair about 1° east-southeast of Phi (ϕ) Velorum. It probably has a deep color, especially with 150-mm or more.
10ʰ 03.2ᵐ 10ʰ 04.5ᵐ	−52° 03′ −51° 48′	h 4282 h 4283	1999 1999	195° 181°	49.5″ 7.8″	7.4 7.3	8.3 8.4	*A4/5IV/V* *A0V*	 binary	These pairs are about 20′ apart in a group of four 7th-magnitude stars. Gould, 175-mm, 100×: h 4282 is a "very wide, fairly bright white pair"; h 4283 is also a "fairly bright white pair."
10ʰ 05.1ᵐ	−45° 54′	h 4284	2000	66°	6.6″	7.4	9.5	*K0III*	binary	Gould, 175-mm, 100×: An "easy uneven pair, orange and ashy; the field is moderately starry."
10ʰ 06.2ᵐ	−47° 22′	I 173	1997	1°	0.9″	5.3	7.1	*K1IV+G5V*	p=232 yr	This binary should be resolvable with 120-mm; its separation is increasing.
10ʰ 19.0ᵐ	−56° 01′	R 140	1992	281°	3.1″	7.5	8.2	*A2*	binary	Gould, 175-mm, 65×: "A beautiful fairly bright white pair, in a very good field. The triple Rmk 13 [J Velorum, below] is in the field."
10ʰ 20.9ᵐ	−56° 03′	J Vel AB AC	 2000 2000	 102° 191°	 7.1″ 36.2″	 4.5 4.5	 7.2 9.2	 *B3III*	 binary	Showcase pair. Hartung, referring to the pair as Rmk 13: "This bright white pair shines like twin gems in a rich field; the third star is wider . . . even with 105-mm this is a beautiful object."
10ʰ 31.4ᵐ	−53° 43′	h 4329	2000	103°	73.9″	5.0	8.6	*F6V*		Gould, 350-mm, 120×: A "very wide pair, pale yellow and dull purple/brown; west is a small clump of several stars."
10ʰ 32.0ᵐ	−45° 04′	Pz 3	1999	219°	13.6″	5.6	6.0	*B8*	optical	Showcase pair. Gould, 175-mm, 100×: A "fine, bright and easy pair of white stars in a moderately starry field." [It forms a shallow arc, about 1° in size, with three other bright stars.]
10ʰ 32.9ᵐ 10ʰ 33.5ᵐ	−47° 00′ −46° 59′	h 4330 h 4332	1999 1999	163° 162°	40.4″ 28.4″	5.2 7.1	8.6 9.8	*K4III* *A*	 optical	Gould, 175-mm, 100×: "Curious effect. A bright orange star with an easy wide companion [h 4330] . . . that forms a very long rectangle with a parallel pair [h 4332]."

Vela *(continued)*

RA	Dec.	Name	Year	P.A.	Sep.	m_1	m_2	Spec.	Status	Comments
10h 33.3m	−55° 23′	Δ 89								Gould, 175-mm, 100×: This "bright yellow star has a tiny pair [BC] wide beside it . . . which is nearly even in brightness. The field is fairly rich." [x Velorum, below, is about 1° to the east-southeast.]
		AB	2000	30°	25.5″	6.8	7.8	*A1V*		
		BC	1991	253°	1.4″	7.8	8.1			
10h 39.3m	−55° 36′	x Vel								Gould, 350-mm, 120×: "The field is dominated by this bright yellow star with a wide and less bright, deep yellow companion. The good starry field includes two small asterisms."
		AB	2000	105°	51.7″	4.4	6.1	*G2-3I*		
10h 46.8m	−49° 25′	μ Vel	2001	52°	2.0″	2.8	5.7	*G5III+G2V*	p=138 yr	A 160-mm should be able to split this fast-moving binary through 2025. Hartung: A "bright close yellow pair."

Virgo

RA	Dec.	Name	Year	P.A.	Sep.	m_1	m_2	Spec.	Status	Comments
11h 38.4m	−02° 26′	Σ1560	1991	280°	5.0″	6.4	9.4	*G9III*	binary	Smyth: "A fine but very delicate double star . . . pale orange; reddish . . . This beautiful object [is] far too delicate for metrical observation with a small instrument."
11h 52.0m	+08° 50′	Σ1575	2003	210°	29.9″	7.4	7.9	*K0*		This pair is about 1° west-northwest of 6 Virginis. Phillips, 225-mm: A "nice neat pair, white in color. There is a third star nearby [which forms a straight line with the pair]." Harshaw, 200-mm: "Orange, white."
12h 14.5m	+08° 47′	Σ1616	2002	296°	23.2″	7.6	9.7	*G0*		Phillips, 225-mm: This pair shows a "wonderful color contrast – the stars are white and purplish blue. Nearby is a third star, in a [straight] line with this double."
12h 15.1m	−07° 15′	Σ1619	2002	269°	6.9″	8.0	8.3	*G5*	p. binary	Phillips, 225-mm: This double star is a "beautiful, close, perfectly equal little double. White." Harshaw, 200-mm: "Orange, blue."
12h 18.2m	−03° 57′	Σ1627	2003	196°	19.8″	6.6	6.9	*F2V F3V*	optical	60-mm: An attractively close pair that seems to jump out of a nearly black field. It's a pair of nearly equal stars, pure white and blue-white, split by a tiny gap at 25×.
12h 22.5m	+05° 18′	17 Vir	2004	339°	20.5″	6.6	10.5	*F8V*		60-mm, 25×: Striking contrast! A bright white star with a tiny dot of light beside it, modestly wide apart. This is among the handful of guide stars to the Virgo Cluster of galaxies.
12h 30.6m	+03° 31′	Σ1648	2000	40°	7.9″	7.5	9.8	*G8III*	binary	Nice color contrast. Phillips, 225-mm: A "white star with a hint of yellow, which has a small blue companion." Harshaw: "Yellow and (possibly) red."
12h 31.6m	−11° 04′	Σ1649	2003	194°	15.8″	8.0	8.4	*A7*		The data and location suggest a conspicuous wide pair in a desolate field. Phillips, 225-mm: A "light bluish pair, with a faint third star nearby." Harshaw: "White, white."
12h 38.3m	−11° 31′	Σ1664								Mullaney, 200-mm: "This is an arrow shaped multiple, called the 'Little Sagitta' [because it looks like the constellation]; the arrow points to the Sombrero Galaxy (M104)."
		AB	2003	224°	37.2″	7.8	9.2	*K0*		
		BC	2002	340°	66.5″	9.4	11.6	*G5*		
		CD	2002	259°	30.7″	11.6	11.6	*K0*		

RA	Dec.	Name	Year	P.A.	Sep.	m_1	m_2	Spec.	Status	Comments
12ʰ 38.7ᵐ	−04° 22′	S 639	2002	110°	56.0″	6.8	10.0	*M*0III		Phillips, 225-mm: "Surprisingly nice pair, given its large separation. The colors are yellow and blue, and they are seen *vividly.*" Smyth: "Pale yellow; greenish; several small stars in the field."
12ʰ 41.7ᵐ	−01° 27′	γ Vir	2004	179°	0.4″	3.5	3.5	*F*0V *F*0V	p=169 yr	Showcase pair *(Porrima).* This binary is among the best known. 60-mm, 120×: A pair of identical stars, brilliantly bright and stunning. They were easily apart in 1985, a figure-8 in 1995, and a single star in 2000. Both are yellow-white. Smyth: "Silvery white; pale yellow."
12ʰ 45.3ᵐ	−03° 53′	Σ1677	2004	349°	15.6″	7.3	8.1	*A*9IV	optical	60-mm, 25×: A pleasant combination. A deep white star with a smaller gray companion; these stars are attractively close while wide enough to be easy.
12ʰ 51.4ᵐ	−10° 20′	Σ1682	2003	298°	28.9″	6.6	9.7	*G*8III		60-mm, 35×: Striking contrast! A pearly white star with a ghostly, ethereal presence close beside it. Smyth: "Topaz yellow; lucid purple, the colors finely contrasted."
12ʰ 55.5ᵐ	+11° 30′	Σ1689	2003	221°	30.1″	7.1	9.1	*M*4III		125-mm, 50×: An easy pair with pretty colors. It's a peach-white star with a small blue companion very wide part. Smyth: "Pale white; sky blue." Webb: "Yellowish, bluish."
12ʰ 56.3ᵐ	−04° 52′	Σ1690	1999	148°	5.8″	7.2	9.0	*A*0V	binary	60-mm, 45×: Striking contrast for the separation! An ash-white star almost touching a tiny nebulous dot, and the little companion can be seen only with averted vision.
13ʰ 00.6ᵐ	−03° 22′	46 Vir	1991	167°	0.8″	6.2	8.8	*K*1IV	binary	The data suggest a bright star with a very close dim companion. Webb: Dembowski calls it "yellow, ash or olive."
13ʰ 03.8ᵐ	−20° 35′	β 341	2001	312°	0.8″	6.5	6.3	*F*8V	binary	This binary should be splittable with a 150-mm. The magnitudes suggest a fairly bright pair.
13ʰ 03.9ᵐ	−03° 40′	48 Vir	2004	197°	0.6″	7.1	7.7	*F*0V	p. binary	Hartung calls this double an "almost equal yellow pair . . . 200-mm shows the stars clearly apart." Webb: Both stars are "green white."
13ʰ 07.3ᵐ	+00° 35′	Σ1719	2000	359°	7.0″	7.6	8.2	*F*5V *F*9V	p. binary	Phillips, 225-mm: This double is a "perfectly equal little double star . . . both stars white. An excellent object." Harshaw, 200-mm: "Very fine object. White, lilac."
13ʰ 09.9ᵐ	−05° 32′	θ Vir	1999	342°	6.9″	4.4	9.4	*A*1IV+*A*		125-mm, 83×: Fantastic contrast for the separation! A brilliant yellow star with a vivid little drop of bluish turquoise beside it — close but wide enough to be easy. Smyth: "Pale white; violet."
13ʰ 12.1ᵐ	−16° 12′	53 Vir	1908	7°	79.3″	5.0	12.5	*F*5III-IV		Smyth: "A wide and very delicate double star . . . yellowish white; bluish . . . I could only catch a sight of B [the companion] by gleams."

RA	Dec.	Name	Year	P.A.	Sep.	m_1	m_2	Spec.	Status	Comments
$13^h 13.4^m$	−18° 50′	54 Vir	2003	33°	5.3″	6.8	7.2	A0V A1V	binary	Showcase pair. 60-mm, 45×: An attractively close little pair — two gloss-white stars, nearly alike, that are just kissing at 45×. Webb: Franks calls it "yellow, blue [in] contrast."
$13^h 14.9^m$	−11° 22′	Rst 3829								125-mm, 50×: The very wide Aa-B pair is interesting — the stars look similar in brightness, but one is white and the other pale red. It's easily noticed about 1½° west of 62 Virginis.
		Aa	2001	132°	0.6″	7.4	9.1	G3	p=168 yr	
		SHJ 162								
		Aa-B	2002	46°	107.6″	7.1	8.2	K0		
$13^h 23.7^m$	+02° 43′	Σ1740	2000	74°	26.3″	7.1	7.4	G5V G5V	optical	60-mm, 35×: Splendid view — a pair of orange-red eyes in the field with two bright white stars. It's a very wide pair of identical stars.
$13^h 24.0^m$	−20° 55′	β 610	1991	17°	4.2″	6.6	10.1		binary	Ferguson saw the close little companion of this bright star with 150-mm. He calls it an "interesting pair; white and faint blue."
$13^h 25.2^m$	−11° 10′	α Vir								*(Spica)* Smyth: "Brilliant white; bluish tinge. This beautiful bright star is in a clear dark field . . . for it has no companion nearer than the one here described [probably referring to AC]."
		AB	1999	33°	151.4″	1.0	12.0	B1V		
		AC	1879	62°	359.8″	1.0	10.5			
$13^h 28.2^m$	+09° 28′	Σ1746	2003	246°	23.0″	7.6	9.8	K0		Phillips, 225-mm: "Very nice colors — white and purple-hinting blue."
$13^h 28.4^m$	+13° 47′	70 Vir	2002	127°	268.6″	5.0	8.7	G2.5V		60-mm, 25×: This pair is included because of its pretty primary star. It's a bright grapefruit-orange star with a little red star beside it. They're super-wide apart but look like a pair.
$13^h 30.4^m$	−06° 28′	72 Vir	1999	17°	29.9″	6.1	10.7	F2V		Smyth: "A very delicate double star . . . yellowish white, violet tint; a third star is in the south preceding [meaning southwest]." W. Herschel "Large white, small red."
$13^h 34.3^m$	−00° 19′	Σ1757	2004	129°	2.0″	7.8	8.8	K4III	p=461 yr	This binary is 18′ north of Zeta (ζ) Virginis, and should splittable with 103-mm (5.1-inch). Smyth: "Pale white; yellowish; and the two point to a telescopic star."
$13^h 34.7^m$	−13° 13′	β 932	2004	59°	0.4″	6.3	7.3	A0V	p=190 yr	A 300-mm should be able to split this binary; the separation is slowly widening.
$13^h 35.6^m$	+10° 12′	h 228	2002	15°	70.1″	6.6	9.0	K0		The data suggest a bright star with a very wide companion. Phillips, 225-mm: "A nice, colorful pair. Yellow, bluish."
$13^h 37.6^m$	−07° 52′	81 Vir	1998	40°	2.6″	7.8	8.1	K0III	binary	Phillips, 225-mm: "Fantastic pair. Yellowish twins, beautiful at 171×." Smyth: "A close double star . . . bright white, yellowish . . . [a] fine object."
$13^h 37.7^m$	+02° 23′	Σ1764	2003	32°	16.1″	6.8	8.6	K2III	optical	Phillips, 225-mm, 171×: "Excellent object. A yellow and blue pair with vivid colors, in a beautiful field."
$13^h 43.1^m$	+03° 32′	84 Vir	2003	228°	2.8″	5.6	8.3	K1III	binary	125-mm, 200×: A pretty star that gives a hint of having a companion. It's a bright grapefruit-orange star with an uneven outer ring — a fragment of it is slightly bulged.

Virgo *(continued)*

RA	Dec.	Name	Year	P.A.	Sep.	m_1	m_2	Spec.	Status	Comments
13ʰ43.5ᵐ	−04° 16′	Σ1775	2002	335°	27.7″	7.2	10.1	K2III F7V		The data suggest a fairly bright star with a companion. Harshaw, 200-mm: "Very fine pair; strong orange, white." Webb: "Yellowish, no color given." Phillips, 225-mm: "White, faint bluish."
13ʰ46.1ᵐ	+05° 07′	Σ1781	2004	186°	0.8″	7.9	8.1	F8V+F0	p=274 yr	This binary should be splittable with 300-mm; its separation is increasing.
13ʰ55.0ᵐ	−08° 04′	Σ1788	2004	100°	3.6″	6.7	7.3	F8V+G0	p=2,613 yr	60-mm, 77×: A bright pair of hair-split stars, nearly alike, that are lucid white with a touch of yellow. Hartung: A "fine yellow pair . . . a good object for small apertures."
14ʰ01.6ᵐ	+01° 33′	τ Vir	2002	290°	81.0″	4.3	9.5	A3V		60-mm, 25×: Grand sight! A bright white star with a tiny pinpoint of light beside it. The two stars are easily seen, fantasic in contrast, and look very much closer than the data suggest.
14ʰ08.1ᵐ	−12° 56′	Σ1802	2002	279°	5.9″	8.0	9.0	G8/K0V	binary	Phillips, 225-mm, 171×: "The stars of this binary are yellowish and purple hinting blue. Almost missed [seeing resolution] at 62×, but superb at 171×."
14ʰ15.3ᵐ	+03° 08′	Σ1819	2004	191°	0.9″	7.7	7.9	G0V	p=220 yr	This binary should be splittable with 270-mm until at least 2025. Webb: "Both stars are yellowish."
14ʰ22.6ᵐ	−07° 46′	Σ1833	1999	175°	5.8″	7.5	7.5	G0V G0V	p. binary	60-mm: Ideal brightness for the separation! A pair of pearly white stars, split by a hair at 45×, that are modestly dim but razor-sharp in definition. They're very slightly unequal.
14ʰ28.2ᵐ	−02° 14′	φ Vir	1998	112°	5.3″	4.9	10.0	G2III		Smyth: "A most delicate double star . . . pale yellow; fine blue. The companion is clear and distinct, but [is] too minute for illumination [by a filar micrometer]."
14ʰ47.1ᵐ	+00° 58′	Σ1881	2003	1°	3.5″	6.7	8.8	B9.5V	binary	The data suggest a fairly bright star with a companion. Webb: "Very white, ash."
14ʰ48.9ᵐ	+05° 57′	Σ1883	2004	280°	0.9″	7.0	9.0	F6V	p=216 yr	A 270-mm should be able to separate this binary; the gap is slowly widening. Webb, listed under Boötes: Both stars are "yellowish."
15ʰ01.8ᵐ	−00° 08′	β 348	2004	108°	0.5″	6.1	7.5	M0.5I	binary	Hartung: A "bright orange star with [a] companion very close . . . which 300-mm shows only in good conditions."
15ʰ04.1ᵐ	+05° 30′	Σ1904	2002	348°	9.9″	7.2	7.4	F0V	binary	125-mm, 50×: This binary is attractively close but wide enough to be easy. Both stars look white with flashes of orange and yellow. They're nearly equal, and just a small gap apart.

Volans

RA	Dec.	Name	Year	P.A.	Sep.	m_1	m_2	Spec.	Status	Comments
06ʰ41.0ᵐ	−71° 47′	I 351	1998	333°	16.4″	6.5	10.0	K1III		Gould, 350-mm, 120×: This "bright orange star has an easy, well separated, fainter companion. The field is patchy starry."

Volans *(continued)*

RA	Dec.	Name	Year	P.A.	Sep.	m1	m2	Spec.	Status	Comments
07h08.7m	−70° 30′	γ Vol	1999	298°	13.7″	3.9	5.4	K0III F2V	optical	Showcase pair. Gould, 350-mm, 120×: "Beautiful, bright, and easy double. The stars are deep yellow and dull yellow, and the field is moderately starry."
07h35.4m	−74° 17′	h 3997	1996	126°	1.9″	7.1	7.0	B9IV B9IV	p. binary	The data suggest a bright and easy double star. Gould, 350-mm, 120×: A "fine close and even pair, pale yellow in color. The field is thin."
07h41.8m	−72° 36′	ζ Vol	1999	117°	16.5″	4.0	9.7	K0III		Hartung: "A star sprinkled field is dominated by this brilliant orange star with a white companion, well separated . . . an attractive object for small apertures."
08h07.9m	−68° 37′	ε Vol	1999	24°	6.0″	4.4	7.3	B6IV	binary	Showcase pair. Hartung: "This star is like a bright white gem in a field sprinkled with stars; it has a yellowish companion well clear . . . easily seen by 75-mm."
08h19.8m	−71° 31′	κ Vol								Gould, 175-mm, 100×: This "very broad triple is a little arc of three stars: two very bright, one fairly bright. The AB pair is visible in an 8 × 50 finder."
		AB	1999	58°	64.8″	5.3	5.6	B9III-IV		
		BC	1999	31°	37.6″	5.6	7.7			
08h57.4m	−66° 12′	h 4164	2000	145°	10.7″	7.7	9.5	K0III	optical	This pair is about 1/2° west-northwest of Alpha (α) Volantis. Gould, 175-mm, 100×: A "fairly bright deep yellow-orange star with a fainter companion, well separated and easy. Faint field."

Vulpecula

RA	Dec.	Name	Year	P.A.	Sep.	m1	m2	Spec.	Status	Comments
19h04.6m	+23° 20′	Σ2445	2002	262°	12.3″	7.3	8.6	B2V	optical	125-mm, 50×: Nice effect. An attractively close pair, peach white and nebulous, which forms a fishhook shape with a close field star. Webb: "Very white, ashy."
19h06.9m	+22° 10′	Σ2455	2003	29°	8.8″	7.4	9.4	F4IV	binary	125-mm, 50×: Remarkable — these two pairs are a short distance apart, look nearly identical, and all four stars are in a straight line! They're white stars with small gray companions.
19h07.1m	+22° 35′	Σ2457	2000	201°	10.2″	7.5	9.5	A7IV	optical	
19h17.7m	+23° 02′	2 Vul	2001	131°	1.6″	5.4	8.8	B0.5IV	binary	The data suggest a bright star with a close companion. Hartung: "I have not found this close pair easy, even with 200-mm."
19h25.5m	+19° 48′	4 Vul								This is the brightest star in the well-known Coathanger asterism (Cr 399). With 150-mm, Ferguson sees a triple star. He calls it "white and two faint blues."
		AB	1957	100°	18.9″	5.2	10.0	G9III		
		AC	1997	209°	51.6″	5.2	11.7			
19h26.5m	+19° 53′	Σ2521	1997	33°	27.9″	5.8	10.5	K5III		This star forms the bend of the hook in the Coathanger asterism. Smyth, listed under Anse [sic]: "A very delicate double star . . . topaz yellow; deep blue."
19h26.8m	+21° 10′	Σ2523	1997	148°	6.5″	8.0	8.1	B3V B7V	binary	125-mm: Splendid view. A close pair of twins beside a matching field star. The edge of the Coathanger asterism is in the field. All three stars are lovely deep white.

Vulpecula *(continued)*

RA	Dec.	Name	Year	P.A.	Sep.	m_1	m_2	Spec.	Status	Comments
$19^h 28.7^m$	+24° 40′	6-8 Vul	2002	30°	424.5″	4.6	5.9	*M0III*		Showcase pair. 60-mm, 25×: A bright pair of stars in fantastic contrast – a brilliant citrus orange beside a bright azure white. They're super-wide apart. Webb: "Deep and pale yellow."
$19^h 33.3^m$	+20° 25′	Σ2540	2004	147°	5.2″	7.5	9.2	*A3*	binary	125-mm, 50×: Striking contrast for the separation! A fairly bright white star with a little nebulous globe on its edge, and a bright white star (9 Vulpeculae) wide in the field.
$19^h 34.6^m$	+19° 46′	9 Vul	1968	30°	8.0″	5.0	13.4	*B8III*		The data suggest a bright star with an extremely faint companion. Webb gives its position angle and separation, so he must have seen the tiny companion.
$19^h 40.7^m$	+23° 43′	Σ2560	2003	298°	14.2″	6.6	10.5	*B6IV*		125-mm, 200×: Fantastic contrast! A bright white star with a ghostly, ethereal presence within its glow.
$19^h 52.4^m$	+25° 51′	OΣ 388								125-mm, 50×: This triple is almost a miniature Beta (β)
		AB	2003	137°	3.9″	8.3	8.5	*B9.5V*	binary	Monocerotis! It's a pair of hair-split white twins and a
		AB-C	1929	134°	29.3″	7.5	9.6			wide gray star, all in a straight line.
$19^h 53.4^m$	+20° 20′	Σ48	2003	148°	42.5″	7.1	7.3	*A0*		60-mm, 83×: A fine sight for low power. A wide easy pair of bluish white stars, nearly alike, that dominate a beautiful field packed with stars.
$20^h 00.1^m$	+25° 57′	h 1462	1999	23°	37.0″	7.6	9.9	*K2*		125-mm, 50×. Splendid view at low power. A lovely tangerine-orange star with a tiny wide companion that dominates a small cloud of stars.
$20^h 02.0^m$	+24° 56′	16 Vul	2003	124°	0.8″	5.8	6.2	*F2III*	p. binary	The data suggest a close pair of bright stars. Hartung: "In 1961 [when the separation was 1.0″], 150-mm showed the deep yellow stars just in contact."
$20^h 13.7^m$	+24° 14′	Σ2653	2003	274°	2.9″	6.7	9.2	*A1*	binary	Hartung: "This dainty unequal yellow pair lies in an attractive starry field and is quite clear with 75-mm."
$20^h 14.1^m$	+22° 13′	Σ2655								125-mm: Interesting triple. It's a close bright binary and a
		AB	2003	3°	6.2″	7.9	8.0	*A2V*	binary	wide third star in a fishhook shape. The bright AB couple
		AC	2000	154°	60.4″	7.9	10.1			is a pair of pearly white twins split by a tiny gap at 50×.
$20^h 18.3^m$	+25° 39′	h 1499								125-mm, 83×: Striking contrast. A gloss-white star with
		AC	2003	355°	21.2″	7.0	10.9	*B2V*		a drop of silvery vapor within its glow.
$21^h 10.5^m$	+22° 27′	Σ2769	2003	299°	17.9″	6.7	7.4	*A1V*	optical	60-mm, 25×: An easy wide pair that seems to jump out of a black background. It's a yellow-white star and a blue-white star, slightly unequal, that are split by a modest gap.

RESOURCES

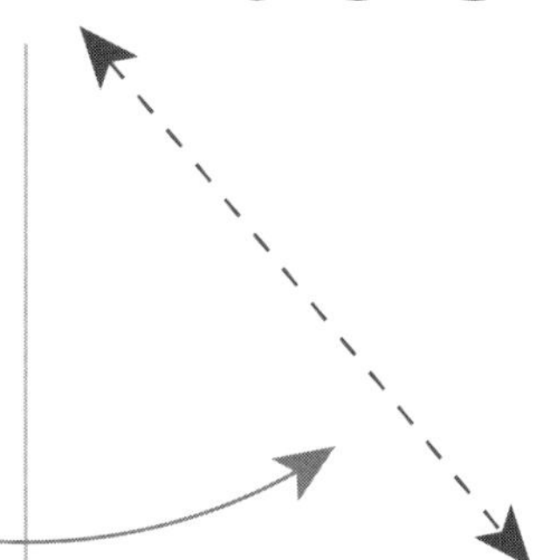

ASTRONOMY CLASSICS

Dawes, William. "Micrometrical Measurements of Double Stars." *Memoirs of the Royal Astronomical Society,* Vol. XXXV (1895): 159.

Hartung, Ernst J. *Astronomical Objects for Southern Telescopes.* Cambridge University Press, 1968.

Herschel, Sir John F. W. *Treatise on Astronomy, 3rd edition.* Carey, Lea and Blanchard, 1835.

Herschel, Sir John F. W. *Outlines of Astronomy.* 1869. Reprint, Vol. 2. P. F. Collier & Son, 1901

Sagot, Robert, and Texereau, Jean. *Revue des Constellations.* Société Astronomique de France, 1958.

Smyth, William Henry. *A Cycle of Celestial Objects. Vol. 2 The Bedford Catalogue.* 1844. Reprint, Willmann-Bell, 1986.

Treanor, Patrick J. "On the Telescopic Resolution of Unequal Binaries." The Observatory, April 1946, 255–258.

Webb, Thomas William. *Celestial Objects for Common Telescopes.* 1917. Revised and edited by Margaret W. Mayall. Dover Publications, 1962.

RECENT ARTICLES AND BOOKS

Argüelles, Luis. "Finding Your Double-Star Limit." *Sky & Telescope,* January 2002, 63–67.

Argyle, Bob (editor). *Observing and Measuring Visual Double Stars.* Springer-Verlag, 2004.

Haas, Sissy. "Enjoying Unequal Double Stars." *Sky & Telescope,* January 2002, 118–121.

Inglis, Mike. "A Basket of Multiple Stars." *Sky & Telescope,* December 2003, 113–115.

Inglis, Mike. "Seeking Colorful Double Stars." *Sky & Telescope,* June 2003, 105–108.

Jaworski, Richard. "Southern Double-Star Gems." *Sky & Telescope,* December 2002, 111–113.

Mullaney, James. *Double and Multiple Stars.* Springer-Verlag, 2005.

Mullaney, James. "Seeing Double." *Night Sky,* September/October 2005, 40–44

CATALOGS AND ATLASES

ESA. *Double and Multiple Systems Annex + Solar System Objects.* Vol. 10 of The Hipparcos and Tycho Catalogues. ESA SP-1200. European Space Agency, 1997.

Jones, Kenneth Glyn, ed. *Double Stars.* Vol. 1 of *Webb Society Deep-Sky Observer's Handbook.* Enslow Publishers, 1979.

Sinnott, Roger W. *Sky & Telescope's Pocket Sky Atlas.* Sky Publishing Corporation, 2006.

Sinnott, Roger W., and Perryman, Michael A. C. *Millennium Star Atlas*, softcover edition. Sky Publishing Corporation (with the European Space Agency), 2006.

Tirion, Wil. *Cambridge Star Atlas, 2nd edition.* Cambridge University Press, 1996.

Tirion, Wil, and Sinnott, Roger W. *Sky Atlas 2000.0, 2nd edition.* Sky Publishing Corporation, 1998.

USEFUL WEB SITES

Astronomical League's Double Star Club
www.astroleague.org/al/obsclubs/dblstar/dblstar1.html

Double Star Library
http://ad.usno.navy.mil/wds/dsl.html

Hipparcos Space Astrometry Mission
www.rssd.esa.int/Hipparcos

Hipparchos and Tycho Catalogs Online
http://cdsweb.u-strasbg.fr/hipparcos.html

Journal of Double Star Observations
www.southalabama.edu/physics/jdso

Sixth Catalog of Orbits of Visual Binary Stars
http://ad.usno.navy.mil/wds/orb6.html

Washington Double Star Catalog
http://ad.usno.navy.mil/wds

Webb Society Double Star Section
www.webbsociety.freeserve.co.uk/notes/doublest01.html

INDEX

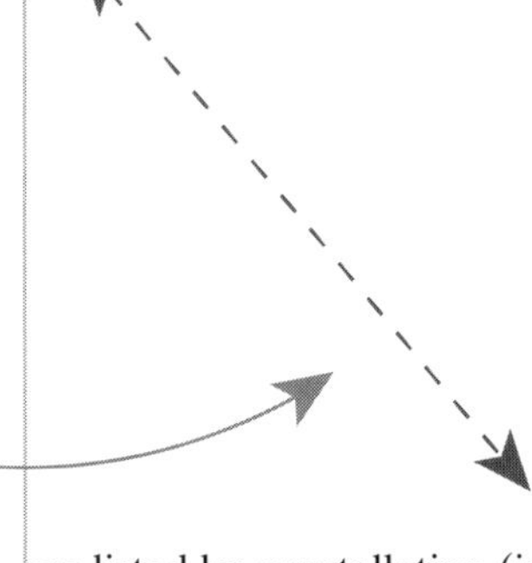

THE DOUBLE STARS included in the catalog (beginning on page 11) are listed by constellation (in alphabetical order) and then by increasing right ascension within each constellation. Page numbers in boldface refer to photographs or illustrations.